Aerospace Engineering

Aerospace Engineering

Edited by **Ted Dunham**

Clanrye International
New Jersey

Published by Clanrye International,
55 Van Reypen Street,
Jersey City, NJ 07306, USA
www.clanryeinternational.com

Aerospace Engineering
Edited by Ted Dunham

International Standard Book Number: 978-1-63240-057-4 (Hardback)

Printed in the United States of America.

Contents

Preface

Space has always been the final frontier. And such a challenge drives people with curiosity in their minds to try and decode such an intriguing and infinitely mysterious field. While such a leap towards understanding the universe might seem like a purely cerebral science, there is of course a highly specialized part of space and its study that is essential to our existence. Aerospace Engineering is the primary branch of engineering that is related to space and all its wonders. It is basically concerned with designing, research, construction, testing, development, specially the science and technology of aircraft and spacecrafts. Aerospace Engineering has two sub branches itself, which are in no way small fields of study and application, and both often overlapping with one another- Astronautical Engineering and Aeronautical Engineering. While one, that is aeronautics, deals with aircrafts that will operate within the atmosphere of the Earth, Astronautics deals with the aircraft, or more appropriately, the spacecraft that are able to operate outside the atmosphere of the Earth. Working in such constantly innovative fields, one does need to have a vivid imagination and of course a sense of wonder and adventure. The temperament of a scientific and technical know-how with the deep sense of wonder of the unknown needed in this field is no easy task. No wonder then that Aerospace Engineering is still regarded one the most intriguing fields of study and work, encompassing diverse areas from fluid mechanics to aero acoustics.

This collection of preliminary and advanced research and data on such a vast area of study has not been easy and so I would like to thank those whose hard work made this possible.

Editor

1

Satellite Attitude Control Using Analytical Solutions to Approximations of the Hamilton-Jacobi Equation

Stefan LeBel and Christopher J. Damaren

University of Toronto Institute for Aerospace Studies, 4925 Dufferin Street, Toronto, ON, Canada M3H 5T6

Correspondence should be addressed to Christopher J. Damaren; damaren@utias.utoronto.ca

Academic Editors: A. Desbiens, C. Meola, and S. Simani

The solution to the Hamilton-Jacobi equation associated with the nonlinear $\mathscr{H}_\infty$ control problem is approximated using a Taylor series expansion. A recently developed analytical solution method is used for the second-, third-, and fourth-order terms. The proposed controller synthesis method is applied to the problem of satellite attitude control with attitude parameterization accomplished using the modified Rodrigues parameters and their associated shadow set. This leads to kinematical relations that are polynomial in the modified Rodrigues parameters and the angular velocity components. The proposed control method is compared with existing methods from the literature through numerical simulations. Disturbance rejection properties are compared by including the gravity-gradient and geomagnetic disturbance torques. Controller robustness is also compared by including unmodeled first- and second-order actuator dynamics, as well as actuation time delays in the simulation model. Moreover, the gap metric distance induced by the unmodeled actuator dynamics is calculated for the linearized system. The results indicated that a linear controller performs almost as well as those obtained using higher-order solutions for the Hamilton-Jacobi equation and the controller dynamics.

1. Introduction

The attitude control problem is critical for most satellite applications and has thus attracted extensive interest. While many control methods have been developed to address this problem, most of them are concerned primarily with the optimality of attitude maneuvers [1–4]. In the present work, we shall focus on robust nonlinear control systems. We note that, throughout this paper, by nonlinear $\mathscr{H}_\infty$ we mean the $\mathscr{L}_2$-gain of a nonlinear system.

Control laws are generally developed based on mathematical models that are, at best, a close approximation of real-world phenomena. For such control methods to have any real practical value, they must be made robust with regard to unmodeled dynamics and disturbances that may act on the system. The study of robust control is therefore an essential part of the application of control theory to physical systems. In general, the development of an optimal nonlinear state feedback control law is characterized by the solution to a Hamilton-Jacobi partial differential equation (HJE) [5], while a robust nonlinear controller is obtained from the solution of one or more Hamilton-Jacobi equations [6–9]. However, no general analytical solution has yet been obtained to solve this optimization problem. Solutions have thus far only been obtained under certain conditions: in the case of linear systems with a quadratic performance index, the HJE reduces to the well-known algebraic Riccati equation (ARE). It is noted that the concept of dissipativity, which is closely related to optimal and robust control, is characterized by a Hamilton-Jacobi inequality [10–12].

Extensive work has been carried out to approximate the solution of Hamilton-Jacobi equations through a Taylor series expansion [13–17]. Although such a series expansion results in an infinite-order polynomial, finite-order approximations can be used to obtain suboptimal solutions to an HJE. We also note the work in [18, 19] which uses series solution methods for nonlinear optimal control problems. It has been shown that a local solution to an HJE can be obtained by solving the ARE for the linear approximation of the system [6, 7, 20]. Methods that have been developed over the past decades to attempt to solve this problem include the Zubov procedure [21, 22], the state-dependent Riccati equation [23, 24], the Galerkin method for the equivalent sequence of first-order partial differential equations [25, 26], and the use of symplectic geometry to examine the associated Hamiltonian system [27]. However, one aspect that is lacking in all the above methods is an analytical solution to the approximate equations.

The primary purpose of this paper is to develop robust nonlinear controllers based on analytical expressions for approximate solutions to the Hamilton-Jacobi equation. In particular, we shall provide analytical expressions for the second-, third-, and fourth-order terms of the approximation solution. These controllers are then compared through numerical simulations with existing methods from the literature for spacecraft attitude regulation [1–4]. Our objective is to examine the effects of different disturbances and uncertainties on the performance and robustness of the various controllers. More specifically, we include gravity-gradient and geomagnetic torques, as well as unmodeled actuator dynamics and actuation time delays. Moreover, we make use of the gap metric [28] to characterize the difference in the input-output (IO) map of the system induced by the unmodeled actuator dynamics. However, since we cannot calculate the gap between two nonlinear systems, we calculate the gap metric distance for the linearized system only. In contrast with some of the methods used for comparison, which were developed specifically to address the attitude control problem, the method presented in this paper is a general controller synthesis method and has also been applied to spacecraft formation flying [29].

The outline of the paper is as follows. In Section 2, a detailed description of the general class of systems is given, along with the controller synthesis method that is proposed. This controller is the solution of an appropriate nonlinear $\mathscr{H}_\infty$ problem and is taken from the work of James et al. [9]. Then, the nonlinear equations of motion for the satellite attitude dynamics are given in Section 3. Section 4 presents simulation results using the proposed controller and comparisons are made with existing methods. Finally, some conclusions and suggestions for future work are stated in Section 5. We note here that some of these results also appear in past conference proceedings [30, 31] with the present paper containing improvements to the overall presentation.

2. Nonlinear Controller Synthesis Approach

This section provides the main results of our approach to robust nonlinear controller synthesis. We begin by describing the class of nonlinear systems with which we are concerned and define robustness in the gap metric. Then, the nonlinear $\mathscr{H}_\infty$ control problem is presented. Finally, the analytic solutions for the second-, third-, and fourth-order terms in the Taylor series approximation to the solution of the Hamilton-Jacobi equation (HJE) are stated.

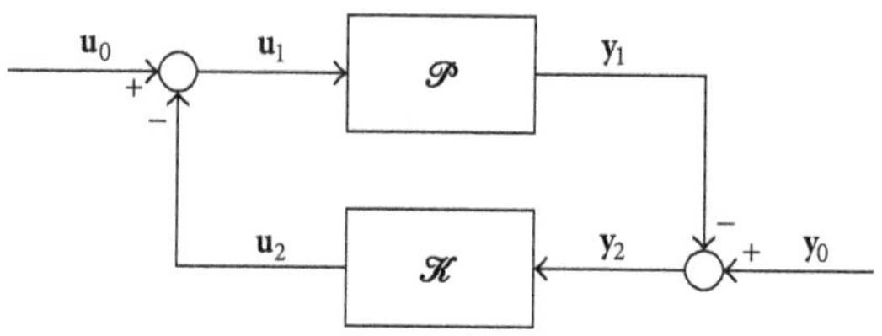

FIGURE 1: Block diagram of system (1) with controller (2).

2.1. Class of Nonlinear Systems. Consider the nonlinear system shown in Figure 1. The plant is given by

$$\mathscr{P}: \begin{cases} \dot{\mathbf{x}}(t) = \mathbf{a}(\mathbf{x}) + \mathbf{b}(\mathbf{x})\,\mathbf{u}_1(t), \\ \mathbf{y}_1(t) = \mathbf{c}(\mathbf{x}), \end{cases} \tag{1}$$

where $\mathbf{a}(\mathbf{x})$, $\mathbf{b}(\mathbf{x})$, and $\mathbf{c}(\mathbf{x})$ are assumed to be smooth (i.e., C^∞) nonlinear functions of the plant states with $\mathbf{a}(\mathbf{0}) = \mathbf{0}$ and $\mathbf{c}(\mathbf{0}) = \mathbf{0}$ (i.e., $\mathbf{x}(t) = \mathbf{0}$ is an equilibrium). The controller is described by

$$\mathscr{K}: \begin{cases} \dot{\mathbf{x}}_c(t) = \mathbf{a}_c(\mathbf{x}_c) + \mathbf{b}_c(\mathbf{x}_c)\,\mathbf{y}_2(t), \\ \mathbf{u}_2(t) = \mathbf{c}_c(\mathbf{x}_c), \end{cases} \tag{2}$$

where $\mathbf{a}_c(\mathbf{x}_c)$, $\mathbf{b}_c(\mathbf{x}_c)$, and $\mathbf{c}_c(\mathbf{x}_c)$ are smooth nonlinear functions of the controller states. Additionally, the following relations hold:

$$\begin{aligned} \mathbf{u}_1(t) &= \mathbf{u}_0(t) - \mathbf{u}_2(t), \\ \mathbf{y}_2(t) &= \mathbf{y}_0(t) - \mathbf{y}_1(t). \end{aligned} \tag{3}$$

In the above, $\mathbf{x}(t) \in \mathbb{R}^N$ is the plant state vector, $\mathbf{x}_c(t) \in \mathbb{R}^{N_c}$ is the controller state vector, $\mathbf{u}_2(t) \in \mathbb{R}^{N_u}$ is the control signal, $\mathbf{u}_0(t) \in \mathbb{R}^{N_u}$ is the exogenous (disturbance) input, $\mathbf{u}_1(t) \in \mathbb{R}^{N_u}$ is the actual plant input, $\mathbf{y}_1(t) \in \mathbb{R}^{N_y}$ is the actual plant output, $\mathbf{y}_0(t) \in \mathbb{R}^{N_y}$ is the reference and/or sensor noise signal, and $\mathbf{y}_2(t) \in \mathbb{R}^{N_y}$ is the tracking error.

The plant in (1) can be written as

$$\mathscr{P}^\star: \begin{cases} \dot{\mathbf{x}}(t) = \mathbf{a}(\mathbf{x}) + \begin{bmatrix}\mathbf{b}(\mathbf{x}) & \mathbf{0}\end{bmatrix}\mathbf{w}(t) + [-\mathbf{b}(\mathbf{x})]\,\mathbf{u}(t), \\ \begin{bmatrix}\mathbf{z}_1(t) \\ \mathbf{z}_2(t)\end{bmatrix} = \begin{bmatrix}\mathbf{0} \\ -\mathbf{c}(\mathbf{x})\end{bmatrix} + \begin{bmatrix}\mathbf{0} & \mathbf{0} \\ \mathbf{0} & \mathbf{1}\end{bmatrix}\mathbf{w}(t) + \begin{bmatrix}\mathbf{1} \\ \mathbf{0}\end{bmatrix}\mathbf{u}(t), \\ \mathbf{y}(t) = [-\mathbf{c}(\mathbf{x})] + \begin{bmatrix}\mathbf{0} & \mathbf{1}\end{bmatrix}\mathbf{w}(t), \end{cases} \tag{4}$$

where

$$\mathbf{w}(t) = \begin{bmatrix}\mathbf{u}_0(t) \\ \mathbf{y}_0(t)\end{bmatrix}, \qquad \mathbf{z}(t) = \begin{bmatrix}\mathbf{z}_1(t) \\ \mathbf{z}_2(t)\end{bmatrix} = \begin{bmatrix}\mathbf{u}_2(t) \\ \mathbf{y}_2(t)\end{bmatrix},$$
$$\mathbf{u}(t) = \mathbf{u}_2(t), \qquad \mathbf{y}(t) = \mathbf{y}_2(t). \tag{5}$$

The generalized system in (4) is shown in Figure 2. The first equation in (4) defines the plant dynamics with state variable

$\mathbf{x}(t)$, control input $\mathbf{u}(t)$, and subject to a set of exogenous inputs $\mathbf{w}(t)$, which includes disturbances (to be rejected), references (to be tracked), and/or noise (to be filtered). The second equation defines the penalty variable $\mathbf{z}(t)$ representing the outputs of interest, which may include a tracking error, a function of some of the exogenous variables $\mathbf{w}(t)$, and a cost of the input $\mathbf{u}(t)$ needed to achieve the prescribed control goal. The third equation defines the set of measured variables $\mathbf{y}(t)$, which are functions of the plant state $\mathbf{x}(t)$ and the exogenous inputs $\mathbf{w}(t)$. As we shall see next, we will be concerned with defining an upper bound on the $\mathscr{L}_2$-gain from the disturbance inputs $\mathbf{w}(t)$ to the outputs $\mathbf{z}(t)$ of the system in (4).

2.2. Robustness in the Gap Metric. In general, the plant and controller are assumed to be causal mappings from their respective inputs and outputs; that is, $\mathscr{P} : \mathscr{U} \to \mathscr{Y}$ and $\mathscr{K} : \mathscr{Y} \to \mathscr{U}$, which satisfy $\mathscr{P}\mathbf{0} = \mathbf{0}$ and $\mathscr{K}\mathbf{0} = \mathbf{0}$, where $\mathscr{U}$ and $\mathscr{Y}$ are appropriate signal spaces. In particular, these mappings can be represented by (1) and (2). Thus, in the feedback configuration of Figure 1, the signals $\mathbf{u}_i$ belong to $\mathscr{U}$ and the signals $\mathbf{y}_i$ belong to $\mathscr{Y}$, where $i \in \{0, 1, 2\}$. The input-output relations describing the plant and controller can be represented by their respective graphs:

$$\mathscr{G}_{\mathscr{P}} := \left\{ \begin{bmatrix} \mathbf{u} \\ \mathscr{P}\mathbf{u} \end{bmatrix} : \mathbf{u} \in \mathscr{U},\ \mathscr{P}\mathbf{u} \in \mathscr{Y} \right\},$$
$$\mathscr{G}_{\mathscr{K}} := \left\{ \begin{bmatrix} \mathscr{K}\mathbf{y} \\ \mathbf{y} \end{bmatrix} : \mathscr{K}\mathbf{y} \in \mathscr{U},\ \mathbf{y} \in \mathscr{Y} \right\}, \tag{6}$$

where $\mathscr{G}_{\mathscr{P}}, \mathscr{G}_{\mathscr{K}} \subset \mathscr{W} = \mathscr{U} \times \mathscr{Y}$. For a plant $\mathscr{P}_0$ with $\mathscr{M}_0 := \mathscr{G}_{\mathscr{P}_0}$ and a plant $\mathscr{P}_1$ with $\mathscr{M}_1 := \mathscr{G}_{\mathscr{P}_1}$, the ρ-gap between these two plants is defined as [9, 28]

$$\rho_g(\mathscr{P}_0, \mathscr{P}_1) = \max\left\{ \vec{\rho}_g(\mathscr{P}_0, \mathscr{P}_1), \vec{\rho}_g(\mathscr{P}_1, \mathscr{P}_0) \right\}, \tag{7}$$

where

$$\vec{\rho}_g(\mathscr{P}_0, \mathscr{P}_1) = \sup_{0 \neq \mathbf{v}_0 \in \mathscr{M}_0} \inf_{\mathbf{v}_1 \in \mathscr{M}_1} \frac{\left(\|\mathbf{v}_0 - \mathbf{v}_1\|^2 + \|\mathscr{P}_0\mathbf{v}_0 - \mathscr{P}_1\mathbf{v}_1\|^2 \right)^{1/2}}{\left(\|\mathbf{v}_0\|^2 + \|\mathscr{P}_0\mathbf{v}_0\|^2 \right)^{1/2}} \tag{8}$$

and $\vec{\rho}_g(\mathscr{P}_1, \mathscr{P}_0)$ is similarly defined. The norm used here, $\|(\cdot)\|$, is defined in [9].

It has been shown [9, 32] that, for a stable feedback pair $(\mathscr{P}_0, \mathscr{K})$, if a system $\mathscr{P}_1$ is such that

$$\rho_g(\mathscr{P}_0, \mathscr{P}_1) < \left\| \mathbf{\Pi}_{\mathscr{N}\|\mathscr{M}_0} \right\|_\infty^{-1}, \tag{9}$$

then the feedback system $(\mathscr{P}_1, \mathscr{K})$ is also stable. The symbol $\mathbf{\Pi}_{\mathscr{N}\|\mathscr{M}}$ defines a parallel projection operator [32] and represents the closed-loop mapping from the disturbances $\mathbf{w}(t)$ to the outputs $\mathbf{z}(t)$ shown in Figure 2. The $\mathscr{L}_2$-gain of $\mathbf{\Pi}_{\mathscr{N}\|\mathscr{M}}$ is the induced norm defined by

$$\|\mathbf{\Pi}_{\mathscr{N}\|\mathscr{M}}\|_\infty = \sup_{0 \neq \mathbf{w} \in \mathscr{L}_2} \frac{\|\mathbf{z}\|_2}{\|\mathbf{w}\|_2}, \tag{10}$$

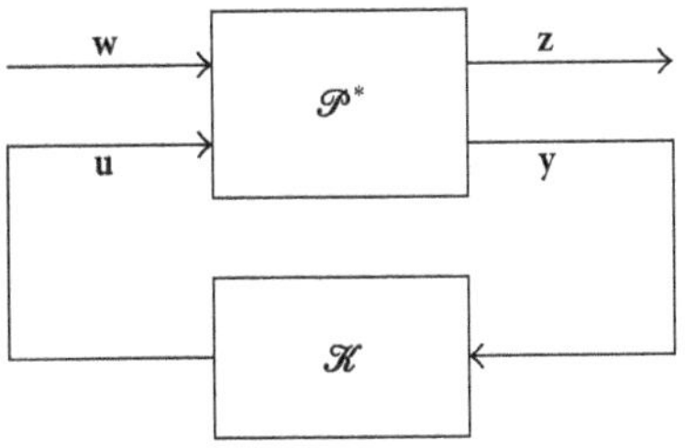

FIGURE 2: Block diagram of generalized system (4) with controller (2).

where the $\mathscr{L}_2$-norm is defined by $\|\mathbf{z}\|_2 = \sqrt{\int_0^\infty \mathbf{z}^T\mathbf{z}\, dt}$.

Thus, to minimize the effects of the disturbances $\mathbf{w}(t)$ on the outputs $\mathbf{z}(t)$, which simultaneously achieves optimal robustness, the objective is to design a controller $\mathscr{K}$ such that $\|\mathbf{\Pi}_{\mathscr{N}\|\mathscr{M}_0}\|_\infty$ is minimized. However, no closed-form solution exists to this optimal control problem. Instead, we shall be concerned with the suboptimal problem: given some constant scalar γ, design $\mathscr{K}$ such that $\|\mathbf{\Pi}_{\mathscr{N}\|\mathscr{M}_0}\|_\infty < \gamma$. Note that a γ-iteration procedure can be applied to obtain the optimal solution within an arbitrary tolerance. It will be shown that the controller that satisfies this objective can be obtained from the solution to a single Hamilton-Jacobi equation.

2.3. Nonlinear $\mathscr{H}_\infty$ Control Problem. As indicated in the previous subsection, we are concerned with rendering $\|\mathbf{\Pi}_{\mathscr{N}\|\mathscr{M}_0}\|_\infty < \gamma$. In other words, we wish to bound the $\mathscr{L}_2$-gain from the disturbance inputs $\mathbf{w}(t)$ to the outputs $\mathbf{z}(t)$ of the system in (4). Without providing the details, it is noted that the HJE corresponding to (4) is dependent on γ and the quadratic term is sign-indefinite (i.e., is neither positive definite nor negative definite). However, by performing a certain transformation [9, 33], the generalized system can be written in a form where the outputs $\mathbf{z}(t)$ are not explicit functions of the disturbances $\mathbf{w}(t)$. The resulting HJE is no longer dependent on γ and is sign-definite. The system resulting from this transformation is described by

$$\mathscr{P}^\star : \begin{cases} \dot{\mathbf{x}}(t) = \mathbf{a}(\mathbf{x}) + \left[-\mathbf{b}(\mathbf{x})\ \ \mathbf{0}\right]\mathbf{w}(t) + \left[-\mathbf{b}(\mathbf{x})\right] + \mathbf{u}(t). \\ \begin{bmatrix} \mathbf{z}_1(t) \\ \mathbf{z}_2(t) \end{bmatrix} = \begin{bmatrix} \mathbf{0} \\ -\beta^{-1}\mathbf{c}(\mathbf{x}) \end{bmatrix} + \begin{bmatrix} \mathbf{1} \\ \mathbf{0} \end{bmatrix}\mathbf{u}(t), \\ \mathbf{y}(t) = \left[-\beta^{-1}\mathbf{c}(\mathbf{x})\right] + \left[\mathbf{0}\ \ {-\mathbf{1}}\right]\mathbf{w}(t), \end{cases} \tag{11}$$

where $\beta = \sqrt{1 - \gamma^{-2}}$. Now, the HJE corresponding to the system in (11) is given by

$$\nabla V(\mathbf{x})\, \mathscr{A}(\mathbf{x}) - \frac{1}{2}\nabla V(\mathbf{x})\, \mathscr{R}(\mathbf{x})\, \nabla V^T(\mathbf{x}) + \frac{1}{2}\mathscr{Q}(\mathbf{x}) = 0, \tag{12}$$

where $\nabla V(\mathbf{x})$ is the Jacobian matrix of the storage function $V(\mathbf{x})$ and is defined as

$$\nabla V(\mathbf{x}) = \frac{\partial V}{\partial \mathbf{x}^T} = \text{row}_i \left\{ \frac{\partial V}{\partial x_i} \right\} \tag{13}$$

with $\text{row}_i\{\cdot\}$ denoting a row matrix in the index $i \in \{1, 2, \ldots, N\}$. For the system in (11), we have the parameters [9]

$$\begin{aligned} \mathcal{A}(\mathbf{x}) &= \mathbf{a}(\mathbf{x}), \\ \mathcal{R}(\mathbf{x}) &= \mathbf{b}(\mathbf{x})\mathbf{b}^T(\mathbf{x}), \\ \mathcal{Q}(\mathbf{x}) &= \mathbf{c}^T(\mathbf{x})\mathbf{c}(\mathbf{x}). \end{aligned} \tag{14}$$

The name used in the literature for the HJE in (12) is not uniform. In the context of robust control, it is sometimes referred to as the Hamilton-Jacobi-Isaacs equation and when used in nonlinear optimal control it is often referred to as the Hamilton-Jacobi-Bellman equation. In [9] it is referred to as the Hamilton-Jacobi-Bellman-Isaacs equation. We will simply call it the HJE.

It is shown in [9] that the controller $\mathcal{K}$ that solves the suboptimal $\mathcal{H}_\infty$ problem for the plant $\mathcal{P}^\star$ (i.e., renders $\|\Pi_{\mathcal{N}\|\mathcal{M}_0}\|_\infty < \gamma$) is related to that (call it $\overline{\mathcal{K}}$) which solves the same problem for the modified plant $\overline{\mathcal{P}}^\star$ as follows: $\mathcal{K} = \overline{\mathcal{K}}\beta$. We now construct this $\mathcal{K}$. Denote by $V_+(\mathbf{x})$ the unique smooth solution to the HJE (12) that satisfies $V_+(\mathbf{x}) \geq 0$ and $V_+(\mathbf{0}) = 0$, with

$$+\left[\mathcal{A}(\mathbf{x}) - \mathcal{R}(\mathbf{x})\nabla V_+^T(\mathbf{x})\right] \tag{15}$$

asymptotically stable. Similarly, we denote by $V_-(\mathbf{x})$ the unique smooth solution to the HJE that satisfies $V_-(\mathbf{x}) \leq 0$ and $V_-(\mathbf{0}) = 0$, with

$$-\left[\mathcal{A}(\mathbf{x}) - \mathcal{R}(\mathbf{x})\nabla V_-^T(\mathbf{x})\right] \tag{16}$$

asymptotically stable.

Following the approach of James et al. [9], a local solution to the nonlinear $\mathcal{H}_\infty$ control problem for the plant in (4) is obtained if the following conditions are satisfied.

(1) There exists a C^3 positive-definite function $V_+(\mathbf{x})$ defined in a neighbourhood of the origin with $V_+(\mathbf{0}) = 0$ that satisfies the HJE (12).

(2) There exists a C^3 negative-definite function $V_-(\mathbf{x})$ defined in a neighbourhood of the origin with $V_-(\mathbf{0}) = 0$ that satisfies the HJE (12), and additionally satisfies

$$\begin{aligned} \nabla^2 \Big\{ &-\nabla r_{2e}\left(\mathcal{A} + \gamma^{-2}\beta^{-2}\mathcal{R}\nabla V_+^T\right) \\ &+ \frac{1}{2}\gamma^{-2}\nabla r_{2e}\mathcal{R}\nabla r_{2e}^T + \frac{1}{2}\beta^{-4}\nabla V_+\mathcal{R}\nabla V_+^T \\ &- \frac{1}{2}\gamma^2\beta^{-2}\mathcal{Q}\Big\} < 0, \end{aligned} \tag{17}$$

where

$$r_{2e}(\mathbf{x}) = \beta^{-2}V_+(\mathbf{x}) + \gamma^2 V_-(\mathbf{x}). \tag{18}$$

(3) There exists a C^2 function $\mathbf{G}_2(\mathbf{x})$ such that

$$\nabla r_{2e}(\mathbf{x})\mathbf{G}_2(\mathbf{x}) = \gamma^2\beta^{-1}\mathbf{c}^T(\mathbf{x}) \tag{19}$$

in a neighbourhood of the origin.

The resulting nonlinear controller is given by

$$\mathcal{K}: \begin{cases} \dot{\mathbf{x}}_c = \mathbf{a}(\mathbf{x}_c) - \mathbf{b}(\mathbf{x}_c)\mathbf{b}^T(\mathbf{x}_c)\nabla V_+^T(\mathbf{x}_c) \\ \qquad + \mathbf{G}_2(\mathbf{x}_c)\left[\beta^{-1}\mathbf{c}(\mathbf{x}_c) + \beta\mathbf{y}_2\right], \\ \mathbf{u}_2(\mathbf{x}_c) = \beta^{-2}\mathbf{b}^T(\mathbf{x}_c)\nabla V_+^T(\mathbf{x}_c). \end{cases} \tag{20}$$

In the next subsection, we shall examine how to obtain analytical expressions for the approximate solution to the HJE in (12) required for this control law.

2.4. Analytical Solutions to HJE Approximation. Our approach for the synthesis of nonlinear controllers is based on the Taylor series approximation of the solution to the Hamilton-Jacobi equation, where each order of the controller is built using the previous orders. The following notation will be used: $\text{row}_k\{\cdot\}$ denotes a row matrix with index k, $\text{col}_k\{\cdot\}$ denotes a column matrix with index k, and $\text{mat}_{mn}\{\cdot\}$ denotes a matrix with row index m and column index n. It should be emphasized that the symbols k, m, and n used here are dummy indices. In general, $A^{(i,j)}$ refers to the (i, j)th entry of the matrix $\mathbf{A}$. We will denote the positive-semidefinite solution $V_+(\mathbf{x})$ of the HJE (12) simply by $V(\mathbf{x})$.

Consider the nonlinear system in (1). We begin by making the assumptions that $\mathbf{b}(\mathbf{x}) = \mathbf{B}$ and $\mathbf{c}(\mathbf{x}) = \mathbf{C}\mathbf{x}$, where $\mathbf{B}$ and $\mathbf{C}$ are constant matrices. From these assumptions, $\mathcal{R}(\mathbf{x})$ is a constant matrix, which we will denote simply by $\mathbf{R} = \mathbf{B}\mathbf{B}^T$, and $\mathcal{Q}(\mathbf{x})$ is a quadratic form, which we will write as $\mathbf{x}^T\mathbf{Q}\mathbf{x}$, where $\mathbf{Q} = \mathbf{C}^T\mathbf{C}$. Thus, the only nonlinearities present are in the system $\mathbf{a}(\mathbf{x})$ matrix. For the purpose of the results to be presented, this nonlinear function will be approximated to fourth order. It should be noted that some of these terms may be zero, depending on the system considered. Therefore, we have the following approximation:

$$\mathbf{a}(\mathbf{x}) = \mathbf{a}_1(\mathbf{x}) + \mathbf{a}_2(\mathbf{x}) + \mathbf{a}_3(\mathbf{x}) + \mathbf{a}_4(\mathbf{x}), \tag{21}$$

where

$$\begin{aligned} \mathbf{a}_1(\mathbf{x}) &= \mathbf{A}_1\mathbf{x} = \text{col}_k\left\{\sum_n A_1^{(k,n)}x_n\right\}, \\ \mathbf{a}_2(\mathbf{x}) &= \text{col}_k\left\{\mathbf{x}^T\mathbf{A}_{2k}\mathbf{x}\right\} = \text{col}_k\left\{\sum_i\sum_j x_ix_jA_{2k}^{(i,j)}\right\}, \\ \mathbf{a}_3(\mathbf{x}) &= \text{mat}_{mn}\left\{\mathbf{x}^T\mathbf{A}_{3mn}\mathbf{x}\right\}\mathbf{x} \\ &= \text{col}_m\left\{\sum_n\left(\sum_i\sum_j x_ix_jA_{3mn}^{(i,j)}\right)x_n\right\}, \\ \mathbf{a}_4(\mathbf{x}) &= \text{col}_k\left\{\mathbf{x}^T\text{mat}_{mn}\left\{\mathbf{x}^T\mathbf{A}_{4kmn}\mathbf{x}\right\}\mathbf{x}\right\} \\ &= \text{col}_k\left\{\sum_m\sum_n x_mx_n\left(\sum_i\sum_j x_ix_jA_{4kmn}^{(i,j)}\right)\right\} \end{aligned} \tag{22}$$

and the summations run from 0 to N. Here, $\mathbf{A}_1$, $\mathbf{A}_{2k}$, $\mathbf{A}_{3mn}$, and $\mathbf{A}_{4kmn}$ are families of $N \times N$ square matrices. We shall also find it useful to define the column matrices $\mathbf{a}_{2ij} = \text{col}_k\{A_{2k}^{(i,j)}\}$ and $\mathbf{a}_{3njk} = \text{col}_m\{A_{3mn}^{(j,k)}\}$ whose entries can be used to form $\mathbf{A}_{2k}$ and $\mathbf{A}_{3mn}$, respectively.

Additionally, consider a storage function $V(\mathbf{x})$ for which

$$\nabla V(\mathbf{x}) = \nabla V_1(\mathbf{x}) + \nabla V_2(\mathbf{x}) + \nabla V_3(\mathbf{x}) + \nabla V_4(\mathbf{x}), \quad (23)$$

where

$$\begin{aligned}
\nabla V_1(\mathbf{x}) &= \mathbf{x}^T\mathbf{P}_1 = \text{row}_n\left\{\sum_m x_m P_1^{(m,n)}\right\},\\
\nabla V_2(\mathbf{x}) &= \text{row}_k\left\{\mathbf{x}^T\mathbf{P}_{2k}\mathbf{x}\right\} = \text{row}_k\left\{\sum_i\sum_j x_i x_j P_{2k}^{(i,j)}\right\},\\
\nabla V_3(\mathbf{x}) &= \mathbf{x}^T\text{mat}_{mn}\left\{\mathbf{x}^T\mathbf{P}_{3mn}\mathbf{x}\right\}\\
&= \text{row}_n\left\{\sum_m x_m\left(\sum_i\sum_j x_i x_j P_{3mn}^{(i,j)}\right)\right\},\\
\nabla V_4(\mathbf{x}) &= \text{row}_k\left\{\mathbf{x}^T\text{mat}_{mn}\left\{\mathbf{x}^T\mathbf{P}_{4kmn}\mathbf{x}\right\}\mathbf{x}\right\}\\
&= \text{row}_k\left\{\sum_m\sum_n x_m x_n\left(\sum_i\sum_j x_i x_j P_{kmn}^{(i,j)}\right)\right\}.
\end{aligned} \quad (24)$$

Here, $\mathbf{P}_1$, $\mathbf{P}_{2k}$, $\mathbf{P}_{3mn}$, and $\mathbf{P}_{4kmn}$ are families of $N \times N$ square matrices. In general, $P^{(i,j)}$ refers to the (i, j)th entry of the matrix $\mathbf{P}$. We shall also find it useful to define the column matrices $\mathbf{p}_{2ij} = \text{col}_k\{P_{2k}^{(i,j)}\}$ and $\mathbf{p}_{3mij} = \text{col}_n\{P_{3mn}^{(i,j)}\}$ whose entries can be used to form $\mathbf{P}_{2k}$ and $\mathbf{P}_{3mn}$, respectively.

It is important to recognize that, while $\mathbf{a}_i(\mathbf{x})$ may be zero for some i, the corresponding $\nabla V_i(\mathbf{x})$ is not necessarily zero as well. This is because the present method involves a Taylor series expansion of the nonlinear solution to the HJE. Substituting (21) and (23) into the HJE in (12) and grouping terms of the same order yields

$$\mathcal{O}\left[\|\mathbf{x}\|^2\right] : \nabla V_1(\mathbf{x})\,\mathbf{a}_1(\mathbf{x}) - \frac{1}{2}\nabla V_1(\mathbf{x})\,\mathbf{R}\nabla V_1^T(\mathbf{x}) + \frac{1}{2}\mathbf{x}^T\mathbf{Q}\mathbf{x} = 0, \quad (25)$$

$$\mathcal{O}\left[\|\mathbf{x}\|^k\right] : \nabla V_{(k-1)}\mathbf{a}_{c1} + \nabla V_1\mathbf{a}_{(k-1)} + \sum_{i=2}^{k-2}\nabla V_i\mathbf{a}_{c(k-i)} = 0 \quad (26)$$

for $k \geq 3$, where $\mathbf{a}_{c1} = \mathbf{a}_1(\mathbf{x}) - \mathbf{R}\nabla V_1^T(\mathbf{x}) = \mathbf{A}_{c1}\mathbf{x}$ with $\mathbf{A}_{c1} = \mathbf{A}_1 - \mathbf{B}\mathbf{B}^T\mathbf{P}_1$ and $\mathbf{a}_{ci} = \mathbf{a}_i(\mathbf{x}) - (1/2)\mathbf{R}\nabla V_i^T(\mathbf{x})$ for $i \geq 2$. Thus, at each order k, the objective is to solve for the unknown $\nabla V_{(k-1)}$. Note that in (26) the summation term is equal to zero for $k = 3$, since $i > k - 2$.

We now present the general expressions to compute the unknowns $\nabla V_{(k-1)}$ in (23). The first-order solution, $\mathbf{P}_1 = \mathbf{P}_1^T$, is obtained by solving the ARE corresponding to (25), which is given by

$$\mathbf{P}_1\mathbf{A}_1 + \mathbf{A}_1^T\mathbf{P}_1 - \mathbf{P}_1\mathbf{B}\mathbf{B}^T\mathbf{P}_1 + \mathbf{C}^T\mathbf{C} = \mathbf{0}, \quad (27)$$

where it has been noted that $\mathbf{R} = \mathbf{B}\mathbf{B}^T$ and $\mathbf{Q} = \mathbf{C}^T\mathbf{C}$. We will assume that $(\mathbf{A}_1, \mathbf{B})$ is controllable and $(\mathbf{C}, \mathbf{A}_1)$ is observable so that $\mathbf{P}_1$ is positive definite. Then, the higher-order solutions are obtained by solving (26) recursively for increasing values of k.

The second-order solution is given by

$$\mathbf{p}_{2ij}^T = -\mathbf{a}_{2ij}^T\mathbf{P}_1\mathbf{A}_{c1}^{-1}, \quad i, j = 1, \ldots, N, \quad (28)$$

where the column matrices $\mathbf{p}_{2ij}$ and $\mathbf{a}_{2ij}$ were defined above. We emphasize that the multiplications indicated in (28) are standard matrix multiplications.

The third-order solution is given by

$$\text{mat}_{mn}\left\{P_{3mn}^{(i,j)}\right\} = -\left[\text{mat}_{mn}\left\{\mathbf{p}_{2mi}^T\mathbf{a}_{c2jn}\right\} + \mathbf{P}_1\text{mat}_{mn}\left\{A_{3mn}^{(i,j)}\right\}\right]\mathbf{A}_{c1}^{-1}, \quad i, j = 1, \ldots, N, \quad (29)$$

where $\mathbf{a}_{c2jn} = \mathbf{a}_{2jn} - (1/2)\mathbf{B}\mathbf{B}^T\mathbf{p}_{2jn}$. Here, $\text{mat}_{mn}\{P_{3mn}^{(i,j)}\}$ consists of a square matrix (with row index m and column index n) containing the (i, j)th entries of each $\mathbf{P}_{mn}$ for given i and j. Again, all of the indicated multiplications are standard matrix multiplications.

The fourth-order solution is given by

$$\text{mat}_{mn}\left\{P_{4nmk}^{(i,j)}\right\} = -\left[\text{mat}_{mn}\left\{\mathbf{p}_{3mij}^T\mathbf{a}_{c2kn}\right\} + \text{mat}_{mn}\left\{\mathbf{p}_{2mi}^T\mathbf{a}_{c3njk}\right\} + \text{mat}_{mn}\left\{A_{4nmk}^{(i,j)}\right\}\mathbf{P}_1\right]\mathbf{A}_{c1}^{-1}, \quad i, j, k = 1, \ldots, N, \quad (30)$$

where $\mathbf{a}_{c3njk} = \mathbf{a}_{3njk} - (1/2)\mathbf{B}\mathbf{B}^T\mathbf{p}_{3njk}$. Similar comments on the multiplications involved above apply here.

The negative-semidefinite solution of the HJE, $V_-(\mathbf{x})$, can be determined using (23)–(30) with the proviso that the matrix $\mathbf{P}_1$ is replaced with the negative-definite solution of the Riccati equation in (27) which we will denote by $\mathbf{Q}_1$ and $\mathbf{A}_{c1}$ is replaced by $\mathbf{A}_{u1} = \mathbf{A}_1 - \mathbf{B}\mathbf{B}^T\mathbf{Q}_1$. The matrices corresponding to those in (24) will be denoted by $\mathbf{Q}_1$, $\mathbf{Q}_{2k}$, $\mathbf{Q}_{3mn}$, and $\mathbf{Q}_{4kmn}$ and they are obtained by solving (27)–(30) with $\mathbf{Q}_1$ replacing $\mathbf{P}_1$, $\mathbf{q}_{2ij}$ replacing $\mathbf{p}_{2ij}$, and so forth. In the next section, the satellite attitude dynamics are presented.

3. Attitude Dynamics and Control

3.1. The Attitude Dynamics and Kinematics. The attitude dynamics of a rigid spacecraft are given by Euler's equation:

$$\dot{\boldsymbol{\omega}} = -\mathbf{I}^{-1}\boldsymbol{\omega}^\times\mathbf{I}\boldsymbol{\omega} + \mathbf{I}^{-1}\mathbf{u}, \quad (31)$$

where $\boldsymbol{\omega} = [\omega_1\ \omega_2\ \omega_3]^T$ are the body angular velocities, $\mathbf{I}$ is the moment of inertia matrix, and $\mathbf{u}_1 = [u_1\ u_2\ u_3]^T$ are the body torques. The notation $\boldsymbol{\omega}^\times$ is the matrix representation of the cross product and is defined as

$$\boldsymbol{\omega}^\times = \begin{bmatrix} 0 & -\omega_3 & \omega_2 \\ \omega_3 & 0 & -\omega_1 \\ -\omega_2 & \omega_1 & 0 \end{bmatrix}. \quad (32)$$

While many representations are possible to define the spacecraft attitude kinematics, the modified Rodrigues parameters (MRPs) are chosen here because they are polynomial in the states, which fit nicely with the present controller synthesis approach, and they possess neither singularities nor norm constraints when used in conjunction with the shadow parameters. The MRP vector $\boldsymbol{\sigma} = [\sigma_1 \ \sigma_2 \ \sigma_3]^T$ can be defined in terms of the principal rotation axis $\hat{\mathbf{e}} = [e_1 \ e_2 \ e_3]^T$ and principal rotation angle Φ of Euler's theorem according to

$$\boldsymbol{\sigma} = \hat{\mathbf{e}} \tan\left(\frac{\Phi}{4}\right). \quad (33)$$

The attitude kinematics using MRPs are defined by

$$\dot{\boldsymbol{\sigma}} = \frac{1}{2}\left[\frac{1}{2}\left(1 - \boldsymbol{\sigma}^T\boldsymbol{\sigma}\right)\mathbf{1} + \boldsymbol{\sigma}^\times + \boldsymbol{\sigma}\boldsymbol{\sigma}^T\right]\boldsymbol{\omega}, \quad (34)$$

where $\mathbf{1}$ is the identity matrix.

Upon closer inspection of (33), it is seen that the MRPs encounter a singularity for rotations of $\Phi = \pm 2\pi$ rad. This corresponds to a complete rotation in either direction about the principal axis. To circumvent this, another set of MRPs, called the shadow parameters and denoted by $\boldsymbol{\sigma}_S$, is used in conjunction with the regular MRPs. By switching from one set to the other at rotations of $\Phi = \pm\pi$ rad, it is possible to avoid any singularities. The parameter switching occurs on the surface defined by

$$\boldsymbol{\sigma}^T\boldsymbol{\sigma} = \boldsymbol{\sigma}_S^T\boldsymbol{\sigma}_S = 1. \quad (35)$$

The kinematics are identical for both the regular and the shadow parameters. However, when the switching surface is encountered, both the MRPs and their rates must be converted from one set to the other. This can be accomplished with the following relations:

$$\dot{\boldsymbol{\sigma}}_S = -\frac{\dot{\boldsymbol{\sigma}}}{(\boldsymbol{\sigma}^T\boldsymbol{\sigma})} + \frac{1 + (\boldsymbol{\sigma}^T\boldsymbol{\sigma})}{2(\boldsymbol{\sigma}^T\boldsymbol{\sigma})^2}\boldsymbol{\sigma}\boldsymbol{\sigma}^T\boldsymbol{\omega}, \quad \boldsymbol{\sigma}_S = -\frac{\boldsymbol{\sigma}}{(\boldsymbol{\sigma}^T\boldsymbol{\sigma})}. \quad (36)$$

More details can be found in the text by Schaub and Junkins [34].

Defining the state vector $\mathbf{x} = [\omega_1 \ \omega_2 \ \omega_3 \ \sigma_1 \ \sigma_2 \ \sigma_3]^T$ and grouping terms of the same order, the attitude dynamics in (31) and kinematics in (34) can be expressed as

$$\begin{aligned}\dot{\mathbf{x}}(t) &= \mathbf{a}_1(\mathbf{x}) + \mathbf{a}_2(\mathbf{x}) + \mathbf{a}_3(\mathbf{x}) + \mathbf{B}\mathbf{u}_1(t),\\ \mathbf{y}_1(t) &= \mathbf{C}\mathbf{x},\end{aligned} \quad (37)$$

where

$$\mathbf{a}_1(\mathbf{x}) = \mathbf{A}_1\mathbf{x} = \begin{bmatrix} \mathbf{0} & \mathbf{0} \\ \frac{1}{4}\mathbf{1} & \mathbf{0}\end{bmatrix}\begin{bmatrix}\boldsymbol{\omega} \\ \boldsymbol{\sigma}\end{bmatrix}, \qquad \mathbf{B} = \begin{bmatrix}\mathbf{I}^{-1} \\ \mathbf{0}\end{bmatrix}, \quad \mathbf{C} = \mathbf{1}, \quad (38)$$

and the second- and third-order terms are given by

$$\mathbf{a}_2(\mathbf{x}) = \begin{bmatrix} -\mathbf{I}^{-1}\boldsymbol{\omega}^\times\mathbf{I}\boldsymbol{\omega} \\ \frac{1}{2}\boldsymbol{\sigma}^\times\boldsymbol{\omega}\end{bmatrix}, \qquad \mathbf{a}_3(\mathbf{x}) = \begin{bmatrix} \mathbf{0} \\ \left(\frac{1}{2}\boldsymbol{\sigma}\boldsymbol{\sigma}^T - \frac{1}{4}\boldsymbol{\sigma}^T\boldsymbol{\sigma}\mathbf{1}\right)\boldsymbol{\omega}\end{bmatrix}, \quad (39)$$

respectively. It should be noted that these third-order attitude dynamics are exact so that we may take $\mathbf{a}_4(\mathbf{x}) = \mathbf{0}$ and hence $\mathbf{A}_{4kmn} = \mathbf{0}$.

3.2. The Attitude Controller. In this section, the proposed controller synthesis methods will be applied to the satellite attitude dynamics given by (37)–(39). For simplicity, we shall assume that the dynamics are formulated in principal axes so that the inertia matrix is given by $\mathbf{I} = \mathrm{diag}\{I_1, I_2, I_3\}$. Let us begin by comparing the definitions of $\mathbf{a}_2$ and $\mathbf{a}_3$ in (22) with the specific ones given in (39). From this, we identify the nonzero elements of the matrices $\mathbf{A}_{2k}$ and $\mathbf{A}_{3mn}$ as follows:

$$\begin{aligned} &A_{2,1}^{23} = A_{2,1}^{32} = \frac{k_1}{2}, \qquad A_{2,2}^{13} = A_{2,2}^{31} = \frac{k_2}{2},\\ &A_{2,3}^{12} = A_{2,3}^{21} = \frac{k_3}{2},\\ &A_{2,4}^{53} = -A_{2,4}^{62} = -A_{2,5}^{43} = A_{2,5}^{61}\\ &= A_{2,6}^{43} = -A_{2,6}^{51} = \frac{1}{2},\end{aligned} \quad (40)$$

where $k_1 = (I_2 - I_3)/I_1$, $k_2 = (I_3 - I_1)/I_2$ and $k_3 = (I_1 - I_2)/I_3$. In addition,

$$\begin{aligned} &2A_{3,41}^{44} = A_{3,41}^{55} = A_{3,41}^{66} = 2A_{3,42}^{45}\\ &= 2A_{3,42}^{54} = 2A_{3,43}^{64} = 2A_{3,43}^{46} = \frac{1}{2},\\ &2A_{3,51}^{54} = 2A_{3,51}^{45} = A_{3,52}^{44} = 2A_{3,52}^{55}\\ &= A_{3,52}^{66} = 2A_{3,53}^{56} = 2A_{3,53}^{65} = \frac{1}{2},\\ &2A_{3,61}^{64} = 2A_{3,61}^{46} = 2A_{3,62}^{65} = 2A_{3,56}^{45}\\ &= A_{3,63}^{44} = A_{3,63}^{55} = 2A_{3,63}^{66} = \frac{1}{2}.\end{aligned} \quad (41)$$

Given the definitions of $\mathbf{A}_1$, $\mathbf{B}$, and $\mathbf{C}$ in (38), the positive-definite solution of the algebraic Riccati equation in (27) is easily determined. The nonzero elements are given by

$$\begin{aligned} P_1^{(i,i)} &= I_i\sqrt{1+\left(\frac{I_i}{2}\right)}, \\ P_1^{(i,3+i)} &= P_1^{(3+i,i)} = I_i, \\ P_1^{(3+i,3+i)} &= 4\sqrt{1+\left(\frac{I_i}{2}\right)}, \\ i &= 1,2,3. \end{aligned} \tag{42}$$

The negative-definite solution (nonzero elements) is given by

$$\begin{aligned} Q_1^{(i,i)} &= -I_i\sqrt{1+\left(\frac{I_i}{2}\right)}, \\ Q_1^{(i,3+i)} &= Q_1^{(3+i,i)} = I_i, \\ Q_1^{(3+i,3+i)} &= -4\sqrt{1+\left(\frac{I_i}{2}\right)}, \\ i &= 1,2,3. \end{aligned} \tag{43}$$

The corresponding closed-loop matrices $\mathbf{A}_{c1} = \mathbf{A}_1 - \mathbf{B}\mathbf{B}^T\mathbf{P}_1$ and $\mathbf{A}_{u1} = \mathbf{A}_1 - \mathbf{B}\mathbf{B}^T\mathbf{Q}_1$ are readily determined:

$$\begin{aligned} A_{c1}^{(i,i)} &= -I_i^{-1}\sqrt{1+\left(\frac{I_i}{2}\right)}, \\ A_{c1}^{(i,3+i)} &= I_i^{-1}, \\ A_{c1}^{(3+i,i)} &= \frac{1}{4}, \\ A_{c1}^{(3+i,3+i)} &= 0, \\ i &= 1,2,3, \\ A_{u1}^{(i,i)} &= I_i^{-1}\sqrt{1+\left(\frac{I_i}{2}\right)}, \\ A_{u1}^{(i,3+i)} &= I_i^{-1}, \\ A_{u1}^{(3+i,i)} &= \frac{1}{4}, \\ A_{u1}^{(3+i,3+i)} &= 0, \\ i &= 1,2,3. \end{aligned} \tag{44}$$

Using the above quantities, the entries in $\mathbf{P}_{2k}$ (via $\mathbf{p}_{2ij}$), $\mathbf{P}_{3mn}$, and $\mathbf{P}_{4knm}$ can be calculated using (28), (29), and (30). The same equations can be used to determine $\mathbf{Q}_{2k}$ (via $\mathbf{q}_{2ij}$), $\mathbf{Q}_{3mn}$, and $\mathbf{Q}_{4knm}$ with $\mathbf{P}_1$ replaced with $\mathbf{Q}_1$ and $\mathbf{A}_{c1}$ replaced with $\mathbf{A}_{u1}$. The dynamic controller in (20) can be made specific to the attitude control problem:

$$\begin{aligned} \dot{\mathbf{x}}_c &= \mathbf{a}(\mathbf{x}_c) - \mathbf{B}\mathbf{B}^T\nabla V_+^T(\mathbf{x}_c) \\ &\quad + \mathbf{G}_2(\mathbf{x}_c)\left[\beta^{-1}\mathbf{C}\mathbf{x}_c + \beta\mathbf{y}_2\right], \\ \mathbf{u}_2(\mathbf{x}_c) &= \beta^{-2}\mathbf{B}^T\nabla V_+^T(\mathbf{x}_c), \end{aligned} \tag{45}$$

where $\mathbf{a}(\mathbf{x}_c)$ is determined using (22) in conjunction with (40) and (41) and $\nabla V_+^T(\mathbf{x}_c)$ is determined using (24) and the solutions in (27)–(30). The observer gain $\mathbf{G}_c(\mathbf{x}_c)$ which is defined by (19) can be determined as follows. Since, $r_{2e}(\mathbf{x}) = \beta^{-2}V_+(\mathbf{x}) + \gamma^2 V_-(\mathbf{x})$, we have used (24):

$$\begin{aligned} \nabla r_{2e}(\mathbf{x}) &= \mathbf{x}^T\mathbf{R}_{e2}(\mathbf{x}) \\ &= \mathbf{x}^T\left[\left(\beta^{-2}\mathbf{P}_1 + \gamma^2\mathbf{Q}_1\right)\right. \\ &\quad \left. + \operatorname{row}_k\left(\beta^{-2}\mathbf{P}_{2k}\mathbf{x} + \gamma^2\mathbf{Q}_{2k}\mathbf{x}\right) + \cdots\right]. \end{aligned} \tag{46}$$

Using this in (19) yields the following expression for the observer gain:

$$\begin{aligned} \mathbf{G}(\mathbf{x}_c) &= \gamma^2\beta^{-1}\left[\left(\beta^{-2}\mathbf{P}_1 + \gamma^2\mathbf{Q}_1\right)\right. \\ &\quad \left. + \operatorname{row}_k\left(\beta^{-2}\mathbf{P}_{2k}\mathbf{x}_c + \gamma^2\mathbf{Q}_{2k}\mathbf{x}_c\right) + \cdots\right]^{-1}\mathbf{C}^T. \end{aligned} \tag{47}$$

The condition in (17) needs to be satisfied in a region of the origin. Using the lowest order terms in the expansions for $\mathscr{A} = \mathbf{a}(\mathbf{x})$, ∇V_+, ∇V_-, and ∇r_{2e}, the Hessian matrix defined by (17) is given by

$$\begin{aligned} &-\left(\beta^{-2}\mathbf{P}_1 + \gamma^2\mathbf{Q}_1\right)\left(\mathbf{A}_1 + \gamma^{-2}\beta^{-2}\mathbf{B}\mathbf{B}^T\mathbf{P}_1\right) \\ &-\left(\mathbf{A}_1 + \gamma^{-2}\beta^{-2}\mathbf{B}\mathbf{B}^T\mathbf{P}_1\right)^T\left(\beta^{-2}\mathbf{P}_1 + \gamma^2\mathbf{Q}_1\right) \\ &+\gamma^{-2}\left(\beta^{-2}\mathbf{P}_1 + \gamma^2\mathbf{Q}_1\right)\mathbf{B}\mathbf{B}^T\left(\beta^{-2}\mathbf{P}_1 + \gamma^2\mathbf{Q}_1\right) \\ &+\beta^{-4}\mathbf{P}_1\mathbf{B}\mathbf{B}^T\mathbf{P}_1 - \gamma^2\beta^{-2}\mathbf{C}^T\mathbf{C} \end{aligned} \tag{48}$$

which must be negative definite. This condition limits the chosen value of γ. In the sequel, we shall refer to the controller in (45) as the $\mathscr{H}_\infty$ controller of order k, where k is the order of the approximation adopted for $\mathbf{a}(\mathbf{x})$, ∇V_+, ∇V_-, and ∇r_{2e} in determining $\mathbf{G}_2$.

4. Numerical Example and Comparisons

In this section, the nonlinear controller presented above is compared through numerical simulations with existing methods from the literature for spacecraft attitude regulation. The purpose of these comparisons is to examine the effects of different disturbances and uncertainties on the performance and robustness of the various controllers. In particular, we include gravity-gradient and geomagnetic torques, as well as unmodeled actuator dynamics and actuation time delays. In addition to these comparisons, we make use of the gap metric [28] to characterize the difference in the input-output (IO) map of the system induced by the unmodeled actuator dynamics.

The simulation parameters are as follows. The satellite is in a circular orbit with an inclination of $i = 87$ degrees and a longitude of the ascending node of $\Omega = 0$. The initial value of the argument of latitude is zero. The altitude is 700 km and we take $R_e = 6378.14$ km for the Earth's radius. These orbital parameters will be used to determine the gravity-gradient and geomagnetic disturbance torques acting on the spacecraft. The spacecraft position in the geocentric inertial frame, $\mathbf{r}_i(t)$, is determined using a simple Keplerian model. The gravity-gradient torque is then given by $\mathbf{u}_{0g} = 3(\mu/R^5)\mathbf{r}_b^{\times}\mathbf{I}\mathbf{r}_b$, where μ is the geocentric gravitational parameter, $R = \sqrt{\mathbf{r}_i^T\mathbf{r}_i}$, and $\mathbf{r}_b = \mathbf{C}_{bi}\mathbf{r}_i$, where the rotation matrix relating the body-fixed frame to the inertial frame may be expressed in terms of the MRPs as $\mathbf{C}_{bi} = (\mathbf{1} - \boldsymbol{\sigma}^{\times})^2(\mathbf{1} + \boldsymbol{\sigma}^{\times})^2$.

For the purposes of the geomagnetic disturbance torque, the satellite is assumed to generate a magnetic dipole of $\mathbf{m} = [0.1\ \ 0.1\ \ 0.1]^T$ A·m^2. The magnetic field model is the tilted dipole model, $\mathbf{B}_i(\mathbf{r}_i)$, presented in [35], where $\mathbf{B}_i$ are the geomagnetic field components expressed in the geocentric inertial frame. The geomagnetic disturbance torque is given by $\mathbf{u}_{0m} = \mathbf{m}^{\times}\mathbf{C}_{bi}\mathbf{B}_i$. The satellite inertia matrix is given by $\mathbf{I} = \text{diag}\{10.0, 6.3, 8.5\}$ kg·m^2. In all comparisons, we consider the regulation problem only, hence $\mathbf{y}_0 = \mathbf{0}$ and the input to the controller is $\mathbf{y}_2 = -\mathbf{y}_1 = -\mathbf{C}\mathbf{x} = -\mathbf{x}$. Hence, we assume perfect measurements of the state.

In the case of the disturbance rejection comparisons, the simulations start from the desired attitude and we compare the ability of the different controllers to maintain that attitude. For all other comparisons, the initial states are chosen from Schaub et al. [2] to be $\boldsymbol{\omega}(0) = [1.4\ \ 0.9\ \ 0.8]^T$ rad/s and $\boldsymbol{\sigma}(0) = [0.87\ \ 0\ \ 0]^T$. These initial conditions have the satellite oriented almost π rad from the desired attitude with large angular velocities moving it towards this upside-down attitude. All simulations will be performed for one orbit using a 4th-order Runge-Kutta numerical integration method with a step-size $\Delta t = 0.01$ s. The $\mathscr{H}_\infty$ controllers are designed with $\gamma = 4.0$, which was chosen to satisfy the linear version of the condition in (17); that is, such that the matrix in (48) is negative definite.

The methods to be used, in addition to the $\mathscr{H}_\infty$ controllers of the previous section, are the linear and nonlinear proportional-derivative (PD) laws of Tsiotras [1], the open-loop (OL) optimal control method by Schaub et al. [2], the closed-loop (CL) optimal nonlinear method of Tewari [3], and the sum of squares (SOS) approach of Gollu and Rodrigues [4]. In order to gauge the performance and properly compare the different controllers, we will use the following metrics:

$$E_{\text{rms}} = \left[\frac{1}{T}\int_0^T \mathbf{y}_2^T(t)\,\mathbf{y}_2(t)\,dt\right]^{(1/2)},$$
$$T_{\text{rms}} = \left[\frac{1}{T}\int_0^T \mathbf{u}_2^T(t)\,\mathbf{u}_2(t)\,dt\right]^{(1/2)}, \tag{49}$$

where $\mathbf{y}_2 = \mathbf{y}_0 - \mathbf{y}_1 = -\mathbf{y}_1 = -\mathbf{x} = -[\omega_1\ \ \omega_2\ \ \omega_3\ \ \sigma_1\ \ \sigma_2\ \ \sigma_3]^T$, T is the orbital period, E_{rms} is the RMS tracking error, and T_{rms} is the RMS control torque (note that $-\mathbf{u}_2$ are the control torques).

4.1. Existing Control Methods. We now provide a brief overview of the controller synthesis methods to be used for comparisons with our method. The interested reader is referred to the appropriate literature for a more detailed exposition of these methods.

4.1.1. Proportional-Derivative Controllers. Tsiotras [1] developed two proportional-derivative (PD) control laws for the attitude control problem. The linear PD controller is given by

$$\mathbf{u}_2(t) = \boldsymbol{\omega} + 2\boldsymbol{\sigma}, \tag{50}$$

while the nonlinear version is given by

$$\mathbf{u}_2(t) = \boldsymbol{\omega} + 2\left(1 + \boldsymbol{\sigma}^T\boldsymbol{\sigma}\right)\boldsymbol{\sigma}. \tag{51}$$

4.1.2. Open-Loop Optimal Control. The optimal control method presented by Schaub et al. [2] is designed to minimize the cost function

$$J = \underbrace{\frac{1}{2}K_1 g\left(\boldsymbol{\sigma}\left(t_f\right)\right) + \frac{1}{2}\boldsymbol{\omega}^T\left(t_f\right)\mathbf{K}_2\boldsymbol{\omega}\left(t_f\right)}_{\phi(t_f)} + \int_{t_0}^{t_f}\underbrace{\frac{1}{2}\left[K_3 g(\boldsymbol{\sigma}) + \boldsymbol{\omega}^T\mathbf{K}_4\boldsymbol{\omega} + \mathbf{u}_2^T\mathbf{R}\mathbf{u}_2\right]}_{p(\mathbf{x},\mathbf{u}_2,t)}dt, \tag{52}$$

where K_1 and K_3 are scalar weights, $\mathbf{K}_2$, $\mathbf{K}_4$, and $\mathbf{R}$ are weighting matrices, and

$$g(\boldsymbol{\sigma}) = 4\frac{\boldsymbol{\sigma}^T\boldsymbol{\sigma}}{\left(1 + \boldsymbol{\sigma}^T\boldsymbol{\sigma}\right)^2}. \tag{53}$$

The Hamiltonian relating to this optimal control problem is defined as

$$H = p(\mathbf{x}, \mathbf{u}, t) + \boldsymbol{\Lambda}^T\left[\mathbf{a}(\mathbf{x}) - \mathbf{B}\mathbf{u}_2(t)\right], \tag{54}$$

where we have used the plant model in (1) and the second relation of (3) with $\mathbf{y}_1 = \mathbf{0}$. The costates, denoted by $\boldsymbol{\Lambda}$, have dynamics

$$\dot{\boldsymbol{\Lambda}} = -\frac{\partial H}{\partial \mathbf{x}}, \qquad \boldsymbol{\Lambda}\left(t_f\right) = \left.\frac{\partial \phi}{\partial \mathbf{x}}\right|_{t=t_f}. \tag{55}$$

Note that the costates are specified at some final time t_f, not the initial time. The optimal control law for this problem is determined from

$$\frac{\partial H}{\partial \mathbf{u}_2} = \mathbf{0}, \tag{56}$$

which yields

$$\mathbf{u}_2(t) = \mathbf{R}^{-1}\mathbf{B}^T\boldsymbol{\Lambda}. \tag{57}$$

The primary disadvantage of this method from a practical point of view is that it requires the solution of a two-point boundary value problem (TPBVP) and results in an open-loop control strategy. For the simulation results presented below, the following weighting parameters are used: $K_1 = 5.0$, $\mathbf{K}_2 = 5.01$, $K_3 = 1.0$, $\mathbf{K}_4 = \mathbf{1}$, and $\mathbf{R} = \mathbf{1}$. Moreover, the maneuver is optimized for a final time $t_f = 60$ s.

4.1.3. Closed-Loop Optimal Controller. The optimal control method presented by Tewari [3] is based on obtaining an exact analytical solution to the Hamilton-Jacobi equation. Consider the HJE in (12) with parameters

$$\begin{aligned}\mathcal{A}(\mathbf{x}) &= \mathbf{a}(\mathbf{x}),\\ \mathcal{R}(\mathbf{x}) &= \mathbf{B}\mathbf{R}^{-1}\mathbf{B}^T,\\ \mathcal{Q}(\mathbf{x}) &= \begin{bmatrix}\mathbf{x}\\ \mathbf{x}^{[2]}\end{bmatrix}^T \underbrace{\begin{bmatrix}\mathbf{Q}_{11} & \mathbf{Q}_{12}\\ \mathbf{Q}_{21} & \mathbf{Q}_{22}\end{bmatrix}}_{\mathbf{Q}} \begin{bmatrix}\mathbf{x}\\ \mathbf{x}^{[2]}\end{bmatrix},\end{aligned} \tag{58}$$

where the inertia matrix is defined as $\mathbf{I} = \mathrm{diag}_i\{I_i\}$, $\mathbf{R} = \mathrm{diag}_i\{R_i\}$ is symmetric and positive definite, $\mathbf{Q}$ is symmetric and positive semidefinite, and $\mathbf{x}^{[2]} = [\omega_1^2\ \omega_2^2\ \omega_3^2\ \sigma_1^2\ \sigma_2^2\ \sigma_3^2]^T$. It is assumed that $V(\mathbf{x})$ has the same form as $\mathcal{Q}(\mathbf{x})$; thus,

$$V(\mathbf{x}) = \frac{1}{2}\begin{bmatrix}\mathbf{x}\\ \mathbf{x}^{[2]}\end{bmatrix}^T \underbrace{\begin{bmatrix}\mathbf{P}_{11} & \mathbf{P}_{12}\\ \mathbf{P}_{21} & \mathbf{P}_{22}\end{bmatrix}}_{\mathbf{P}} \begin{bmatrix}\mathbf{x}\\ \mathbf{x}^{[2]}\end{bmatrix}, \tag{59}$$

where $\mathbf{P}$ is symmetric and positive definite. The state feedback controller is given by

$$\mathbf{u}_2(t) = \mathbf{R}^{-1}\mathbf{B}^T\nabla V^T(\mathbf{x}), \tag{60}$$

where

$$\begin{aligned}\nabla V^T(\mathbf{x}) &= \mathbf{P}_{11}\mathbf{x} + 2\,\mathrm{diag}_i\{x_i\}\,\mathbf{P}_{12}\mathbf{x} + \mathbf{P}_{12}\mathbf{x}^{[2]}\\ &\quad + 2\,\mathrm{diag}_i\{x_i\}\,\mathbf{P}_{22}\mathbf{x}^{[2]}.\end{aligned} \tag{61}$$

The matrix $\mathbf{P}$ is obtained as follows. First, $\mathbf{P}_{11}$ is calculated from the algebraic Riccati equation in (27) corresponding to the parameters in (58), that is, with $\mathbf{R}$ in (27) replaced by $\mathbf{B}\mathbf{R}^{-1}\mathbf{B}^T$ and $\mathbf{Q}$ in (27) replaced by $\mathbf{Q}_{11}$. Then, the equations in (26) with $k = 3, \ldots, 6$, are solved simultaneously for the remaining unknowns. In particular, the nonzero elements of the matrices $\mathbf{P}_{12}$ and $\mathbf{P}_{22}$ are

$$\begin{aligned}
P_{12}^{11} &= \frac{I_1^2R_1\left(2Q_{12}^{11} + k_2P_{11}^{12} + k_3P_{11}^{13}\right)}{\left(6P_{11}^{11}\right)},\\
P_{12}^{16} &= \frac{I_1^2R_1Q_{12}^{16}}{\left(2P_{11}^{11}\right)},\\
P_{12}^{22} &= \frac{I_2^2R_2\left(Q_{12}^{12} - 2k_2P_{11}^{22}\right)}{\left(3P_{11}^{22}\right)},\\
P_{12}^{25} &= \frac{I_2^2R_2\left(2Q_{12}^{25} - 3P_{12}^{22}P_{11}^{25}\right)}{\left(2P_{11}^{22}\right)},\\
P_{12}^{33} &= \frac{I_3^2R_3Q_{12}^{33}}{\left(3P_{11}^{33}\right)},\\
P_{22}^{11} &= \frac{\left[I_1^2R_1Q_{22}^{11} - 9\left(P_{12}^{11}\right)^2\right]}{\left(2P_{11}^{11}\right)},\\
P_{22}^{12} &= \frac{Q_{22}^{12}}{\left[P_{11}^{11}/\left(I_1^2R_1\right) + P_{11}^{22}/\left(I_2^2R_2\right)\right]},\\
P_{22}^{13} &= \frac{Q_{22}^{13}}{\left[P_{11}^{11}/\left(I_1^2R_1\right) + P_{11}^{33}/\left(I_3^2R_3\right)\right]},\\
P_{22}^{22} &= \frac{\left[I_2^2R_2Q_{22}^{22} - 9\left(P_{12}^{22}\right)^2\right]}{\left(2P_{11}^{22}\right)},\\
P_{22}^{23} &= \frac{Q_{22}^{23}}{\left[P_{11}^{33}/\left(I_3^2R_3\right) + P_{11}^{22}/\left(I_2^2R_2\right)\right]},\\
P_{22}^{33} &= \frac{\left[I_3^2R_3Q_{22}^{33} - 9\left(P_{12}^{33}\right)^2\right]}{\left(2P_{11}^{33}\right)},
\end{aligned} \tag{62}$$

where, as before, $k_1 = (I_2 - I_3)/I_1$, $k_2 = (I_3 - I_1)/I_2$ and $k_3 = (I_1 - I_2)/I_3$.

For the simulation results presented below, the following weighting parameters are used [3]:

$$\begin{aligned}
\mathbf{R} &= \mathrm{diag}_i = \left\{I_i^{-2}\right\}, \qquad \mathbf{Q}_{11} = (0.01)\,\mathbf{1},\\
\mathbf{Q}_{12} &= \left[\begin{array}{c|c}(0.1)\,\mathbf{1} & \mathbf{0}\\ \hline \mathbf{0} & \mathbf{0}\end{array}\right],\\
\mathbf{Q}_{22} &= \left[\begin{array}{ccc|c} P_{11}^{11} & P_{11}^{12} & P_{11}^{13} & \\ P_{11}^{21} & P_{11}^{22} & P_{11}^{23} & \mathbf{0}\\ P_{11}^{31} & P_{11}^{32} & P_{11}^{33} & \\ \hline & \mathbf{0} & & \mathbf{0}\end{array}\right].
\end{aligned} \tag{63}$$

We now make a few remarks regarding the characteristics of the synthesis method of Tewari [3]. Like the controller presented in this paper, Tewari's closed-loop optimal method results in a polynomial feedback controller. However, his solution is derived specifically for the attitude control problem. In contrast, the method developed in the present paper is applicable to a wider class of systems.

4.1.4. Sum of Squares Controller. A multivariate polynomial $p(\mathbf{x})$ is a sum of squares (SOS) if there exist some polynomials $f_i(\mathbf{x})$, $i = 1, \ldots, M$, such that

$$p(\mathbf{x}) = \sum_{i=1}^{M} f_i^2(\mathbf{x}). \tag{64}$$

The SOS controller synthesis approach relaxes the search for positive- or negative-definite functions to a search for SOS functions. It should be noted, however, that the use of sums of squares is conservative, since $p(\mathbf{x})$ being SOS implies that $p(\mathbf{x}) \geq 0$, while the converse is not true in general. Also, it can be shown that if $\mathbf{v}^T\mathbf{P}(\mathbf{x})\mathbf{v}$ is SOS for $\mathbf{v} \in \mathbb{R}^n$, then $\mathbf{P}(\mathbf{x}) \geq 0$ for all $\mathbf{x} \in \mathbb{R}^N$ [36].

In applying the SOS controller synthesis method, we first rewrite the system in (1) as

$$\dot{\mathbf{x}} = \mathbf{A}(\mathbf{x})\mathbf{x} - \mathbf{B}\mathbf{u}_2(t), \tag{65}$$

where we have used the second relation of (3) with $\mathbf{y}_1 = \mathbf{0}$. Note that the matrix $\mathbf{A}(\mathbf{x})$ is not unique. The SOS state feedback controller for this problem is given by [4]

$$\mathbf{u}_2(t) = -\mathbf{K}(\mathbf{x})\mathbf{P}^{-1}\mathbf{x}, \tag{66}$$

where we note that $\mathbf{P}$ is now taken to be constant (state independent).

Consider the Lyapunov function

$$V(\mathbf{x}) = \mathbf{x}^T\mathbf{P}^{-1}\mathbf{x}. \tag{67}$$

Taking the time derivative of this function along the trajectories of the system in (65) with the controller of (66) yields

$$\begin{aligned}\dot{V}(\mathbf{x}) = {} & \left[\mathbf{A}(\mathbf{x})\mathbf{x} + \mathbf{B}\mathbf{K}(\mathbf{x})\mathbf{P}^{-1}\mathbf{x}\right]^T\mathbf{P}^{-1}\mathbf{x} \\ & + \mathbf{x}^T\mathbf{P}^{-1}\left[\mathbf{A}(\mathbf{x})\mathbf{x} + \mathbf{B}\mathbf{K}(\mathbf{x})\mathbf{P}^{-1}\mathbf{x}\right].\end{aligned} \tag{68}$$

Using the change of variables $\mathbf{x} = \mathbf{P}\mathbf{v}$, this last expression can be written as

$$\dot{V}(\mathbf{x}) = \mathbf{v}^T\left[\mathbf{P}\mathbf{A}^T(\mathbf{x}) + \mathbf{A}(\mathbf{x})\mathbf{P} + \mathbf{K}^T(\mathbf{x})\mathbf{B}^T + \mathbf{B}\mathbf{K}(\mathbf{x})\right]\mathbf{v}. \tag{69}$$

The conditions $V(\mathbf{x}) > 0$ and $\dot{V}(\mathbf{x}) < 0$ can be replaced by the conditions that $\mathbf{P}$ is positive definite and $-\dot{V}(\mathbf{x})$ is SOS. The second of these conditions will be strengthened to $-[\dot{V}(\mathbf{x}) + \epsilon(\mathbf{x})]$ being SOS, where $\epsilon(\mathbf{x})$ is some SOS function. This semidefinite programming (SDP) problem can then be written as follows:

$$\begin{aligned}& \text{find } \mathbf{P}, \mathbf{K}(\mathbf{x}), \epsilon(\mathbf{x}) \\ & \text{s.t. } \mathbf{P} = \mathbf{P}^T > 0 \\ & \qquad \epsilon(\mathbf{x}) \text{ is SOS} \\ & \qquad -\mathbf{v}^T\left[\mathbf{P}\mathbf{A}^T(\mathbf{x}) + \mathbf{A}(\mathbf{x})\mathbf{P} + \mathbf{K}^T(\mathbf{x})\mathbf{B}^T \right. \\ & \qquad \left. + \mathbf{B}\mathbf{K}(\mathbf{x}) + \epsilon(\mathbf{x})\mathbf{1}\right]\mathbf{v} \text{ is SOS.}\end{aligned} \tag{70}$$

This optimization problem can be solved using the SOSTOOLS software package [37].

4.2. Disturbance Rejection. We begin by comparing the controllers presented in this paper with the methods from the literature with regard to disturbance rejection. The two disturbances considered here are the gravity-gradient and geomagnetic torques. For the purposes of this comparison, the simulations start from the desired attitude (i.e., $\mathbf{y}_0 = \mathbf{0}$) and we compare the ability of the different controllers to maintain that attitude over one complete orbit. Table 1 presents values of the performance metrics for the different controllers.

The optimal control method of Schaub et al. [2] is not included in the table because it is unable to reject any disturbances, which is entirely due to its open-loop nature. There is no apparent difference in performance between the linear and nonlinear PD controllers. There is also very little difference between any of the nonlinear $\mathscr{H}_\infty$ controllers developed using the present method. This is due to the very small magnitude of these disturbance torques, which the linear term in the controller can effectively overcome. With regard to disturbance rejection, the present $\mathscr{H}_\infty$ controllers do not perform as well as the existing methods.

4.3. Response to Initial Conditions. In this and the following subsections, the disturbance torques are set to zero and the nonzero initial conditions noted at the beginning of this section are applied. All of the resulting controllers described previously are stable.

The resulting attitude and control torques when the fourth-order $\mathscr{H}_\infty$ controller is applied are given in Figures 3 and 4. The MRP switching can clearly be seen in Figure 3, where we also note that it requires several rotations for the controller to sufficiently slow down the spacecraft. This is due to the small torques applied, as seen in Figure 4. It is important to note that the control torques are continuous, which is a result of the dynamic aspect of the controller developed in this paper. Once again, there is little to distinguish the behaviour of the $\mathscr{H}_\infty$ controllers of varying order of approximation. This will be discussed further after the robustness of these controllers has been assessed.

4.4. Robustness to Actuation Time Delay. The robustness properties of the different controllers are now examined with regard to a time delay in the actuation. Such a delay could represent the finite time required by a satellite on-board computer to take the sensor measurements and calculate the required control signal. The time delay is made equal to an integer number of the numerical integration step-size Δt. Table 2 indicates the maximum allowable time delay, h_{max}, for each controller such that the desired attitude maneuver is achieved within one orbit. As can be seen from these results, the four $\mathscr{H}_\infty$ controllers are more robust with regard to this effect than the other control methods. In particular, the present fourth-order $\mathscr{H}_\infty$ controller is the most robust. We also note that the open-loop method of Schaub et al. [2] has no robustness to this effect; the time delay results in a

TABLE 1: Controller disturbance rejection.

	$\mathscr{H}_\infty$ (1st-order)	$\mathscr{H}_\infty$ (2nd-, 3rd-, and 4th-order)	PD (both)	Optimal (CL)	SOS
E_{rms} ($\times 10^{-6}$)	10.809686	10.809683	1.989095	5.328614	1.319846
T_{rms} (N·m) ($\times 10^{-6}$)	3.979559	3.979559	3.977874	3.977763	3.976904

TABLE 2: Robustness to actuation time delay.

	$\mathscr{H}_\infty$ (1st-order)	$\mathscr{H}_\infty$ (2nd-order)	$\mathscr{H}_\infty$ (3rd-order)	$\mathscr{H}_\infty$ (4th-order)	
$h_{\max}$ (s)	3.93	3.94	3.97	3.98	
	PD (linear)	PD (nonlinear)	Optimal (OL)	Optimal (CL)	SOS
$h_{\max}$ (s)	1.69	1.09	0.00	1.41	0.81

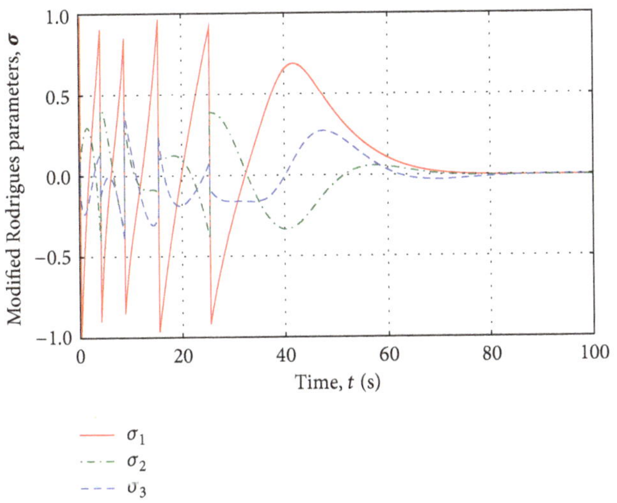

FIGURE 3: MRP trajectories for $\mathbf{I} = \text{diag}\{10, 6.3, 8.5\}$ kg·m^2 with 4th-order $\mathscr{H}_\infty$ controller.

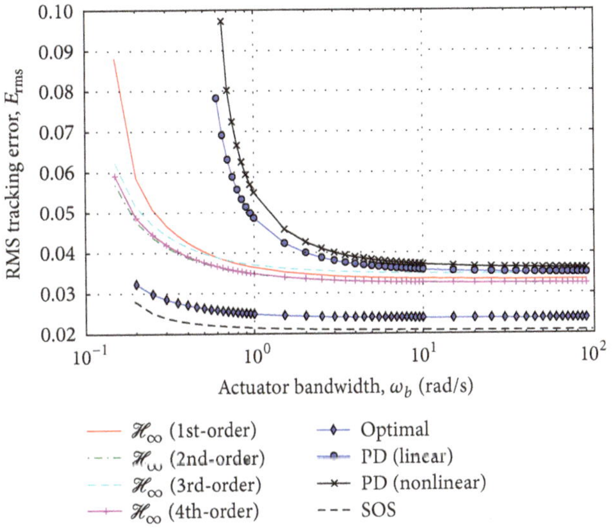

FIGURE 5: RMS tracking error with respect to 1st-order actuator bandwidth.

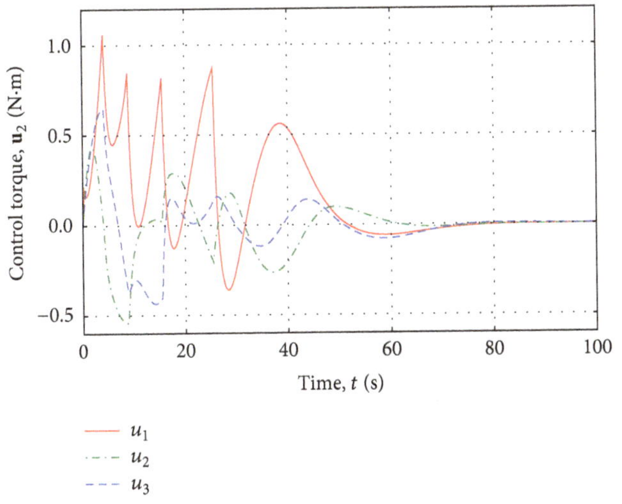

FIGURE 4: Control torques for $\mathbf{I} = \text{diag}\{10, 6.3, 8.5\}$ kg·m^2 with 4th-order $\mathscr{H}_\infty$ controller.

nonzero final angular velocity such that the system will rotate endlessly.

4.5. Robustness to Unmodeled Actuator Dynamics. The robustness properties of the different controllers are now examined with regard to unmodeled actuator dynamics. For the purposes of this study, we make use of first- and second-order actuator models. In practice the actuator dynamics may be far more complex than the ones used here. However, we use these simple models here for the purposes of studying the capabilities of the different control methods. Evidently, as the actuator bandwidth decreases, it becomes harder for the controller to stabilize the system. Thus, we are able to infer the relative robustness properties of the different controllers by examining Figures 5–8. In particular, the farther a line reaches towards the left-hand side of the graph, the more robust that controller is with regard to the unmodeled actuator dynamics. As we shall see, the present $\mathscr{H}_\infty$ controllers are always more robust than the other methods.

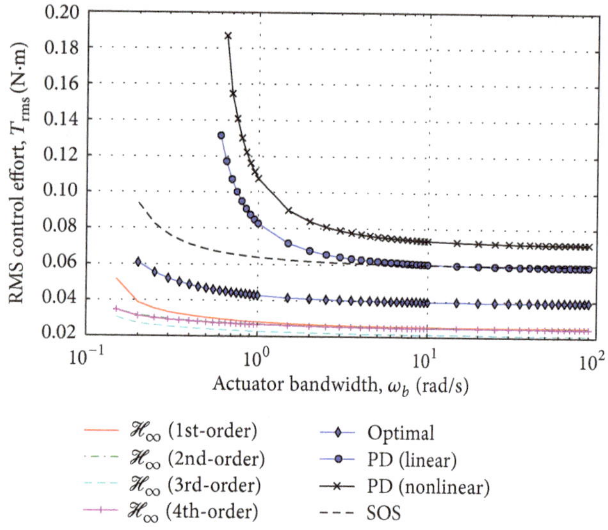

FIGURE 6: RMS control effort with respect to 1st-order actuator bandwidth.

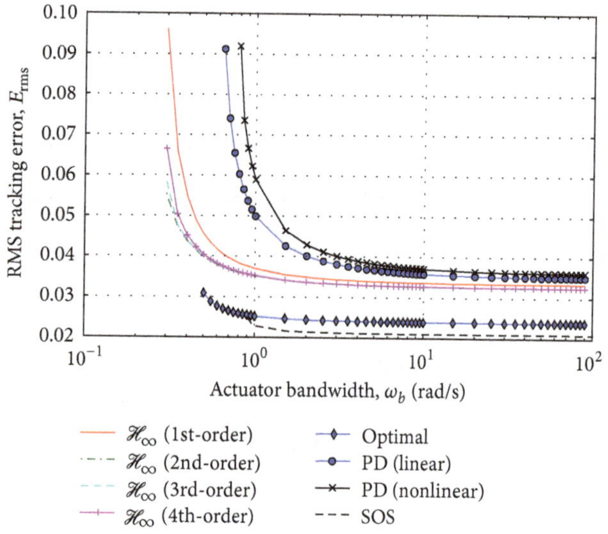

FIGURE 7: RMS tracking error with respect to 2nd-order actuator bandwidth.

4.5.1. Unmodeled 1st-Order Actuator Dynamics. We begin by examining the robustness of the controllers with regard to unmodeled first-order actuator dynamics. This shall be accomplished by including the following first-order dynamics model in each component ($i = 1, 2, 3$) of the controller output:

$$\dot{x} = -\omega_b x + \omega_b u_{2,i},$$
$$y_i = x, \tag{71}$$

where ω_b is the actuator bandwidth. Figures 5 and 6 show E_{rms} and T_{rms}, respectively, as a function of the actuator bandwidth. It is noted from these figures that all four $\mathscr{H}_\infty$ controllers provide nearly the same tracking error and control effort. Moreover, while the closed-loop optimal control law

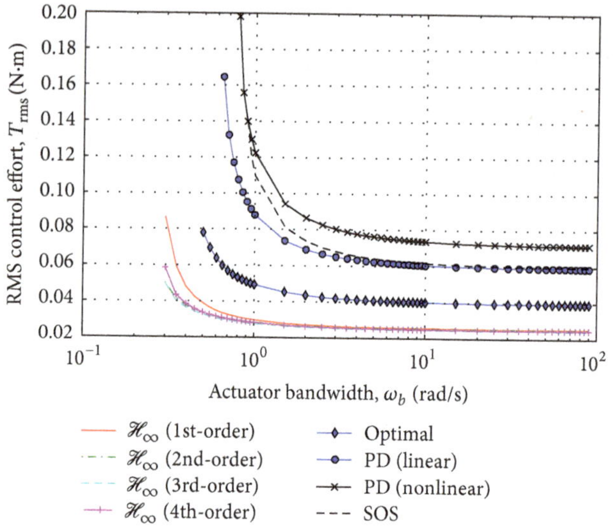

FIGURE 8: RMS control effort with respect to 2nd-order actuator bandwidth.

provides better tracking error compared with the $\mathscr{H}_\infty$ controllers, the trade-off is that it requires more control effort. Similarly, the SOS controller yields very low tracking error at the expense of greater control effort. The two PD laws, on the other hand, have a higher tracking error and control effort than the other methods.

4.5.2. Unmodeled 2nd-Order Actuator Dynamics. The robustness properties of the different controllers are now examined with regard to unmodeled second-order actuator dynamics. This shall be accomplished by including the following second-order dynamics model in each component ($i = 1, 2, 3$) of the controller output:

$$\begin{bmatrix} \dot{x}_1 \\ \dot{x}_2 \end{bmatrix} = \begin{bmatrix} 0 & 1 \\ -\omega_b^2 & -2\zeta\omega_b \end{bmatrix} \begin{bmatrix} x_1 \\ x_2 \end{bmatrix} + \begin{bmatrix} 0 \\ \omega_b^2 \end{bmatrix} u_{2,i},$$
$$y_i = x_1, \tag{72}$$

where ζ is the actuator damping ratio and ω_b is the actuator bandwidth. All simulations were performed with a damping ratio $\zeta = 0.5$. Figures 7 and 8 show E_{rms} and T_{rms}, respectively, as a function of the actuator bandwidth. It is seen from these figures that the various control methods follow the same trends as in the case of the first-order actuator dynamics.

4.6. Robustness in the Gap Metric (Revisited). We now make use of the gap metric [28] to characterize the difference in the input-output (IO) map of the system induced by the unmodeled actuator dynamics. However, since we cannot calculate the gap between two nonlinear systems, we calculate the gap metric distance for the linearized system only. As the actuator bandwidth and damping ratio change, the value of the gap metric will also vary. Figure 9 shows the gap metric value with respect to the first-order actuator bandwidth. Figure 10 shows the gap metric value with respect to the second-order actuator bandwidth for a damping ratio $\zeta = 0.5$

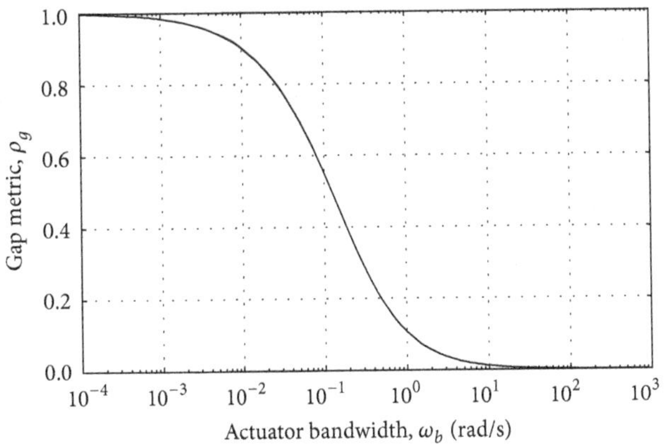

FIGURE 9: Gap metric with respect to 1st-order actuator bandwidth.

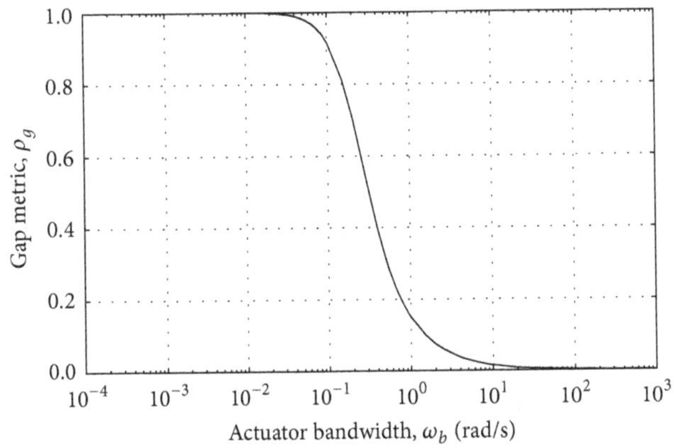

FIGURE 10: Gap metric with respect to 2nd-order actuator bandwidth.

as used in the above numerical simulations. As the damping ratio varies, the curve of this figure moves to the left or right slightly, although there does not appear to be any discernible trend. Moreover, it should be noted that as the actuator bandwidth approaches infinity, the effects of the actuator in the controller input-output map become negligible. This is to be expected and is seen in both Figures 9 and 10, where the gap metric approaches zero with increasing actuator bandwidth.

We now return for a moment to the question of controller robustness in the gap metric. From (9) we have the following small-gain-type stability criterion:

$$\rho_g(\mathcal{P}_0, \mathcal{P}_1) \cdot \underbrace{\left\|\Pi_{\mathcal{N}\|\mathcal{M}_0}\right\|_\infty}_{\le\gamma} < 1. \tag{73}$$

In the numerical simulations performed with $\gamma = 4$, it was determined that the closed-loop system is stable in the presence of the unmodeled second-order actuator with bandwidth as low as $\omega_b = 0.3$ rad/s. With this actuator bandwidth, the calculation of the gap between the linear attitude dynamics with and without the actuator results in a value of $\rho_g = 0.522$, which does not satisfy the condition of (73). This could be explained by the conservativeness of the small-gain criterion and the fact that it is a sufficient (but not necessary) condition for stability. However, it can also be attributed to the nonlinear effects of the dynamics not taken into account in the gap calculation here. This emphasizes the need for a method to calculate the gap metric for nonlinear systems.

4.7. Overall Assessment of the $\mathcal{H}_\infty$ Controllers. The previous sections show that the methodology presented here can be used to develop very robust attitude controllers. However, their performance was not as good as some of the other techniques to which it was compared. We have not demonstrated significant benefits for the higher-order terms in the controller which, while disappointing, is an important contribution to the literature. The fact that a linear controller can perform very closely to the higher-order ones is a strong vindication of the most popular approach used for actual attitude control: linear feedback of angular velocity and attitude information. This is consistent with [38] which showed that a linear combination of angular velocity and quaternion (Euler parameter) feedback solves the state feedback nonlinear (suboptimal)-$\mathcal{H}_\infty$ control problem for rigid spacecraft attitude control. Our results are entirely consistent with this and we strongly suspect that results analogous to [38] can be obtained for the case of angular velocity and MRPs.

5. Conclusions

The results presented here clearly show the trade-off between performance and robustness. Existing methods from the attitude control literature typically focus on performance. In contrast, the method developed in this paper emphasizes robustness. In particular, while the methods from the literature are better at disturbance rejection, the new nonlinear controllers have overall better robustness properties. In general, there is still a need to characterize the trade-off between these two properties. In addition to the results presented in this paper, there is still much room for improvement and many more areas to be explored. Some topics that could be explored in future work include analytical solutions to higher-order terms in the approximation and the effects of varying γ on the closed-loop response. The need for methods to calculate the gap metric between nonlinear systems was also motivated.

Conflict of Interests

The authors declare that there is no conflict of interests regarding the publication of this paper.

Acknowledgment

This work was supported by the Natural Sciences and Engineering Research Council of Canada [Application no. 121947-2010].

References

[1] P. Tsiotras, "Stabilization and optimality results for the attitude control problem," *Journal of Guidance, Control, and Dynamics*, vol. 19, no. 4, pp. 772–779, 1996.

[2] H. Schaub, J. L. Junkins, and R. D. Robinett, "New penalty functions and optimal control formulation for spacecraft attitude control problems," *Journal of Guidance, Control, and Dynamics*, vol. 20, no. 3, pp. 428–434, 1997.

[3] A. Tewari, "Optimal nonlinear spacecraft attitude control through Hamilton-Jacobi formulation," *Journal of the Astronautical Sciences*, vol. 50, no. 1, pp. 99–112, 2002.

[4] N. Gollu and L. Rodrigues, "Control of large angle attitude maneuvers for rigid bodies using sum of squares," in *Proceedings of the American Control Conference (ACC '07)*, pp. 3156–3161, New York, NY, USA, July 2007.

[5] A. P. Sage and C. C. White III, *Optimum System Control*, Prentice-Hall, Englewood Cliffs, NJ, USA, 2nd edition, 1977.

[6] A. J. van der Schaft, "On a state space approach to nonlinear H_∞ control," *Systems and Control Letters*, vol. 16, no. 1, pp. 1–8, 1991.

[7] A. J. van der Schaft, "L_2-gain analysis of nonlinear systems and nonlinear state-feedback H_∞ control," *IEEE Transactions on Automatic Control*, vol. 37, no. 6, pp. 770–784, 1992.

[8] J. A. Ball, J. W. Helton, and M. L. Walker, "H_∞ control for nonlinear systems with output feedback," *IEEE Transactions on Automatic Control*, vol. 38, no. 4, pp. 546–559, 1993.

[9] M. R. James, M. C. Smith, and G. Vinnicombe, "Gap metrics, representations, and nonlinear robust stability," *SIAM Journal on Control and Optimization*, vol. 43, no. 5, pp. 1535–1582, 2005.

[10] J. C. Willems, "Dissipative dynamical systems part I: general theory," *Archive for Rational Mechanics and Analysis*, vol. 45, no. 5, pp. 321–351, 1972.

[11] J. C. Willems, "Dissipative dynamical systems. Part II: linear systems with quadratic supply rates," *Archive for Rational Mechanics and Analysis*, vol. 45, no. 5, pp. 352–393, 1972.

[12] D. Hill and P. Moylan, "The Stability of nonlinear dissipative systems," *IEEE Transactions on Automatic Control*, vol. AC-21, no. 5, pp. 708–711, 1976.

[13] E. G. AlBrekht, "On the optimal stabilization of nonlinear systems," *Prikladnaya Matematika*, vol. 25, no. 5, pp. 836–844, 1961.

[14] D. L. Lukes, "Optimal regulation of nonlinear dynamical systems," *SIAM Journal on Control*, vol. 7, no. 1, pp. 75–100, 1969.

[15] W. L. Garrard, "Suboptimal feedback control for nonlinear systems," *Automatica*, vol. 8, no. 2, pp. 219–221, 1972.

[16] W. L. Garrard and J. M. Jordan, "Design of nonlinear automatic flight control systems," *Automatica*, vol. 13, no. 5, pp. 497–505, 1977.

[17] J. Huang and C.-F. Lin, "Numerical approach to computing nonlinear H_∞ control laws," *Journal of Guidance, Control, and Dynamics*, vol. 18, no. 5, pp. 989–994, 1995.

[18] S. R. Vadali and R. Sharma, "Optimal finite-time feedback controllers for nonlinear systems with terminal constraints," *Journal of Guidance, Control, and Dynamics*, vol. 29, no. 4, pp. 921–928, 2006.

[19] R. Sharma, S. R. Vadali, and J. E. Hurtado, "Optimal nonlinear feedback control design using a waypoint method," *Journal of Guidance, Control, and Dynamics*, vol. 34, no. 3, pp. 698–705, 2011.

[20] A. J. van der Schaft, "Relations between (H_∞) optimal control of a nonlinear system and its linearization," in *Proceedings of the 30th IEEE Conference on Decision and Control*, pp. 1807–1808, Brighton, UK, December 1991.

[21] S. G. Margolis and W. G. Vogt, "Control engineering applications of V. I. Zubovs construction procedure for Lyapunov functions," *IEEE Transactions on Automatic Control*, vol. 8, no. 2, pp. 104–113, 1963.

[22] J. R. Hewit and C. Storey, "Optimization of the Zubov and Ingwerson methods for constructing Lyapunov functions," *IEE Electronics Letters*, vol. 3, no. 5, pp. 211–213, 1967.

[23] J. R. Cloutier, "State-dependent Riccati equation techniques: an overview," in *Proceedings of the American Control Conference*, pp. 932–936, Albuquerque, NM, USA, June 1997.

[24] J. R. Cloutier and D. T. Stansbery, "The capabilities and art of State-dependent Riccati equation-based design," in *Proceedings of the American Control Conference*, pp. 86–91, Anchorage, Alaska, USA, May 2002.

[25] R. Beard, G. Saridis, and J. Wen, "Improving the performance of stabilizing controls for nonlinear systems," *IEEE Control Systems Magazine*, vol. 16, no. 5, pp. 27–35, 1996.

[26] R. W. Beard, T. W. McLain, and J. T. Wen, "Successive Galerkin Approximation of the Isaacs Equation," in *Proceedings of the IFAC World Congress*, Beijing, China, 1999.

[27] N. Sakamoto and A. J. van der Schaft, "Analytical approximation methods for the stabilizing solution of the Hamilton-Jacobi equation," *IEEE Transactions on Automatic Control*, vol. 53, no. 10, pp. 2335–2350, 2008.

[28] A. K. El-Sakkary, "The gap metric: robustness of stabilization of feedback systems," *IEEE Transactions on Automatic Control*, vol. AC-30, no. 3, pp. 240–247, 1985.

[29] S. LeBel and C. Damaren, "Analytical solutions to approximations of the Hamilton-Jacobi equation applied to satellite formation flying," in *Proceedings of the AIAA Guidance, Navigation, and Control Conference and Exhibit*, pp. 1–11, Chicago, Ill, USA, August 2009.

[30] S. LeBel and C. J. Damaren, "A Nonlinear robust control method for the spacecraft attitude problem," in *Proceedings of the 15th CASI Conference on Astronautics (ASTRO '10)*, pp. 1–6, Toronto, Canada, May 2010.

[31] S. LeBel and C. Damaren, "Analytical solutions to approximations of the Hamilton-Jacobi equation applied to satellite formation flying," in *Proceedings of the AIAA Guidance, Navigation, and Control Conference and Exhibit*, Toronto, Canada, August 2009.

[32] T. T. Georgiou and M. C. Smith, "Robustness analysis of nonlinear feedback systems: an input-output approach," *IEEE Transactions on Automatic Control*, vol. 42, no. 9, pp. 1200–1221, 1997.

[33] M. Green and D. J. N. Limebeer, *Linear Robust Control*, Prentice-Hall, Englewood Cliffs, NJ, USA, 1995.

[34] H. Schaub and J. L. Junkins, *Analytical Mechanics of Space Systems*, 2003.

[35] J. R. Wertz, *Spacecraft Attitude Determination and Control*, Reidel, 1978.

[36] S. Prajna, A. Papachristodoulou, and F. Wu, "Nonlinear control synthesis by sum of squares optimization: a Lyapunov-based approach," in *Proceedings of the 5th Asian Control Conference*, pp. 1–9, Melbourne, Australia, July 2004.

[37] A. Papachristodoulou and S. Prajna, "On the construction of Lyapunov functions using the sum of squares decomposition," in *Proceedings of the 41st IEEE Conference on Decision and Control*, pp. 3482–3487, Las Vegas, Nev, USA, December 2002.

[38] M. Dalsmo and O. Egeland, "State feedback H_∞-suboptimal control of a rigid spacecraft," *IEEE Transactions on Automatic Control*, vol. 42, no. 8, pp. 1186–1189, 1997.

2

Attitude and Vibration Control of Flexible Spacecraft Using Singular Perturbation Approach

Morteza Shahravi and Milad Azimi

Space Research Institute, Tehran 15875-1774, Iran

Correspondence should be addressed to Milad Azimi; azimi@mut.ac.ir

Academic Editors: Z. Qin, I. Taymaz, and J. Yao

This paper addresses a composite two-time-scale control system for simultaneous three-axis attitude maneuvering and elastic mode stabilization of flexible spacecraft. By choosing an appropriate time coordinates transformation system, the spacecraft dynamics can be divided into double time-scale subsystems using singular perturbation theory (SPT). Attitude and vibration control laws are successively designed by considering a time bandwidths separation between the oscillatory flexible parts motion describing a fast subsystem and rigid body attitude dynamics as a slow subsystem. A nonlinear quaternion feedback control, based on modified sliding mode (MSM), is chosen for attitude control design and a strain rate feedback (SRF) scheme is developed for suppression of vibrational modes. In the attitude control law, the modification to sliding manifold for slow subsystem ensures that the spacecraft follows the shortest possible path to the sliding manifold and highly reduces the switching action. Stability proof of the overall closed-loop system is given via Lyapunov analysis. The proposed design approach is demonstrated to combine excellent performance in the compensation of residual flexible vibrations for the fully nonlinear system under consideration, as well as computational simplicity.

1. Introduction

In many missions of today's spacecraft with high resolution earth observation payloads and/or large flexible systems, the operation plan requires high precision control capability in order to point at certain area of interest. These missions impose increasingly severe requirements over the modeling and control of spacecraft dynamics. However the flexible structural elements such as solar arrays, antennas, and other light weight parts have received significant focus on providing the control effort for targeting flexible parts such as payloads and tracking maneuver with simultaneous vibration suppression to accomplish mission objectives. Design of such control system poses a challenging problem, including spill-over effects due to the unmodeled dynamics, nonlinear characteristics of rigid-flexible fully coupled dynamics, and unexpected perturbations [1]. From the mathematical point of view, the dynamics of flexible spacecraft involves the coupling of ODEs for attitude dynamics and PDEs for vibration of flexible appendages. This represented by a set of hybrid differential equations (HDE) of motion. Therefore, control strategies have emerged for smoothly shaped maneuvers with vibration excited [2]. Also, the actual performance of controllers is highly sensitive to the error introduced by mathematical model simplification. Therefore the key issues can be classified into modeling error, control/structure interaction, robustness, and so forth [3].

There has been a lot of research and investigation effort for such a problem. Numerical techniques have been reported with analysis and experimental verification. Accordingly, many researchers have surmounted finite dimensional approximation of the original systems. Simultaneous attitude maneuver with vibration suppression has been considered by Vadali [4] and Vadali et al. [5].

The design of robust and practical controllers such as sliding mode control (SMC) which is well known for its powerful robustness and ease of implementation have been presented in some previous studies [6, 7]. Also in some approaches, the spacecraft flexible dynamics are considered as an external perturbation which affected the rigid body motion [8, 9].

All the same, in previous works, for the case of three-axis attitude maneuver with fully nonlinear coupled rigid-flexible dynamics, the VSC approach has been modified by ignoring these nonlinear terms. The disadvantage of these modification techniques lies in shifting of the calculated parameters away from what has happened in reality. Hu proposed a robust nonlinear VSC control theory for 3-axis attitude control and vibration suppression of a flexible spacecraft simultaneously with parameter uncertainty and control saturation nonlinearity [10, 11]. Elsewhere the mentioned researcher Hu et al. [12] used a control technique which incorporated both SMC and command input shaping for the vibration suppression of a flexible spacecraft in single axis maneuver without proof of global stability of the system. The traditional sliding mode theory with disturbance accommodating control is combined for attitude tracking maneuver of spacecraft [13]. A modified version of classical SMC called smoothing model-reference control is proposed by Lo and Chen in which the attitude tracking performance is increased [14].

The problem of control of residual vibrations has received tremendous interest and poses a challenge task for spacecraft designers. They suggested using smart materials such as shape memory alloys (SMA) and piezoelectric material (PZT), for this problem. The piezoelectric materials have the advantages of high stiffness, light weight, low power consumption, high frequency response, and easy implementation. Using piezoelectric material as actuator (compensator) in the case of surface bounded layers with VSC during attitude maneuver for vibration reduction of flexible appendages in single axis maneuver is proposed by Hu and Ma [15]. Azadi et al. [16] studied attitude maneuver control and vibration suppression of a flexible satellite in three-axis rotation using adaptive robust control. The global stability of the fully coupled nonlinear system has not been reported in these researches.

This paper presents a method for degrading the induced vibration and limiting the control action during the slew maneuver based on fully nonlinear dynamic model of the system and using singular perturbation approach. The control of spacecraft for high precision pointing is formulated incorporating control of attitude by modified SMC and SRF techniques simultaneously. Global stability of the complete system has been guaranteed. Numerical simulations show the effectiveness of the proposed controller.

The rest of the paper is organized as follows. Section 2 describes the mathematical modeling of three-axis flexible satellite with embedded piezoelectric materials using physical characteristics of the coupled motion and the singular perturbation theory. The next section states attitude control design based on modified SMC and active vibration suppression based on SRF method using piezoelectric patches. The results of numerical simulations are presented to verify the controller performance in Section 5. Finally concluding remarks are given in Section 6.

2. Dynamic Modeling

Figure 1 demonstrates the schematic of the hub with two flexible appendages. A spacecraft model and rigid main body with two clamped loaded Euler-Bernoulli beams bounding with PZT layers are considered to model the elastic deformations of the flexible parts in multiaxis attitude maneuvers. The coordinates used are shown in Figure 1. By choosing the center of the mass of the spacecraft as the body fixed reference frame origin $(O\ X\ Y\ Z)|_b$, the attitude motions may be decoupled from the translational motions. The beams have the same length L_b, thickness t_b, mass per unit length ρ_b, bending moment of inertia I_b, and Young's modulus E_b. The PZT sensors/actuators patches with the length L_p, thickness t_p, mass per unit length ρ_p, bending moment of inertia I_p, and Young's modulus E_p are bounded in both sides of each panel.

The kinematics between the body angular velocity and attitude parameter need to be established. The orientation of the body fixed coordinate $(O\ X\ Y\ Z)|_b$ with respect to an arbitrary inertial frame $(O\ X\ Y\ Z)|_I$ may be defined using orthonormal direction cosine $\mathbf{C}(t) \in \mathbf{R}^{3\times 3}$ matrix [17]:

$$\{(O\ X\ Y\ Z)|_b(t)\} = \mathbf{C}(t)\{(O\ X\ Y\ Z)|_I(t)\}. \quad (1)$$

The unit quaternion, which is a nonminimal representation of an object attitude, completely avoids singular orientations. A quaternion may be presented as a vector $\mathbf{q} = [q_0\ \ \mathbf{q}_{1:3}] \in R_{4\times1}$; that is,

$$\begin{aligned} \mathbf{q}_{1:3} &= [q_1\ \ q_2\ \ q_3]^T \\ &= e(t)\,\mathrm{Sin}\left(\frac{\Phi(t)}{2}\right) \\ q_0 &= \mathrm{Cos}\left(\frac{\Phi(t)}{2}\right) \\ &\text{with } 0 \le \Phi(t) \le 2\pi, \end{aligned} \quad (2)$$

where $\Phi(t)$ is a rotation of a rigid body about the principle Euler rotation axis $e(t)$. The time derivative of the unit quaternion is derived to calculate attitude at any moment:

$$\dot{\mathbf{q}}(t) = \frac{1}{2}\left[q_0\mathbf{I}_{3\times3} + {}^{\times}\mathbf{q}_{1:3}\ \ -\mathbf{q}_{1:3}^T\right]^T\boldsymbol{\omega}, \quad (3)$$

where ${}^{\times}\mathbf{q}_{1:3}$ is the skew-symmetric matrix of $\mathbf{q}_{1:3}$ and $\boldsymbol{\omega} = \omega_x X_b + \omega_y Y_b + \omega_z Z_b$ the body angular velocity of the spacecraft.

The displacement $r_p(t)$ of any point p on the spacecraft can be defined as

$$\mathbf{r}(P,t) = \mathbf{r}_R^i(P)\,X_b + \mathbf{u}^i(P,t)\,Y_b, \quad (4)$$

where $\mathbf{r}_R^i(P)$ is a vector from center of the mass to the undeformed point p and $\mathbf{u}^i(P,t)$, $i = 1, 2$, represents the elastic deflection on ith appendage with respect to nominal position of point p.

The velocity of a point p with respect to the body fixed reference frame can be obtained by differentiation of (4):

$$\mathbf{v}(p,t) = \dot{\mathbf{u}}^i(p,t) + \boldsymbol{\omega} \times \left(\mathbf{r}_R^i(P) + \mathbf{u}^i(P,t)\right). \quad (5)$$

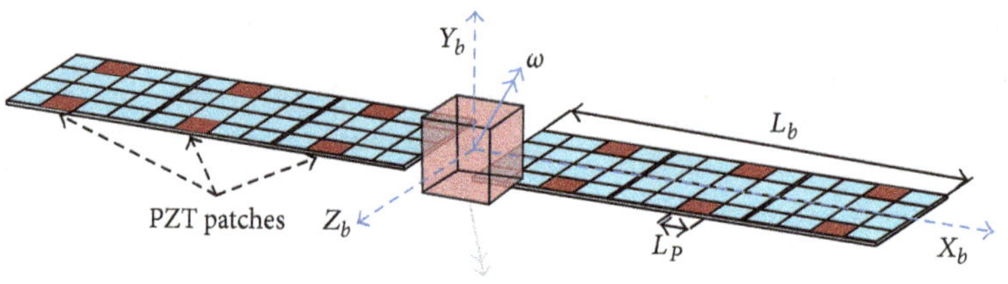

FIGURE 1: Flexible spacecraft model and parameters.

The kinetic energy of the system including PZT patches can be expressed as

$$T = \sum_{i=1}^{2} T_b^i + \sum_{i=1}^{2}\sum_{j=1}^{n_j} {}^{j}T_p^i, \tag{6}$$

where T_b^i, T_P^i and n_j represents the kinetic energy of the main structure, the kinetic energy of the jth sensor/actuator pair, and the number of PZT patches, respectively, and can be expressed as

$$\begin{aligned} T_b^i &= \frac{1}{2}\int_S \rho_b \mathbf{v}(p,t)\cdot\mathbf{v}(p,t)\,dS = \frac{1}{2}\boldsymbol{\omega}^T \mathbf{J}_b \boldsymbol{\omega} \\ &+ \frac{1}{2}\sum_{i=1}^{2}\int_a^{a+L_b} \rho_b^i \dot{\mathbf{u}}^i(p,t)^T \dot{\mathbf{u}}^i(p,t)\,dx \\ &+ \frac{1}{2}\boldsymbol{\omega}\sum_{i=1}^{2}\int_a^{a+L_b} \rho_b^i \left({}^{\times}\mathbf{r}_R^i(P) + {}^{\times}\mathbf{u}^i(P,t)\right)\dot{\mathbf{u}}^i(p,t)\,dx \end{aligned} \tag{7}$$

$$\begin{aligned} T_P^i &= \frac{1}{2}\boldsymbol{\omega}^T \mathbf{J}_p \boldsymbol{\omega} + \frac{1}{2}\sum_{i=1}^{2}\sum_{j=1}^{n_j}\int_{x_i}^{x_i+L_{Pi}} {}^{j}\rho_P^i \dot{\mathbf{u}}^i(p,t)^T \dot{\mathbf{u}}^i(p,t)\,dx \\ &+ \frac{1}{2}\boldsymbol{\omega}\sum_{i=1}^{2}\sum_{j=1}^{n_j}\int_{x_i}^{x_i+L_{Pi}} {}^{j}\rho_P^i \left({}^{\times}\mathbf{r}_R^i(P) + {}^{\times}\mathbf{u}^i(P,t)\right) \\ &\times \dot{\mathbf{u}}^i(p,t)\,dx, \end{aligned}$$

where $\mathbf{J}$ is the hub moment of inertia, a is the distance from the hub center to the root of the beam, L_b is the length of the flexible beam, ρ_b^i is the mass per unit length of the ith appendage, ${}^{j}\rho_P^i$ is the mass per unit length of the jth PZT patch and ith appendage, x_i is starting x-coordinate of PZT patch, and L_P is the length of the PZT patch.

The potential energy of the flexible structure including PZT patches is considered to be

$$V = \sum_{i=1}^{2} V_b^i + \sum_{i=1}^{2}\sum_{j=1}^{n_j} {}^{j}V_p^i, \tag{8}$$

where V_b^i and ${}^{j}V_P^i$ are the potential energy of the ith main structure and the jth sensor/actuator pair, respectively, and can be written as

$$\begin{aligned} V_b^i &= \frac{1}{2}\sum_{i=1}^{2}\int_a^{a+L_b} E_b^i I_b^i \left(\frac{\partial^2 \mathbf{u}^i(P,t)}{\partial x^2}\right)^2 dx, \\ {}^{j}V_P^i &= \frac{1}{2}\sum_{i=1}^{2}\sum_{j=1}^{n_j} {}^{j}E_P^i \left({}^{j}\varpi_p^i\, {}^{j}h_p^i\right)\left({}^{j}y^{i2} + {}^{j}y^i\, {}^{j}h_p^i + \frac{{}^{j}h_p^{i\,2}}{3}\right) \\ &\times \int_{x_i}^{x_i+L_P}\left(\frac{\partial^2 \mathbf{u}^i(p,t)}{\partial x^2}\right)^2 dx, \end{aligned} \tag{9}$$

where E_b^i is the modulus of elasticity for the ith appendage, I_b^i is the moment of inertia for the beam structure, ${}^{j}y^i$ is the starting point of the PZT as measured from the neutral axis of the beam, ${}^{j}\varpi_p^i$ is the width of the jth PZT layer and ith appendage, and ${}^{j}h_p^i$ is the thickness of each PZT element on ith appendage.

The virtual work done by the external torques $\boldsymbol{\tau}$ and PZT patches is given by

$$\delta W_{nc} = \sum_{i=1}^{2}\delta W_\tau + \sum_{i=1}^{2}\sum_{j=1}^{n_j}\delta\, {}^{j}W_p^i, \tag{10}$$

where $\delta^{j}W_p^i$ is the work done by the jth PZT patch on ith appendage. The work done by the external control torque can be expressed as

$$\delta W_\tau = W_T. \tag{11}$$

The work done by the jth PZT patch is the combination of the conservative and nonconservative work terms defined as an integral over the volume of the PZT patches:

$$\delta\, {}^{j}W_p^i = \left.\sum_{i=1}^{2}\sum_{j=1}^{n_j}\delta\, {}^{j}W_p^i\right|_c + \left.\sum_{i=1}^{2}\sum_{j=1}^{n_j}\delta\, {}^{j}W_p^i\right|_{nc}, \tag{12}$$

$$\begin{aligned} {}^{j}W_p^i &= \frac{1}{2}\, {}^{j}\varpi_p^i \sum_{i=1}^{2}\sum_{j=1}^{n_j}\int_{{}^{j}x^i}^{{}^{j}x^i+L_p}\int_{{}^{j}y^i}^{{}^{j}y^i+{}^{j}h_p^i} \begin{Bmatrix} {}^{j}E_3^i \\ {}^{j}S_1^i \end{Bmatrix}^T \\ &\times \begin{bmatrix} 1 & 0 \\ 0 & -1 \end{bmatrix} \begin{Bmatrix} {}^{j}D_3^i \\ {}^{j}T_1^i \end{Bmatrix} dy\,dx \end{aligned} \tag{13}$$

Using constitutive equation of PZT material (See Appendix A) ${}^{j}S_{1}^{i} = -y(\partial^2 \mathbf{u}(p,t)/\partial x^2)$, (13) become

$$\begin{aligned}
{}^{j}W_{p}^{i} &= \frac{1}{2}{}^{j}\varpi_{p}^{i} \times \sum_{i=1}^{2}\sum_{j=1}^{n_j}\int_{{}^{j}x^{i}}^{{}^{j}x^{i}+L_P}\int_{{}^{j}y^{i}}^{{}^{j}y^{i}+{}^{j}h_{p}^{i}}\left\{\left({}^{j}\varepsilon_{3}^{Ti} - {}^{j}d_{31}^{i\,2}\,{}^{j}E_{P}^{i}\right){}^{j}E_{3}^{i\,2} - \left(2\,{}^{j}d_{31}^{i}\,{}^{j}E_{P}^{i}\,{}^{j}E_{3}^{i} \times \left(y\frac{\partial^2 \mathbf{u}^{i}(p,t)}{\partial x^2}\right)\right)\right\}dy\,dx \\
&\quad - \frac{1}{2}{}^{j}\varpi_{p}^{i} \times \sum_{i=1}^{2}\sum_{j=1}^{n_j}\int_{{}^{j}x_{i}}^{{}^{j}x_{i}+L_P}\int_{{}^{j}y_{i}}^{{}^{j}y_{i}+{}^{j}h_{p}^{i}}\left({}^{j}E_{P}^{i}\left(y\frac{\partial^2 \mathbf{u}^{i}(p,t)}{\partial x^2}\right)^2\right)dy\,dx
\end{aligned} \tag{14}$$

The last term of (14) is expressed as PZT potential energy ${}^{j}V_{P}^{i}$; in other words the potential energy expression is composed of conservative works, also introducing

$$\left.{}^{j}W_{P}^{i}\right|_{nc} = \frac{1}{2}{}^{j}\xi_{p}^{i}\,{}^{j}\eta_{p}^{i\,2} - \left\{\underset{\smile k}{\mathbf{q}^{i}}\right\}^{T}\left\{{}^{j}\mathfrak{R}_{p}^{i}\right\}{}^{j}\eta_{p}^{i} \tag{15}$$

with

$$\begin{aligned}
{}^{j}\xi_{p}^{i} &= \sum_{i=1}^{2}\sum_{j=1}^{n_j}\frac{{}^{j}\varpi_{p}^{i}L_{Pi}}{{}^{j}h_{p}^{i}}\left({}^{j}\varepsilon_{3}^{Ti} - {}^{j}d_{31}^{i\,2}\,{}^{j}E_{P}^{i}\right) \\
{}^{j}\eta_{p}^{i} &= \sum_{i=1}^{2}\sum_{j=1}^{n_j}{}^{j}E_{3}^{i} \times {}^{j}h_{p}^{i} \\
{}^{j}\mathfrak{R}_{p}^{i} &= \sum_{i=1}^{2}\sum_{j=1}^{n_j}{}^{j}d_{31}^{i}\,{}^{j}E_{P}^{i}\,{}^{j}\varpi_{p}^{i}\left({}^{j}y^{i} + \frac{{}^{j}h_{p}^{i}}{2}\right) \times \int_{{}^{j}x^{i}}^{{}^{j}x^{i}+L_{Pi}}\left\{\psi''(x)\right\}^{T}dx,
\end{aligned} \tag{16}$$

where η_{p}^{i} is the electrode voltage, $\{\underset{\smile k}{\mathbf{q}^{i}}\} = [q_1 q_2 \cdots q_n]$ is the kth generalized coordinates for ith appendage, and $\{\psi(x)\}$ is the element shape function. Substituting (11) and (15) into (10), the total work can be expressed as

$$W_{nc} = \frac{1}{2}\{\eta\}^{T}[\mathfrak{I}]\{\eta\} - \{q\}^{T}[R]\{\eta\} + W_{T}, \tag{17}$$

where

$$\begin{aligned}
[\mathfrak{I}] &= \operatorname{diag}\left({}^{j}\xi_{p}^{i}\right), \\
[R] &= \left[\left\{{}^{1}\mathfrak{R}_{p}^{i}\right\}\left\{{}^{2}\mathfrak{R}_{p}^{i}\right\}\cdots\left\{{}^{n_j}\mathfrak{R}_{p}^{i}\right\}\right], \\
\eta &= \left[{}^{1}\eta_{p}^{i}\;{}^{2}\eta_{p}^{i}\cdots{}^{n_j}\eta_{p}^{i}\right]^{T}.
\end{aligned} \tag{18}$$

Utilizing assumed mode method (AMM) and defining

$$\mathbf{u}_{P}^{i}(x,t) = \sum_{k=1}^{m}{}^{i}\psi_{k}^{T}(x)\,\underset{\smile k}{\mathbf{q}^{i}}(t) = \{\psi\}\left\{\underset{\smile}{\mathbf{q}}\right\}, \tag{19}$$

where $\mathbf{u}_{P}^{i}(x,t)$ is elastic displacement of ith appendage which can be discretized using m mode by AMM expansion technique. We substitute (6), (8), and (10) into Lagrange's equations of the motion in terms of quasi-coordinate in a vector form; that is,

$$\begin{aligned}
\frac{d}{dt}\left(\frac{\partial L}{\partial \boldsymbol{\omega}}\right) + {}^{\times}\boldsymbol{\omega}\left(\frac{\partial L}{\partial \boldsymbol{\omega}}\right) &= \boldsymbol{\tau} + \mathbf{u} \\
\frac{d}{dt}\left(\frac{\partial L}{\partial \underset{\smile k}{\dot{\mathbf{q}}}}\right) - \frac{\partial L}{\partial \underset{\smile k}{\mathbf{q}}} &= \mathbf{0}.
\end{aligned} \tag{20}$$

Integration over the spatial domains leads to global mass, stiffness and forcing matrixes. In order to account for the structural damping effects in spacecraft dynamics, the Rayleigh's dissipation function may be considered as

$$T_{d} = \frac{1}{2}\underset{\smile k}{\dot{\mathbf{q}}}^{T}\,\mathbf{C}\,\underset{\smile k}{\dot{\mathbf{q}}}. \tag{21}$$

Using (6), (8), (10), and (21) and the extended Hamilton's principle, the attitude dynamic model of a flexible spacecraft can be obtained in the following form:

$$\begin{aligned}
&\left[\begin{array}{cc}\mathbf{M}_{RR} & \mathbf{M}_{RF} \\ \mathbf{M}_{FR} & \mathbf{M}_{FF}\end{array}\right]\cdot\left\{\begin{array}{c}\dot{\boldsymbol{\omega}} \\ \underset{\smile k}{\ddot{\mathbf{q}}}\end{array}\right\} + \left[\begin{array}{cc}\mathbf{C}_{RR} & \mathbf{C}_{RF} \\ \mathbf{C}_{FR} & \mathbf{C}_{FF}\end{array}\right]\left\{\begin{array}{c}\boldsymbol{\omega} \\ \underset{\smile k}{\dot{\mathbf{q}}}\end{array}\right\} \\
&\quad + \left[\begin{array}{cc}\mathbf{0} & \mathbf{0} \\ \mathbf{0} & K_{FF}\end{array}\right]\left\{\begin{array}{c}\boldsymbol{\Phi} \\ \underset{\smile k}{\mathbf{q}}\end{array}\right\} = \left\{\begin{array}{c}\boldsymbol{\tau} + \mathbf{u} \\ -[R][g_a]\{\eta_a\}\end{array}\right\} \\
&\{\eta_S\} = [g_s][\mathfrak{I}]^{-1}[R]^{T}\left\{\underset{\smile k}{\mathbf{q}}\right\},
\end{aligned} \tag{22}$$

where $[g_s]$ and $[g_a]$ represent the sensor and actuator amplifier gains, respectively. The elements of submatrices **M**, **C**, and **K** are given in Appendix B. The PZT patches will be used as sensors and actuators; accordingly, they will have voltage inputs and outputs. As it can be seen from (22), $[R]$ and $[\mathfrak{I}]$ matrices can be decomposed in sensor and actuator parts corresponding to the sensor/actuator voltages, $\{\eta_S\}$ and $\{\eta_a\}$.

3. Implementation of SPT

The fundamental idea of this approach is to separate the system dynamics into the slow and fast subsystems. Control design may then proceed for each lower-order subsystem, and the results are combined to yield a hybrid stabilized controller for the overall system. For this case, a new variable $\mathbf{Z}$ may be considered as follows:

$$\begin{gathered}\left[\underline{\underline{\mathbf{K}_{FF}}}\right]\left\{\underset{\sim k}{\mathbf{q}}\right\}=k_{\text{Min}}\left[\underline{\underline{\mathbf{K}_{FF}^{R}}}\right]\left\{\underset{\sim k}{\mathbf{q}}\right\}\\ =\{\mathbf{Z}\}\Longrightarrow\left[\underline{\underline{\mathbf{K}_{FF}^{R}}}\right]\left\{\underset{\sim k}{\mathbf{q}}\right\}=\frac{1}{k_{\text{Min}}}\{\mathbf{Z}\},\end{gathered} \tag{23}$$

where k_{Min} is the smallest coefficient of the stiffness matrix $\underline{\underline{\mathbf{K}_{FF}}}$. Introducing a new parameter as a singular perturbation parameter, $\varepsilon=(1/k_{\text{Min}})^{0.5}$ [18], (23) become

$$\left[\underline{\underline{\mathbf{K}_{FF}^{R}}}\right]\{\mathbf{q}_k\}=\varepsilon^2\{\mathbf{Z}\}. \tag{24}$$

Comparing ${}^{j}\mathfrak{R}_{p}^{i}$ matrix, with $\sqrt{\varepsilon}$, it can be concluded that $O({}^{j}\mathfrak{R}_{p}^{i})=O(\varepsilon)$; therefore, ${}^{j}\mathfrak{R}_{p}^{i}$ matrix can be written as

$${}^{j}\mathfrak{R}_{p}^{i}=\varepsilon\,{}^{j}\mathfrak{R}_{p}^{i\,R}, \tag{25}$$

where the superscript R denotes to reduced magnitude. Substituting the new variables into the equations of motion, the system becomes

$$\begin{aligned}\dot{\boldsymbol{\omega}}=\boldsymbol{\alpha}_1^{-1}\Big(&\boldsymbol{\tau}+\mathbf{u}-\boldsymbol{\alpha}_2\boldsymbol{\omega}-\varepsilon^2\underline{\underline{\mathbf{K}_{FF}^{R}}}^{-1}\boldsymbol{\alpha}_3\dot{\mathbf{Z}}\\ &-\left\{\mathbf{M}_{RF}\mathbf{M}_{FF}^{-1}\left(-\mathbf{Z}-\varepsilon\,[R]\,[g_a]\,\{\eta_a\}\right)\right\}\Big),\end{aligned} \tag{26}$$

$$\begin{aligned}\ddot{\mathbf{Z}}=\varepsilon^{-2}\underline{\underline{\mathbf{K}_{FF}^{R}}}\boldsymbol{\gamma}_1^{-1}\Big(&-\varepsilon\,[R]\,[g_a]\,\{\eta_a\}-\boldsymbol{\gamma}_2\boldsymbol{\omega}-\varepsilon^{-2}\boldsymbol{\gamma}_3\underline{\underline{\mathbf{K}_{FF}^{R}}}^{-1}\dot{\mathbf{Z}}\\ &-\mathbf{M}_{FR}\mathbf{M}_{RR}^{-1}(\boldsymbol{\tau}+\mathbf{u})-\mathbf{Z}\Big).\end{aligned} \tag{27}$$

The slow subsystem can be obtained by setting terms of $O(\varepsilon^2)$ and higher equal to zero in the equations of motion:

$$\mathbf{Z}_{\text{Slow}}=\left(-\mathbf{M}_{FR}\mathbf{M}_{RR}^{-1}(\boldsymbol{\tau}+\mathbf{u})-\varepsilon\,[R]\,[g_a]\,\{\eta_a\}-\boldsymbol{\alpha}_2\boldsymbol{\omega}\right). \tag{28}$$

Substituting (28) in (26) yields:

$$\begin{aligned}\dot{\boldsymbol{\omega}}_{\text{Slow}}=\boldsymbol{\alpha}_1^{-1}\Big(&\boldsymbol{\tau}-\boldsymbol{\alpha}_2\boldsymbol{\omega}\\ &-\mathbf{M}_{RR}\mathbf{M}_{FF}^{-1}\left(-\boldsymbol{\gamma}_2\boldsymbol{\omega}-\mathbf{M}_{FR}\mathbf{M}_{RR}^{-1}(\boldsymbol{\tau}+\mathbf{u})\right)\Big|_{\text{Slow}}\Big)\end{aligned} \tag{29}$$

The term including control voltage of PZT actuators is of $O(\varepsilon)$ and consequently can be ignored since its magnitude is much less than the other terms, which leads to an $O(\varepsilon)$ approximation of the slow subsystem.

Fast subsystem can be obtained by setting $\psi_f=t/\varepsilon$, introducing $\mathbf{Z}_{\text{Fast}}=\mathbf{Z}-\mathbf{Z}_{\text{Slow}}$, lead the slow variables as constant in fast time scale [19] and considering the terms of $O(\varepsilon^2)$ and higher equal to zero so the equation of the motion for fast subsystem can be expressed as

$$\begin{aligned}\frac{d^2}{d\psi_f^2}\mathbf{Z}_{\text{Fast}}=\mathbf{K}_{FF}^{R}\boldsymbol{\gamma}_1^{-1}\Big(&-\left(\varepsilon[R][g_a]\{\eta_a\}\right)\big|_{\text{Fast}}-\boldsymbol{\gamma}_3\frac{d}{d\psi_f}\mathbf{Z}_{\text{Fast}}\\ &-\mathbf{Z}_{\text{Fast}}-\mathbf{M}_{FR}\mathbf{M}_{RR}^{-1}(\boldsymbol{\tau}+\mathbf{u})\big|_{\text{Fast}}\Big).\end{aligned} \tag{30}$$

The equation of the motion of (26) becomes

$$\begin{aligned}&\begin{bmatrix}\underline{\underline{\boldsymbol{\alpha}_1}}&\underline{\underline{0}}\\ \underline{\underline{0}}&\underline{\underline{\boldsymbol{\gamma}_1}}\end{bmatrix}\cdot\begin{Bmatrix}\underline{\underline{\dot{\boldsymbol{\omega}}}}\\ \ddot{\psi}_f\end{Bmatrix}+\begin{bmatrix}\underline{\underline{\boldsymbol{\alpha}_2+\mathbf{M}_{RF}\mathbf{M}_{FF}\boldsymbol{\gamma}_2}}&\underline{\underline{0}}\\ \underline{\underline{0}}&\underline{\underline{\boldsymbol{\gamma}_3}}\end{bmatrix}\\ &\cdot\begin{Bmatrix}\underline{\underline{\boldsymbol{\omega}}}\\ \dot{\psi}_f\end{Bmatrix}+\begin{bmatrix}\underline{\underline{0}}&\underline{\underline{0}}\\ \underline{\underline{0}}&\underline{\underline{\boldsymbol{\kappa}_{FF}}}\end{bmatrix}\cdot\begin{Bmatrix}\underline{\underline{\boldsymbol{\Phi}}}\\ \psi_f\end{Bmatrix}\\ &=\begin{Bmatrix}\underline{\underline{\mathbf{M}_{RR}^{-1}\boldsymbol{\alpha}_1(\boldsymbol{\tau}+\mathbf{u})}}\\ -\varepsilon\,[R]\,[g_a]\,\{\eta_a\}-\mathbf{M}_{FR}\mathbf{M}_{RR}^{-1}(\boldsymbol{\tau}+\mathbf{u})\end{Bmatrix},\end{aligned} \tag{31}$$

where α_i,γ_i $(i=1,2,3)$ are defined as

$$\begin{aligned}\boldsymbol{\alpha}_1&=\mathbf{M}_{RR}-\mathbf{M}_{RF}\mathbf{M}_{FF}^{-1}\mathbf{M}_{FR},\\ \boldsymbol{\alpha}_2&=\mathbf{C}_{RR}-\mathbf{M}_{RF}\mathbf{M}_{FF}^{-1}\mathbf{C}_{FR},\\ \boldsymbol{\alpha}_3&=\mathbf{C}_{RF}-\mathbf{M}_{RF}\mathbf{M}_{FF}^{-1}\mathbf{C}_{FF},\\ \boldsymbol{\gamma}_1&=\mathbf{M}_{FF}-\mathbf{M}_{FR}\mathbf{M}_{RR}^{-1}\mathbf{M}_{RF},\\ \boldsymbol{\gamma}_2&=\mathbf{C}_{FR}-\mathbf{M}_{FR}\mathbf{M}_{RR}^{-1}\mathbf{C}_{RR},\\ \boldsymbol{\gamma}_3&=\mathbf{C}_{FF}-\mathbf{M}_{FR}\mathbf{M}_{RR}^{-1}\mathbf{C}_{RF}.\end{aligned} \tag{32}$$

4. Controller Design

In the present work, three-axis attitude maneuver and vibration control are considered.

The quaternions are chosen for representation of the attitude of the spacecraft. By taking angular velocity and quaternion vectors, a modified sliding manifold is being proposed as

$$\mathbf{S}=\boldsymbol{\omega}_e+\mathbf{K}\tanh\left(q_{0-e}\right)\mathbf{q}_{1:3-e}, \tag{33}$$

where $\boldsymbol{\omega}_e=\boldsymbol{\omega}-\boldsymbol{\omega}_d$ is the spacecraft angular velocity tracking error, $\mathbf{q}_e=\mathbf{q}\langle\times\rangle\mathbf{q}_d^{-1}$ is the quaternion tracking error, in which $\langle\times\rangle$ is the quaternion products, $\mathbf{q}_d$ and $\boldsymbol{\omega}_d$ are the desired quaternion and angular velocity, respectively.

Theorem 1. *The control objective is to stabilize the flexible spacecraft by forcing the rigid body modes to follow some desired trajectories, while simultaneously reducing the elastic modes. The desired attitude maneuver with high mode flexibility can be realized, if the sliding condition $\dot{V}<0$ is satisfied.*

Proof. The desired state that slides on the sliding surface can be shown to be asymptotically stable by choosing the candidate Lyapunov function as

$$V = \frac{1}{2}\mathbf{S}^T \boldsymbol{\alpha}_1 \mathbf{S}. \quad (34)$$

□

The proposed Lyapunov function is valid since it vanishes at equilibrium point $\mathbf{S} = \mathbf{0}$ and is globally positive definite for $\mathbf{S} \neq \mathbf{0}$ since $\boldsymbol{\alpha}_1$ is positive definite. The time derivative of Lyapunov function is given by

$$\dot{V} = \frac{1}{2}\mathbf{S}^T \boldsymbol{\alpha}_1 \dot{\mathbf{S}}, \quad (35)$$

where the derivative of sliding surface is defined as

$$\dot{\mathbf{S}} = \dot{\boldsymbol{\omega}} + \mathbf{K}\tanh(q_0)\,\dot{\mathbf{q}}_{1:3}. \quad (36)$$

From the equation of the motion we have

$$\begin{aligned}\boldsymbol{\alpha}_1\dot{\boldsymbol{\omega}} = &-\left\{\boldsymbol{\alpha}_2 + \mathbf{M}_{RF}\mathbf{M}_{FF}^{-1}\boldsymbol{\gamma}_2\right\}\boldsymbol{\omega} \\ &+ \left\{1 - \mathbf{M}_{RF}\mathbf{M}_{FF}^{-1}\mathbf{M}_{FR}\mathbf{M}_{RR}^{-1}\right\}(\mathbf{u} + \boldsymbol{\tau}),\end{aligned} \quad (37)$$

where $\mathbf{u}$ is the control torque generated by actuators placed on rigid main body. Multiply each side of (36) by $\boldsymbol{\alpha}_1$ combined with (37) leads to the following expression for $\dot{V}$:

$$\begin{aligned}\dot{V} = \mathbf{S}^T\Big(&-\left\{\boldsymbol{\alpha}_2 + \mathbf{M}_{RF}\mathbf{M}_{FF}^{-1}\boldsymbol{\gamma}_2\right\}\boldsymbol{\omega} \\ &+ \left\{\mathbf{M}_{RR}^{-1}\boldsymbol{\alpha}_1\right\}(\mathbf{u} + \boldsymbol{\tau}) + \left(K_1\tanh(q_0)\,\boldsymbol{\alpha}_1\dot{\mathbf{q}}_{1:3}\right)\Big),\end{aligned} \quad (38)$$

which is clearly negative definite provided that $K_1 > 0$. The controller designed by variable structure approach consists of two different tasks. First one is to define an appropriate sliding surface and the other one is to improve the sliding condition, which it commands the states remain on the sliding surface. By solving the above equation for the control input, the external control torque can be derived in such a way that

$$\mathbf{u} = \mathbf{u}_{\text{eq}} + \mathbf{u}_{\text{VS}}, \quad (39)$$

where the variable structure and equivalent parts of controller input are defined as

$$\mathbf{u}_{\text{VS}} = -\left(K_2\mathbf{S}(t)\right) - \left(K_3\tanh\left(\frac{\mathbf{S}(t)}{P^2}\right)\right) \quad (40)$$

$$\begin{aligned}\mathbf{u}_{\text{eq}} = \left(\mathbf{M}_{RR}^{-1}\boldsymbol{\alpha}_1\right)^{-1}\Big(&\left\{\boldsymbol{\alpha}_2 + \mathbf{M}_{RF}\mathbf{M}_{FF}^{-1}\boldsymbol{\gamma}_2\right\}\boldsymbol{\omega} \\ &- \left\{\mathbf{M}_{RR}^{-1}\boldsymbol{\alpha}_1\boldsymbol{\tau}\right\} - \left(K_1\tanh(q_0)\,\boldsymbol{\alpha}_1\dot{\mathbf{q}}_{1:3}\right)\Big).\end{aligned} \quad (41)$$

The equivalent control $\mathbf{u}_{\text{eq}}$ part turns the sliding surface $\mathbf{S}(t)$ into an invariant manifold for the system, to ensure that $\dot{\mathbf{S}} = 0$. Whereas the variable structure part $\mathbf{u}_{\text{VS}}$ is chosen to ensure that the $\mathbf{S} = 0$; thus, the designing surface is attractive and the desired condition can be reached in finite time.

Substitutions of (40) and (41) into the sliding condition yield

$$\dot{V} = -\mathbf{S}^T(t)\left\{\left(K_2\mathbf{S}(t)\right) + \left(K_3\tanh\left(\frac{\mathbf{S}(t)}{P^2}\right)\right)\right\} < 0, \quad (42)$$

where K_n with $n = 1, 2, 3$ are positive definite matrices, and P^2 is a scalar sharpness function that regulates the control action rates. Note that the term $\dot{\mathbf{q}}_{1:3}$ in (41) introduces nonlinear terms in variable-structure controller. This implies from theorem and the K_n values that as $t \longrightarrow \infty$ the control objective $[\mathbf{q}\ \boldsymbol{\omega}]_d^T = [1\ \mathbf{0}_{1\times3}\ \mathbf{0}_{1\times3}]^T$ and the asymptotic global stability can be achieved due to the $\dot{V} = -(1/2)\mathbf{S}^T\mathbf{u}_{\text{VS}}$.

It can be seen from (40) and (41) that the stability and robustness of controller performance are guaranteed if the upper bounds of the perturbations are known. This knowledge may cause the controller to produce the overconservative high gain K_3. This may cause chattering phenomenon. In order to overcome this source of degradation or overaction, the hyperbolic tangent function is used to reduce chattering.

Also, the active vibration suppression system using the PZT sensor and actuator can actively suppress the solar and environmental induced vibration to the flexible appendages during attitude maneuver. Since no external field is applied to the sensor layer, the electric displacement developed on the sensor surface is directly proportional to the strain acting on it. Also PZT materials can be used as strain rate sensors. The output current of the PZT sensor measures the moment rate of the flexible appendages. This current is converted into the open circuit sensor voltage V_S using a signal conditioning device with the gain G_C and applied to an actuator with a suitable controller gain. Thus, the sensor output voltage is obtained as

$$\begin{aligned}V_S(t) &= G_C i(t) \\ &= G_C e_{31}\left(\frac{h_b}{2} + h_p\right)\varpi_p \int_0^{L_p} \frac{\partial^2}{\partial x^2}\psi_k(x)\,\dot{\mathbf{q}}_k(t)\,dx,\end{aligned} \quad (43)$$

where $i(t)$ is circuit current, and the indices b and p explain the beam and PZT structures, respectively. This sensor voltage is given as input to the controller and the output of the controller is the controller gain multiplied by the sensor voltage. Thus, the input voltage to the actuator V_a, in other words the controller input $u(t)$, is given by

$$V_a(t) = u(t) = \mathbf{K}_p \times V_S(t), \quad (44)$$

where $\mathbf{K}_p$ is the controller gain matrix. Note that feedback gain matrix $\mathbf{K}_p$ consists of each feedback gain which is associated with each flexible PZT patch. The actuator equation is derived from the converse PZT equation and the relative control force $\mathbf{f}_{\text{ctrl}}$ produced by the actuator that is applied on the appendages is obtained using bending moment theory:

$$\mathbf{f}_{\text{ctrl}} = E_p d_{31}\varpi_p\left(\frac{h_p + h_b}{2}\right)\int_0^{L_p}\frac{\partial}{\partial x}\psi(x)\,dx\,V_a(t). \quad (45)$$

TABLE 1: Parameters of flexible spacecraft.

Parameters	Flexible appendage	Piezoelectric layer
Young's Modulus (GPa)	$E = 39.72$	$E_p = 68$
Density (kg/m)	$\rho_b = 4$	$\rho_p = 2.31$
Thickness (m)	$t_b = 0.01$	$t_p = 0.003$
Width (m)	$b = 0.5$	$b = 0.1$
Length (m)	$L_b = 2$	$L_p = 0.1$
PZT Strain constant (m/V)	—	$d_{31} = 125 \times 10^{-12}$
PZT Stress constant (Vm/N)	—	$e_{31} = 10.5 \times 10^{-3}$
Hub dimension (m)	$a = 1$	
Spacecraft moment of inertia (kg·m²)	$\begin{bmatrix} 200 & 0 & 0 \\ 0 & 210 & 0 \\ 0 & 0 & 180 \end{bmatrix}$	

5. Numerical Simulations and Results

Simulation of fully nonlinear 3-axis attitude maneuver of a flexible spacecraft has been carried out using MATLAB/SIMULINK software to demonstrate the performance of proposed approach. The proposed control system objective of a flexible spacecraft model is to reduce the induced vibration and tracking a target in sample mission.

The desired maneuver is 160° slew with simultaneous vibration suppression. This is usually a fast and large angle maneuver. The numerical values of the parameters used in the simulation study are presented in Table 1.

The initial conditions for the angular velocity are set to $\boldsymbol{\omega}(t_0) = [0\ 0\ 0]^T$ and for quaternion parameters are given by $\mathbf{q}(t_0) = [0.174\ -0.263\ 0.789\ -0.526]^T$. The first two flexible modes are retained in the model for discretization of elastic deformations. For control implementation, design parameters are considered as $\mathbf{K}_1 = 0.4\mathbf{I}_{3\times3}$, $\mathbf{K}_2 = 15\mathbf{I}_{3\times3}$ and $\mathbf{K}_3 = 0.28\mathbf{I}_{3\times3}$.

Dynamical behavior of the controlled system is shown in Figures 2, 3, 4, 5, 6, 7, 8, and 9. Smoothness and convergence of attitude error in terms of quaternions and angular rate are shown in Figures 2 and 3. Figures 4–6 show the required control torques for different states.

As shown in these figures, using SPT and accounting complete coupling of flexible/rigid dynamics in controller design process causes actually better response of the closed loop system and controller performance. Also, active suppression of structural vibration causes smooth and fine actuation of attitude controller. This is an important characteristic for actual implementation of the controller. As shown in Figure 6, using classical SMC (without modification) causes steady state error arising from flexible modes excitation. Convergence of the flexible body coordinates and PZT actuation voltage are shown in Figures 7–9.

6. Conclusion

A new methodology and control design approach for multi-axis attitude maneuver and vibration suppression of flexible spacecraft has been proposed. The proposed scheme is based on mapping of the fully coupled nonlinear system dynamics into slow and fast subsystem domain using SPT and designing of hybrid control modified SMC/SRF for system. The hybrid controller can obtain asymptotical reference attitude and suppress structural vibrations excited by, for example, rapid maneuvers or other disturbances. Stability proof of the overall system has been proved using Lyapunov stability analysis.

It has been shown that the performance of the resultant closed-loop system being improved compared to those of traditional, while fast targeting, suppression of residual structural vibration and assuring overall stability.

Appendices

A. Piezoelectric Constitutive Equations

The 3D constitutive equation for a piezoelectric element can be shown to have the following standard notation [20] as

$$\begin{bmatrix} \underline{D_i} \\ \underline{S_j} \end{bmatrix} = \begin{bmatrix} \underline{\varepsilon_i^T} & \underline{d_{ij}^1} & \underline{d_{ij}^2} \\ \underline{d_{ij}^{1T}} & \underline{S_{ij}^{E1}} & \underline{0} \\ \underline{d_{ij}^2} & \underline{0} & \underline{S_{ij}^{E2}} \end{bmatrix} \begin{bmatrix} \underline{E_i} \\ \underline{T_j} \end{bmatrix}, \quad \text{(A.1)}$$

where D_i $(i = 1, 2, 3)$ denotes the electric displacement along the ith axis, E_i $(i = 1, 2, 3)$ represents the applied electrical field density, S_i $(i = 1, \ldots, 6)$ represents strain, σ_i $(i = 1, \ldots, 6)$ represents the stress, $\underline{\varepsilon_i^T}$ $(i = 1, 2, 3)$, $\underline{S_{ij}^{Ek}}$ $(i = 1, 5,\ j = 1, 2, 3, 5,\ k = 1, 2)$, and d_{ij}^k $(i = 1, 3,\ j = 1, 3, 5,\ k = 1, 2)$ are permittivity, elastic compliance, and piezoelectricity (strain) coefficient constants of the PZT material, respectively:

$$\underline{\varepsilon_i^T} = \begin{bmatrix} \varepsilon_1^T & 0 & 0 \\ 0 & \varepsilon_1^T & 0 \\ 0 & 0 & \varepsilon_3^T \end{bmatrix}, \qquad \underline{d_{ij}^1} = \begin{bmatrix} 0 & 0 & 0 \\ 0 & 0 & 0 \\ d_{31} & d_{31} & d_{33} \end{bmatrix},$$

$$\underline{d_{ij}^2} = \begin{bmatrix} 0 & d_{15} & 0 \\ d_{15} & 0 & 0 \\ 0 & 0 & 0 \end{bmatrix} \quad \text{(A.2)}$$

$$\underline{S_{ij}^{E1}} = \begin{bmatrix} S_{11}^E & S_{12}^E & S_{13}^E \\ S_{12}^E & S_{11}^E & S_{13}^E \\ S_{13}^E & S_{13}^E & S_{13}^E \end{bmatrix}, \qquad \underline{S_{ij}^{E2}} = \begin{bmatrix} S_{55}^E & 0 & 0 \\ 0 & S_{55}^E & 0 \\ 0 & 0 & S_{55}^E \end{bmatrix} \quad \text{(A.3)}$$

The strain condition based on Euler-Bernoulli beam theory is defined as

$$\varepsilon_x = -y\frac{\partial^2 w}{\partial x^2}, \qquad \varepsilon_y = \varepsilon_z = \gamma_{xy} = \gamma = \gamma_{yx} = 0. \quad \text{(A.4)}$$

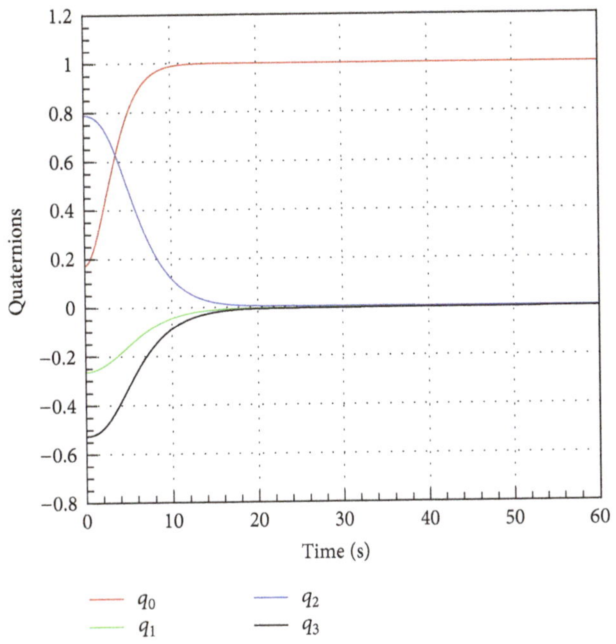

FIGURE 2: Time history of Attitude quaternion.

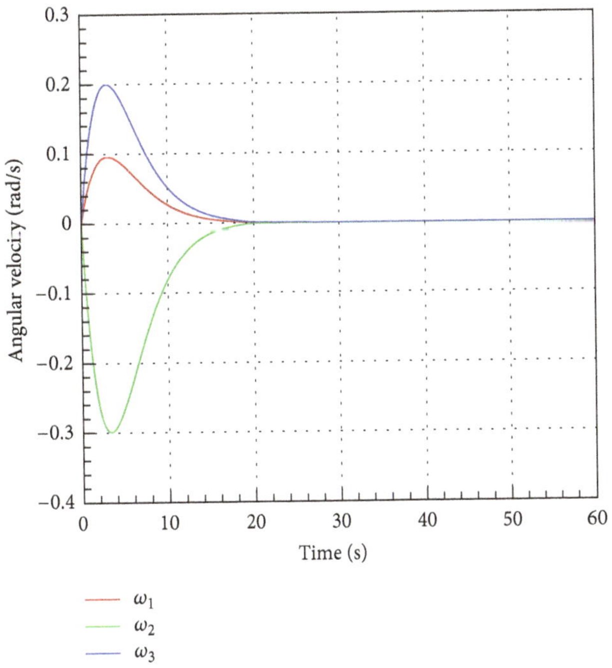

FIGURE 3: Time history of angular velocity.

It can be seen from (A.4) that (A.1) reduced to 1-D Constitutive equation of PZT material and is thereby found to be

$$\begin{Bmatrix} D_3 \\ S_1 \end{Bmatrix} = \begin{bmatrix} \varepsilon_3^T & d_{31} \\ d_{31} & S_{11}^E \end{bmatrix} \begin{Bmatrix} E_3 \\ T_1 \end{Bmatrix} \quad \text{(A.5)}$$

Using the fact that $S_{ij}^E = E_P^{-1}$, where E_P is Young's modulus, (A.3) can be expressed as below:

$$\begin{Bmatrix} D_3 \\ T_1 \end{Bmatrix} = \begin{bmatrix} \varepsilon_3^T - d_{31}^2 E_P & d_{31} E_P \\ -E_P d_{31} & E_P \end{bmatrix} \begin{Bmatrix} E_3 \\ S_1 \end{Bmatrix}. \quad \text{(A.6)}$$

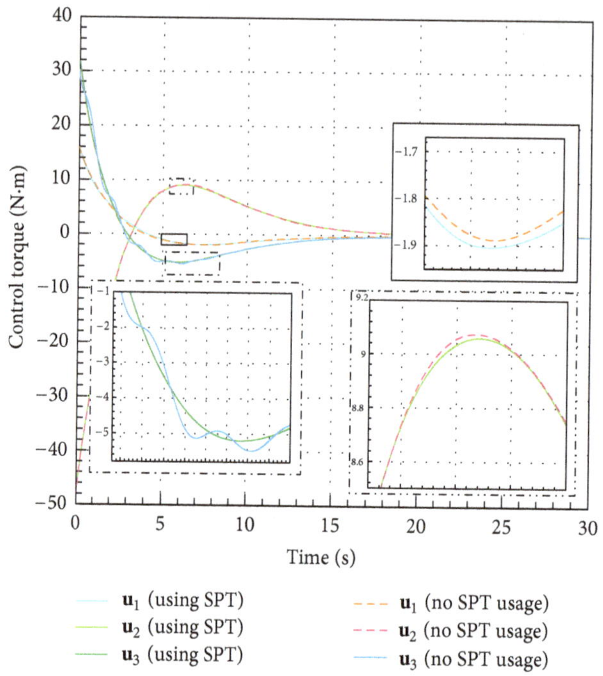

FIGURE 4: Time history of control torque with active vibration suppression.

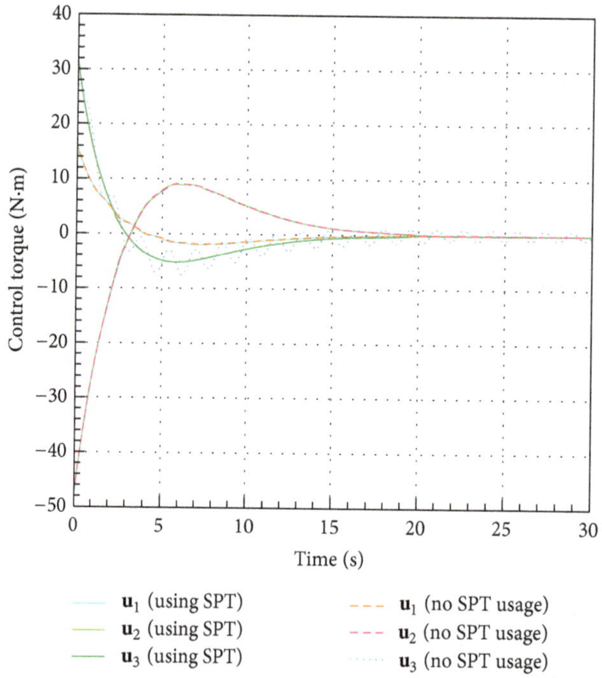

FIGURE 5: Time history of control torque without active vibration suppression.

B. The Elements of the Matrices M, C, K and J

The elements of the sub matrices of the system are:

$$\mathbf{J} = \mathbf{J}_h + \mathbf{J}_b + \mathbf{J}_p = \left[J^{ij}\right]_{3\times 3},$$

$$J^{11} = I_{XX} + \sum_{i=1}^{2} \int_{a}^{a+L_b} \rho_b^i \, {}^i\mathbf{u}^2 dx + \sum_{i=1}^{2} \sum_{j=1}^{n_j} \int_{x_i}^{x_i+L_{Pi}} {}^j\rho_P^i \, {}^i\mathbf{u}^2 dx$$

$$J^{12} = J^{21} = -\sum_{i=1}^{2} \int_{a}^{a+L_b} \rho_b^i \, {}^i\mathbf{u} \, dx + \sum_{i=1}^{2} \sum_{j=1}^{n_j} \int_{x_i}^{x_i+L_{Pi}} {}^j\rho_P^i \, {}^i\mathbf{u} \, dx$$

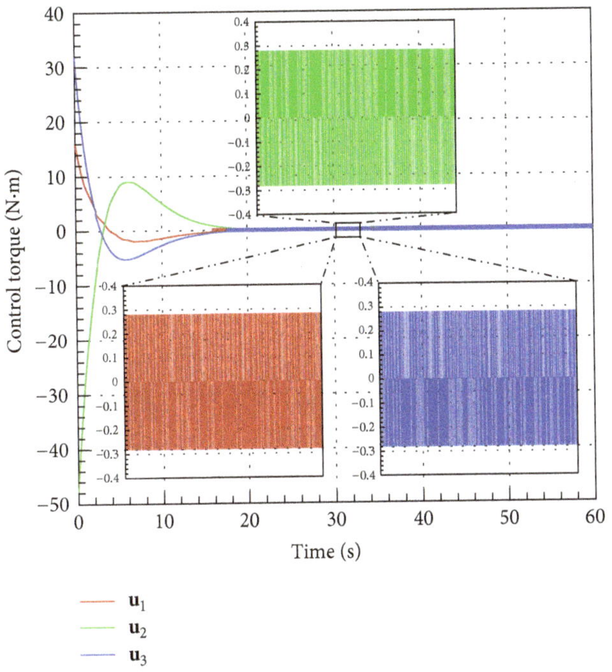

FIGURE 6: Time history of control action using classical (Not-modified) SMC.

$$J^{13} = J^{31} = 0,$$

$$J^{22} = I_{YY},$$

$$J^{23} = J^{32} = 0,$$

$$J^{33} = I_{ZZ} + \sum_{i=1}^{2} \int_{a}^{a+L_b} \rho_b^i \, {}^i\mathbf{u}^2 \, dx + \sum_{i=1}^{2}\sum_{j=1}^{N_j} \int_{x_i}^{x_i+L_{Pi}} {}^j\rho_P^i \, {}^i\mathbf{u}^2 \, dx,$$

$$\mathbf{M}_{RR} = \left[M_{RR}^{ij}\right]_{3\times 3},$$

$$M_{RR}^{11} = I_{XX} + \sum_{i=1}^{2} \left\{ {}^i\underset{\sim}{\mathbf{q}}_k^T \, {}^i\Upsilon_{yy} \, {}^i\underset{\sim}{\mathbf{q}}_k + {}^i a_y \left(\rho_b + \sum_{j=1}^{N_j} {}^j\rho_P \right) + 2 \, {}^i a_y \, {}^i\Upsilon_y^T \, {}^i\underset{\sim}{\mathbf{q}}_k \right\},$$

$$M_{RR}^{12} = M_{RR}^{21} = -\sum_{i=1}^{2} \left\{ {}^i\Upsilon_{yx}^T \, {}^i\underset{\sim}{\mathbf{q}}_k + {}^i a_x \, {}^i a_y + {}^i a_x \, {}^i\Upsilon_{yy}^T \, {}^i\underset{\sim}{\mathbf{q}}_k + {}^i a_y \left(\rho_b \frac{L_b^2}{2} + \sum_{j=1}^{N_j} {}^j\rho_P \frac{{}^jL_P^2}{2} \right) \right\},$$

$$M_{RR}^{13} = M_{RR}^{31} = 0,$$

$$M_{RR}^{22} = I_{YY},$$

$$M_{RR}^{23} = M_{RR}^{32} = 0,$$

$$M_{RR}^{33} = I_{ZZ} + \sum_{i=1}^{2} \left\{ {}^i\underset{\sim}{\mathbf{q}}_k^T \, {}^i\Upsilon_{yy} \, {}^i\underset{\sim}{\mathbf{q}}_k + {}^i a_y^2 \left(\rho_b + \sum_{j=1}^{N_j} {}^j\rho_P \right) \right\} + 2 \, {}^i a_y \, {}^i\Upsilon_y^T \, {}^i\underset{\sim}{\mathbf{q}}_k + \rho_b \int_0^{L_{bi}} x^2 dx + \sum_{j=1}^{N_j} {}^j\rho_P \int_{x_i}^{x_i+L_{Pi}} x^2 dx,$$

$$\mathbf{M}_{RF} = \begin{bmatrix} {}^1a_x & {}^2a_x \end{bmatrix} \begin{bmatrix} {}^1\Upsilon_y^T & \mathbf{0} \\ \mathbf{0} & {}^2\Upsilon_y^T \end{bmatrix} - \begin{bmatrix} {}^1\Upsilon_{yx}^T & {}^2\Upsilon_{yx}^T \end{bmatrix},$$

$$\mathbf{M}_{FR} = \begin{bmatrix} {}^1a_x & {}^2a_x \end{bmatrix} \begin{bmatrix} {}^1\Upsilon_y^T & \mathbf{0} \\ \mathbf{0} & {}^2\Upsilon_y^T \end{bmatrix} - \begin{bmatrix} {}^1\Upsilon_{yx}^T & {}^2\Upsilon_{yx}^T \end{bmatrix},$$

$$\mathbf{M}_{FF} = \begin{bmatrix} {}^1\Upsilon_{yy} & \mathbf{0} \\ \mathbf{0} & {}^2\Upsilon_{yy} \end{bmatrix},$$

$$\mathbf{C}_{RR} = \left[C_R^{ij}\right]_{3\times 3},$$

$$C_{RR}^{11} = \omega_z \sum_{i=1}^{2} \left\{ {}^i\Upsilon_{yx}^T \, {}^i\underset{\sim}{\mathbf{q}}_k + {}^i a_x \, {}^i a_y + {}^i a_x \, {}^i\Upsilon_y^T \, {}^i\underset{\sim}{\mathbf{q}}_k + {}^i a_y \left(\rho_b \frac{L_b^2}{2} + \sum_{j=1}^{N_j} {}^j\rho_P \frac{{}^jL_P^2}{2} \right) \right\} + 2 \sum_{i=1}^{2} \left\{ {}^i\dot{\underset{\sim}{\mathbf{q}}}_k^T \, {}^i\Upsilon_{yy}^T \, {}^i\underset{\sim}{\mathbf{q}}_k + {}^i a_y \, {}^i\Upsilon_y^T \, {}^i\dot{\underset{\sim}{\mathbf{q}}}_k \right\},$$

$$C_{RR}^{21} = \omega_z \left\{ \sum_{i=1}^{2} \left\{ {}^i\underset{\sim}{\mathbf{q}}_k^T \, {}^i\Upsilon_{yy}^T \, {}^i\underset{\sim}{\mathbf{q}}_k + {}^i a_y^2 \left(\rho_b + \sum_{j=1}^{N_j} {}^j\rho_P \right) + 2a_y \left(\rho_b + \sum_{j=1}^{N_j} {}^j\rho_P \right) {}^i\Upsilon_y^T \, {}^i\underset{\sim}{\mathbf{q}}_k \right\}, + \left(I_{ZZ} - I_{yy} \right) \right\},$$

$$C_{RR}^{31} = 0,$$

$$C_{RR}^{21} = -2 \sum_{i=1}^{2} \left\{ {}^i\Upsilon_{yx}^T \, {}^i\dot{\underset{\sim}{\mathbf{q}}}_k + {}^i a_x \, {}^i\Upsilon_y^T \, {}^i\dot{\underset{\sim}{\mathbf{q}}}_k \right\} + \omega_z \left(I_{XX} - I_{yy} \right),$$

$$C_{RR}^{22} = \omega_z \sum_{i=1}^{2} \left\{ {}^i\Upsilon_{yx}^T \, {}^i\underset{\sim}{\mathbf{q}}_k + {}^i a_x \, {}^i a_y + {}^i a_x \, {}^i\Upsilon_y^T \, {}^i\underset{\sim}{\mathbf{q}}_k + {}^i a_y \left(\rho_b \frac{L_b^2}{2} + \sum_{j=1}^{N_j} {}^j\rho_P \frac{{}^jL_P^2}{2} \right) \right\},$$

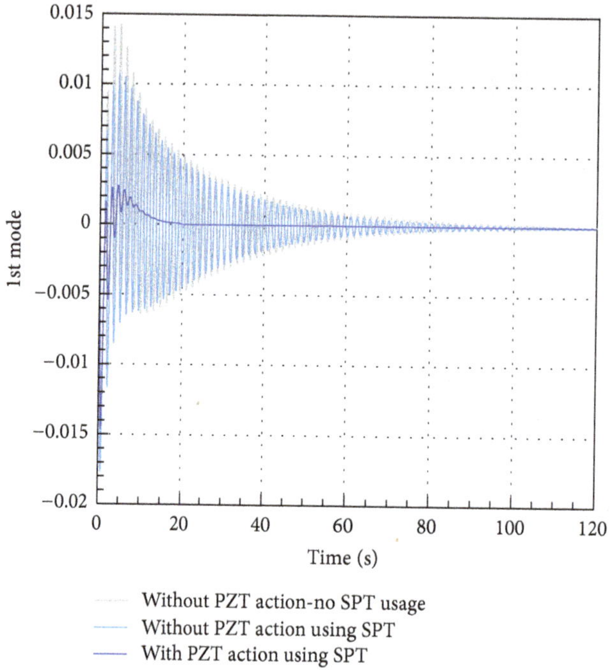

FIGURE 7: Time history of 1st vibrational mode.

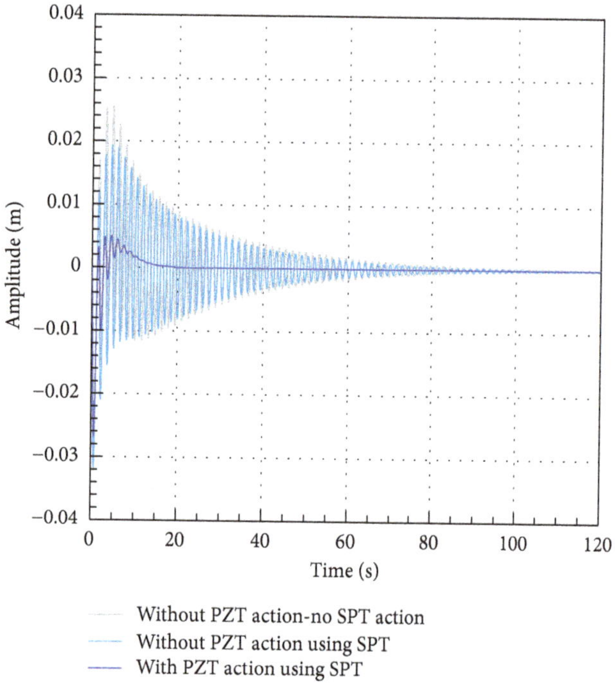

FIGURE 8: Tip deflection response of the appendage.

$$C_{RR}^{32} = 0,$$

$$C_{RR}^{31} = -\omega_x \sum_{i=1}^{2} \left\{ {}^{i}\Upsilon_{yx}^{T}\, {}^{i}\underset{\sim}{\mathbf{q}}_k + {}^{i}a_x\, {}^{i}a_y + {}^{i}a_x\, {}^{i}\Upsilon_y^T\, {}^{i}\underset{\sim}{\mathbf{q}}_k + {}^{i}a_y \left(\rho_b \frac{L_b^2}{2} + \sum_{j=1}^{N_j} {}^{j}\rho_P \frac{{}^{j}L_P^2}{2} \right) \right\}$$

$$- \omega_y \left\{ \sum_{i=1}^{2} \left\{ {}^{i}\underset{\sim}{\mathbf{q}}_k^T\, {}^{i}\Upsilon_{yy}^T\, {}^{i}\underset{\sim}{\mathbf{q}}_k + {}^{i}a_y^2 \left(\rho_b + \sum_{j=1}^{N_j} {}^{j}\rho_P \right) + 2a_y \times \left(\rho_b + \sum_{j=1}^{N_j} {}^{j}\rho_P \right) {}^{i}\Upsilon_y^T\, {}^{i}\underset{\sim}{\mathbf{q}}_k \right\} + (I_{YY} - I_{XX}) \right\},$$

$$C_R^{32} = -\omega_y \sum_{i=1}^{2} \left\{ {}^{i}\Upsilon_{yx}^{T}\, {}^{i}\underset{\sim}{\mathbf{q}}_k + {}^{i}a_x\, {}^{i}a_y + {}^{i}a_x\, {}^{i}\Upsilon_y^T\, {}^{i}\underset{\sim}{\mathbf{q}}_k + {}^{i}a_y \left(\rho_b \frac{L_b^2}{2} + \sum_{j=1}^{N_j} {}^{j}\rho_P \frac{{}^{j}L_P^2}{2} \right) \right\},$$

$$C_R^{33} = 2 \sum_{i=1}^{2} \left\{ {}^{i}\dot{\underset{\sim}{\mathbf{q}}}_k^T\, {}^{i}\Upsilon_{yx}^T\, {}^{i}\underset{\sim}{\mathbf{q}}_k + {}^{i}a_x\, {}^{i}\Upsilon_y^T\, {}^{i}\dot{\underset{\sim}{\mathbf{q}}}_k \right\},$$

$$C_{FR}^{11} = C_{FR}^{21} = C_{FR}^{31},$$

$$C_{FR}^{11} = -\omega_x \left\{ \begin{bmatrix} {}^{1}\Upsilon_{yy} & \mathbf{0} \\ \mathbf{0} & {}^{2}\Upsilon_{yy} \end{bmatrix} {}^{i}\underset{\sim}{\mathbf{q}}_k + \begin{bmatrix} {}^{1}\Upsilon_y & \mathbf{0} \\ \mathbf{0} & {}^{2}\Upsilon_y \end{bmatrix} \left[{}^{1}a_y \;\; {}^{2}a_y \right]^T \right\}$$

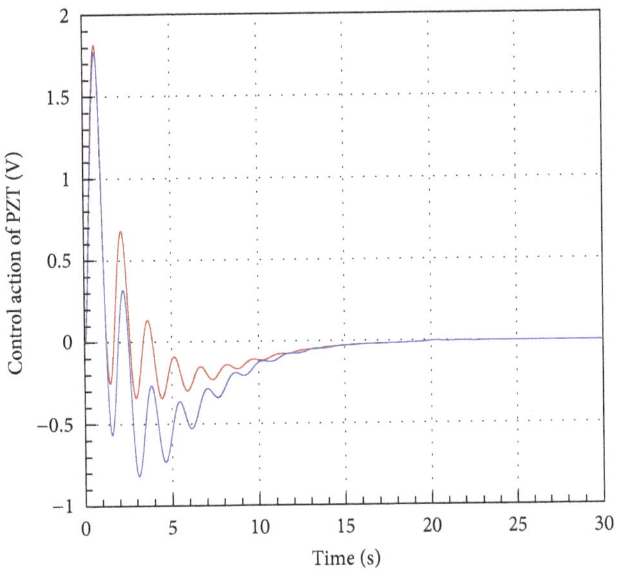

FIGURE 9: Time history of PZT actuation voltage.

$$+\omega_y\left\{\begin{bmatrix}{}^1\Upsilon_{xy}\\ {}^2\Upsilon_{xy}\end{bmatrix}\begin{bmatrix}{}^1\Upsilon_y & \mathbf{0}\\ \mathbf{0} & {}^2\Upsilon_y\end{bmatrix}\begin{bmatrix}{}^1a_x & {}^2a_x\end{bmatrix}^T\right\},$$

$$C_{FR}^{12} = C_{FR}^{22} = C_{FR}^{32} = 0,$$

$$C_{FR}^{13} = C_{FR}^{23}$$

$$= C_{FR}^{33} = -\omega_z\left\{\begin{bmatrix}{}^1\Upsilon_{yy} & \mathbf{0}\\ \mathbf{0} & {}^2\Upsilon_{yy}\end{bmatrix}{}^i\underset{\smile}{\mathbf{q}}_k\right.$$

$$\left.+\begin{bmatrix}{}^1\Upsilon_y & \mathbf{0}\\ \mathbf{0} & {}^2\Upsilon_y\end{bmatrix}\begin{bmatrix}{}^1a_y & {}^2a_y\end{bmatrix}^T\right\},$$

$$\mathbf{C}_{RF} = \mathbf{0},$$

$$\mathbf{K}_{FF} = \begin{bmatrix}{}^1\wp & \mathbf{0}\\ \mathbf{0} & {}^2\wp\end{bmatrix}{}^i\underset{\smile}{\mathbf{q}}_k,$$

$${}^i\Upsilon_{yy} = \left(\int_0^{L_b}\rho_b{}^i\psi(x){}^i\psi^T(x)\,dx\right.$$

$$\left.+\sum_{j=1}^{N_j}\int_{x_i}^{x_i+L_{pi}}{}^j\rho_p{}^i\psi(x){}^i\psi^T(x)\,dx\right),$$

$${}^i\Upsilon_y = \left(\int_0^{L_b}\rho_b{}^i\psi(x)\,dx + \sum_{j=1}^{N_j}\int_{x_i}^{x_i+L_{pi}}{}^j\rho_p{}^i\psi(x)\,dx\right),$$

$${}^i\Upsilon_{yx} = \left(\int_0^{L_b}\rho_b{}^i\psi(x)\cdot x\,dx\right.$$

$$\left.+\sum_{j=1}^{N_j}\int_{x_i}^{x_i+L_{pi}}{}^j\rho_p{}^i\psi(x)\cdot x\,dx\right),$$

$${}^i\wp = \int_0^{L_b}E_b^iI_b^i\left(\frac{\partial^2\psi^i(x)}{\partial x^2}\right)^2dx + \sum_{j=1}^{n_j}{}^jE_P^i\left({}^j\varpi_p^i\,{}^jh_p^i\right)$$

$$\times\left({}^jy^{i2} + {}^jy^i\,{}^jh_p^i + \frac{{}^jh_p^{i\,2}}{3}\right)$$

$$\times\int_{x_i}^{x_i+L_{Pi}}\left(\frac{\partial^2\psi^i(x)}{\partial x^2}\right)^2dx. \tag{B.1}$$

Conflict of Interests

The authors declare that there is no conflict of interests regarding the publishing of this paper.

References

[1] D. C. Hyland, J. L. Junkins, and R. W. Longman, "Active control technology for large space structures," *Journal of Guidance, Control, and Dynamics*, vol. 16, no. 5, pp. 801–821, 1993.

[2] T. Singh and S. R. Vadali, "Input-shaped control of three-dimensional maneuvers of flexible spacecraft," *Journal of Guidance, Control, and Dynamics*, vol. 16, no. 6, pp. 1061–1068, 1993.

[3] Q. Liu and B. Wie, "Robust time-optimal control of uncertain flexible spacecraft," *Journal of Guidance, Control, and Dynamics*, vol. 15, no. 3, pp. 597–604, 1992.

[4] S. Vadali, "Feedback control of flexible spacecraft large angle maneuvers using the Liapunov theory," in *Proceedings of the IEEE American Control Conference*, Piscataway, NJ, USA, 1984.

[5] S. Vadali, J. Junkins, and R. Byers, "Near-minimum time, closed-loop slewing of flexible spacecraft," *Journal of Guidance, Control, and Dynamics*, vol. 13, no. 1, pp. 57–65, 1990.

[6] S. V. Drakunov and V. Utkin, "Sliding mode control in dynamic systems," *International Journal of Control*, vol. 55, no. 4, pp. 1029–1037, 1992.

[7] J. Y. Hung, W. Gao, and J. C. Hung, "Variable structure control. A survey," *IEEE Transactions on Industrial Electronics*, vol. 40, no. 1, pp. 2–22, 1993.

[8] J. D. Bošković, S.-M. Li, and R. K. Mehra, "Robust tracking control design for spacecraft under control input saturation," *Journal of Guidance, Control, and Dynamics*, vol. 27, no. 4, pp. 627–633, 2004.

[9] J. L. Crassidis and F. L. Markley, "Sliding mode control using modified Rodrigues parameters," *Journal of Guidance, Control, and Dynamics*, vol. 19, no. 6, pp. 1381–1383, 1996.

[10] Q. Hu, "Variable structure maneuvering control with time-varying sliding surface and active vibration damping of flexible spacecraft with input saturation," *Acta Astronautica*, vol. 64, no. 11-12, pp. 1085–1108, 2009.

[11] Q. Hu, "Sliding mode attitude control with L_2-gain performance and vibration reduction of flexible spacecraft with actuator dynamics," *Acta Astronautica*, vol. 67, no. 5-6, pp. 572–583, 2010.

[12] Q.-L. Hu, Z. Wang, and H. Gao, "Sliding mode and shaped input vibration control of flexible systems," *IEEE Transactions on Aerospace and Electronic Systems*, vol. 44, no. 2, pp. 503–519, 2008.

[13] J. Kim, J. Kim, and J. L. Crassidis, "Disturbance accommodating sliding mode controller for spacecraft attitude maneuvers," *Advances in the Astronautical Sciences*, vol. 100, pp. 141–154, 1998.

[14] S.-C. Lo and Y.-P. Chen, "Smooth sliding-mode control for spacecraft attitude tracking maneuvers," *Journal of Guidance, Control, and Dynamics*, vol. 18, no. 6, pp. 1345–1349, 1995.

[15] Q. Hu and G. Ma, "Vibration suppression of flexible spacecraft during attitude maneuvers," *Journal of Guidance, Control, and Dynamics*, vol. 28, no. 2, pp. 377–380, 2005.

[16] M. Azadi, S. A. Fazelzadeh, M. Eghtesad, and E. Azadi, "Vibration suppression and adaptive-robust control of a smart flexible satellite with three axes maneuvering," *Acta Astronautica*, vol. 69, no. 5-6, pp. 307–322, 2011.

[17] M. D. Shuster, "Survey of attitude representations," *Journal of the Astronautical Sciences*, vol. 41, no. 4, pp. 439–517, 1993.

[18] B. Siciliano and W. J. Book, "A singular perturbation approach to control of lightweight flexible manipulators," *The International Journal of Robotics Research*, vol. 7, no. 4, pp. 79–90, 1988.

[19] E. Mirzaee, M. Eghtesad, and S. A. Fazelzadeh, "Maneuver control and active vibration suppression of a two-link flexible arm using a hybrid variable structure/Lyapunov control design," *Acta Astronautica*, vol. 67, no. 9-10, pp. 1218–1232, 2010.

[20] A. H. Meitzler, H. F. Tiersten, A. W. Warner et al., *IEEE Standard on Piezoelectricity*, IEEE Ultrasonics, Ferroelectrics, and Frequency Control Society, 1988.

3

Condition Based Maintenance Optimization of an Aircraft Assembly Process Considering Multiple Objectives

J. Li,[1] T. Sreenuch,[2] and A. Tsourdos[3]

[1] *Shanghai Aircraft Manufacturing Co., Ltd., Shanghai 200436, China*
[2] *Integrated Vehicle Health Management Centre, Cranfield University, Bedford MK43 0AL, UK*
[3] *Division of Engineering Sciences, Cranfield University, Bedford MK43 0AL, UK*

Correspondence should be addressed to T. Sreenuch; t.sreenuch@cranfield.ac.uk

Academic Editors: V. G. M. Annamdas, C. Bigelow, R. V. Rao, Y. Shi, and A. Yesildirek

The Commercial Aircraft Cooperation of China (COMAC) ARJ21 fuselage's final assembly process is used as a case study. The focus of this paper is on the condition based maintenance regime for the (semi-) automatic assembly machines and how they impact the throughput of the fuselage assembly process. The fuselage assembly process is modeled and analyzed by using agent based simulation in this paper. The agent approach allows complex process interactions of assembly, equipment, and maintenance to be captured and empirically studied. In this paper, the built network is modeled as the sequence of activities in each stage, which are parameterized by activity lead time and equipment used. A scatter search is used to find multiobjective optimal solutions for the CBM regime, where the maintenance related cost and production rate are the optimization objectives. In this paper, in order to ease computation intensity caused by running multiple simulations during the optimization and to simplify a multiobjective formulation, multiple Min-Max weightings are used to trace Pareto front. The empirical analysis reviews the trade-offs between the production rate and maintenance cost and how sensitive the design solution is to the uncertainties.

1. Introduction

Nowadays, aircraft manufacturers are operating in a global competitive environment. Increasing production rate and reducing costs are the key drivers in aircraft manufacturing. In order to meet the required production rate while meeting high quality requirements, (semi-) automatic assembly machines (e.g., Flexible Drilling Head [1], GRAWDE (Gear Rib Automated Wing Drilling Equipment), and HAWED (Horizontal Automated Wing Drilling Equipment) [2]) are increasingly being used in the aircraft assembly line. These machines can deliver significant productivity gains on the shop floor by reducing the manual multistep processes and overcoming the restricted worker access [3]. This in effect has shifted the production throughput to be now very much dependent on the operational availability of these (semi-) automatic machines [4]. Consequently, machine breakdowns and maintenance are therefore a major cause of bottlenecks in the assembly line. How to manage these machines in an efficient and cost-effective way to maximize the overall product rate is still a key challenge to the aircraft manufacturers [2].

Maintenance involves fixing when equipment becomes out of order (corrective maintenance) and also includes performing routine actions which will keep the equipment working in order or prevent failures from arising (i.e., preventive maintenance) [5, 6]. A maintenance strategy in general includes identification of parameters, inspection methods, plan execution, and repair [7, 8]. In the recent decade, Condition Based Maintenance (CBM) has increasingly been integrated as part of the manufacturing system [4, 9–12]. Its goal is to minimize unscheduled downtime and shift towards a more forward-looking approach by monitoring deterioration of equipment conditions. Examples of integrated CBM in manufacturing system are found in the areas of measurement equipment [13], plastic injection [14], plastic yoghurt pots [15], food and drink industry [16], PBL (performance-based logistics) contracts [17], and generic stochastically deteriorating systems [18]. In these examples, it has been shown that CBM can potentially improve the overall cost and production rate of the manufacturing systems by increasing the machine availability while reducing the maintenance cost [17, 18].

In many cases, manufacturing for an example, where many high-value assets (machines) are part of it, is impractical and economically not feasible to experiment different manufacturing processes based on the real objects [9]. Simulation could allow complex process interactions of assembly, equipment, and maintenance to be captured and empirically studied in a virtual environment without having to build a real manufacturing system. Agent Based Simulation (ABS) and Discrete Event Simulation (DES) have been used in the manufacturing domain. ABS is based on the dynamic interaction of entities involved in the process. Examples of the ABS are autonomic manufacturing execution system [19] and intelligent manufacturing (e.g., enterprise integration and collaboration, manufacturing process planning, and scheduling) [20]. DES is on the other hand based on a fixed sequence of operations or process being performed over entities [9]. DES is more widely adopted in manufacturing as fixed sequence of operations can be naturally captured [9]. However, in a highly complex process it is simpler to model using ABS; complex interactions between entities (e.g., machine, service, and process) can be naturally captured in ABS without having to reformulate a problem into the queue theory framework as required by DES [21, 22]. Both DES and ABS allow the important aspects like quality, cost, and time to be simulated and analyzed which provide the basis in Manufacturing System Development (MSD) and Product Realization Process (PRP) [23].

In simulation, a model comprises several input variables or model parameters such as scheduling properties, process leap time, and machine reliability. The aim of MSD or PRP is to find optimal controllable parameters that will result in the most desirable outputs of the process. In the case of (semi-) automatic assembly lines, examples of performance indicators are maximum production rate and minimum maintenance cost. To find an optimal solution, the simulation is iterated until the most optimal combination of variables is found; at each iteration the controllable variables are adjusted, the model is simulated, and the simulation output is then evaluated against the design objectives [24, 25]. Evolutionary techniques (e.g., scatter search and genetic algorithms) are often applied to solve difficult simulation optimization problems [26–28].

In this paper, CBM is exploited as part of a design solution for a (semi-) automatic aircraft assembly process that demands high production rate and until now there are no studies of CBM in an aircraft assembly process reported in the literature. This and the simulation optimization of a CBM integrated aircraft assembly process model will be the contribution of this paper. In this study, Commercial Aircraft Cooperation of China (COMAC) ARJ21 regional jet final assembly is used as a what-if representative example to illustrate the impact of CBM on the aircraft assembly process.

The outline of this paper is as follows: Section 2 describes an aircraft assembly process and identifies the key performance bottlenecks in the process. Sections 3 and 4 explain CBM and how a CBM enabled aircraft assembly system is modelled using agent concepts. A multiobjective simulation optimization approach is described in Section 5. Section 6 describes performance measures and then evaluates trade-off between competing objectives and their relation to design parameters. Finally, the concluding remarks are made in Section 7.

2. Aircraft Assembly Process

2.1. ARJ21 Structure Assembly. ARJ21 is one of two ongoing COMAC's regional jet development programs [29]. It is a new type of turbofan short/medium range 78–90 seat regional aircraft. The ARJ21 program is in the ongoing certification process and is currently in transition from development stage to batch serial production. The ARJ21 has received a total of 309 orders as of 2013. COMAC has planned to increase its production rate to 30 aircrafts per year by 2015. However, at this state, the production of ARJ21 is heavily relying on manual processes and inevitably limited to 1-2 aircrafts per year. To meet the delivery target (i.e., 30 aircrafts per year) while maintaining quality and cost effectiveness, the manufacturing and assembly processes of the ARJ21 have to be less of manual work, but more automated by adopting the concept of (semi-) automatic assembly process.

Similar to other integrated aircraft manufacturing networks like B777, B787, A340, and A380, the main structure components of the ARJ21 are manufactured and assembled across China by three other ARJ21 consortium members (Tier-1) located in Xi'an, Chengdu, and Shenyang. The parts are then transported and finally assembled by COMAC itself in Shanghai. This also means any delay from the Tier-1 airframe component suppliers or in the final assembly will respectively cause holdup in the production rate or accumulation of components from the suppliers. Hence, in order to maximize the overall production rate, it is important that disruptions in each assembly line at different sites will have to be minimized.

In this paper, the subfinal assembly of ARJ21 fuselage joint is used as a case study to illustrate the impact of maintenance on the assembly process performance. This can be subsequently extended to cover the whole final assembly process or applied to the other Tier-1 component-level assemblies. At the ARJ21 final assembly line, each ARJ21 arrives in seven substructures: nose section, front fuselage, central fuselage, aft fuselage, rear fuselage (including tails), and both wings. The components are uploaded to transporters and taken to three specific assembly stations, where in parallel the forward fuselage is constructed of the nose section and front fuselage, the wings are joined to the central fuselage, and the aft and rear fuselages are joined which form the aft fuselage, see Figure 1. The three main fuselage substructures are then transported to the final assembly station where they are joined together into a complete airframe. In this paper, we will focus on the assembly processes (i.e., Stages 200A and 200B) carried out this station.

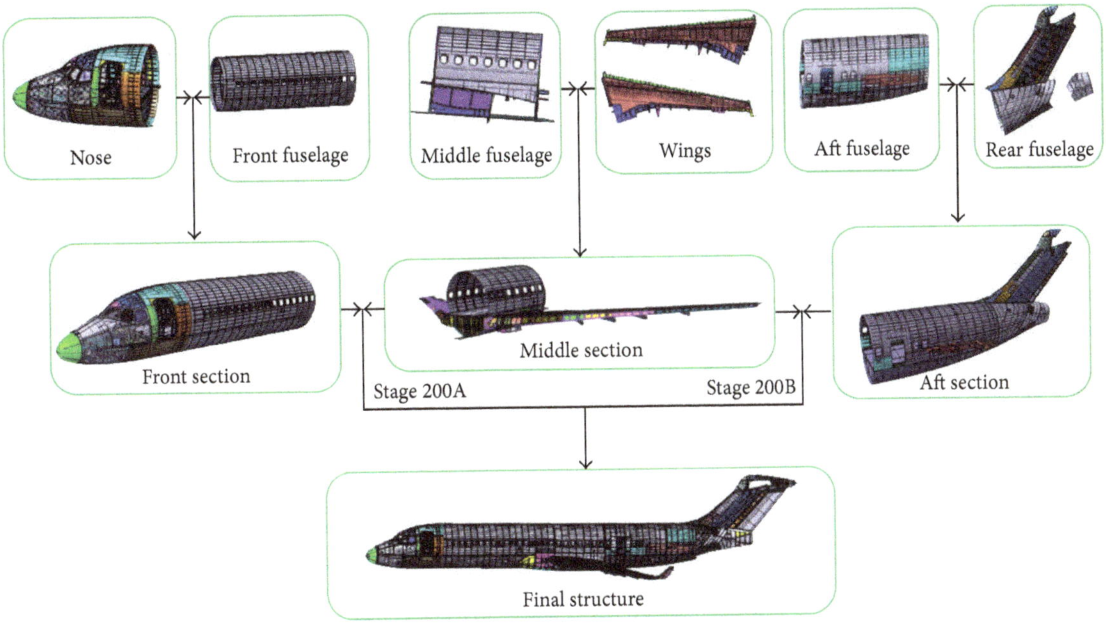

FIGURE 1: COMAC ARJ21 structure assembly.

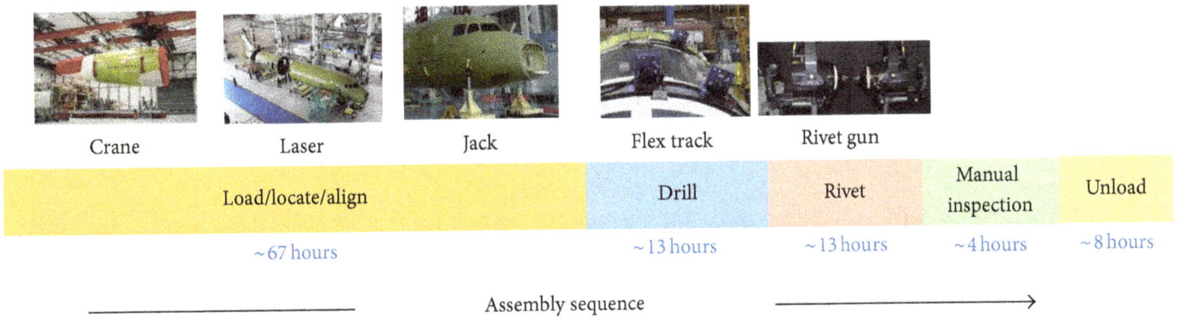

FIGURE 2: ARJ21's main sequence of final joints and related assembly machines.

2.2. Fuselage Joining Process. The main sequence of the final joints (Stages 200A and 200B) can be divided into 5 steps as shown Figure 2. The joint work starts by using an overhead crane to move and prealign the fuselage components into an assembly jig. These components are then aligned with high accuracy by adjusting the supported jacks guided by measurement data from the laser tracking devices. It takes approximately 67 hours to complete this sequence. The next assembly step is to drill the joint structures, that is, front and middle sections and middle and aft sections. In this process, the skins, frames, stringers, and other joint parts of the structure pair are drilled together and then deburred. In manual drilling, this process relies on precision measurements and drilling skills which is time consuming and often imperfect (e.g., oblique holes and excessive countersinks). In an (semi-) automated assembly line, a light weight portable computer numerical control (CNC) drilling machine (a.k.a Flex Track [30]) is used to reduce the lead time while maintaining the required drilling quality. This process is estimated to be 13 hours assuming that 2 Flex Tracks are used. In the third step, the joint fuselage sections are fastened using hammered solid rivets. In this process, "Handheld Electromagnetic Rivet Guns" are used in place of manual riveting to reduce lead time and noise hazard level. For 2 pairs of rivet guns, the lead time of this process is estimated to be 13 hours. The fourth step is to manually inspect the riveting quality such as position, depth, and angle. This process takes approximately 4 hours. Finally, the jig and jacks are removed from the completed fuselage. The fuselage is then towed away from the assembly area. This final unloading process takes about 8 hours.

Table 1 details subprocesses of the main assembly sequence and their estimated lead time. These values were obtained from interviews of COMAC engineer. Note that these estimates do not take into account any of process disruptions which could be caused by part delays, machine breakdowns and maintenance, or other factors.

2.3. Bottlenecks. During the design and development phases, manual work is relevant and sufficient as the assembly process

Table 1: Estimated work process lead time.

Work contents	Assembly machines	Lead time (Hrs)
Load/locate/align		
Load middle section	Crane	9
Prelocate and -align Jig and middle section	Jacks and lasers	19
Load front section	Crane	9
Load aft section	Crane	9
Locate and align Jig and fuselage sections	Jacks and lasers	21
Drill (in parallel)		
Front and middle sections		
Load flex tracks	Flex tracks	2
Drill	Flex tracks	10
Unload flex tracks	Flex tracks	1
Front and middle sections		
Load flex tracks	Flex tracks	2
Drill	Flex tracks	10
Unload flex tracks	Flex tracks	1
Rivet (in parallel)		
Front and middle sections		
Load rivet guns	Rivet guns	2
Rivet	Rivet guns	10
Unload rivet guns	Rivet guns	1
Middle and aft sections		
Load rivet guns	Rivet guns	2
Rivet	Rivet guns	10
Unload rivet guns	Rivet guns	1
Manual inspection		4
Unload		8

is not finalized and the required production rate is limited to 1 or 2 aircrafts per year. To meet the delivery target of 30 aircrafts per year, the assembly process of the ARJ21 has to be of less manual work but to be more automated. However, disruption caused by machine breakdowns and maintenance is one of the key performance bottlenecks in a (semi-) automatic assembly line [2]. From interviews of COMAC engineers, an overhead crane, Flex Tracks, and rivet guns are less reliable in relation to other types of assembly equipment and will likely be the common causes of machine breakdown.

From Table 1, it can be seen that three of the main processes of the ARJ21 fuselage joining process are heavily dependent on the overhead crane, Flex Tracks, and rivet guns. Hence, the downtime of these machines will essentially affect the throughput of the assembly process. Table 2 summarizes estimated maintenance parameters of these assembly machines. In this paper, the crane, Flex Tracks, and rivet guns are the focus of an application of CBM.

3. Condition Based Maintenance

3.1. Degradation Process. Machine failures can be divided into two categories, random failures and those as a consequence of degradation. In this paper, we only consider the degradation failures in which preventive maintenance strategies can be applied. A simplified degradation process is illustrated in Figure 3. R_{PM}, R_F, and $T_M^{(i)}$ are the preventive maintenance threshold, failure threshold, and required duration to perform the ith maintenance (or repair), respectively. The degradation process can be represented by a stochastic process of increasing wear, and hence decreasing in system reliability, finally leading to machine failure. The degradation stages can be modelled using either discrete steps or continuous process in time. The failure occurs when the machine degradation stage reaches a certain reliability level. In Figure 3, maintenances are used to intervene with the degradation process and bring about an improvement to a certain reliability level before failures occur. However,

Table 2: Estimated maintenance parameters.

Assembly machines	MTBM[1] (hrs)	Maintenance time (hrs)	Loss time[2] (hrs)
Overhead crane	720	1	169
Flex track	900	2	506
Rivet gun	1440	1	336

[1]MTBM: mean time before maintenance; [2]Loss time: minimum downtime caused by unexpected breakdown.

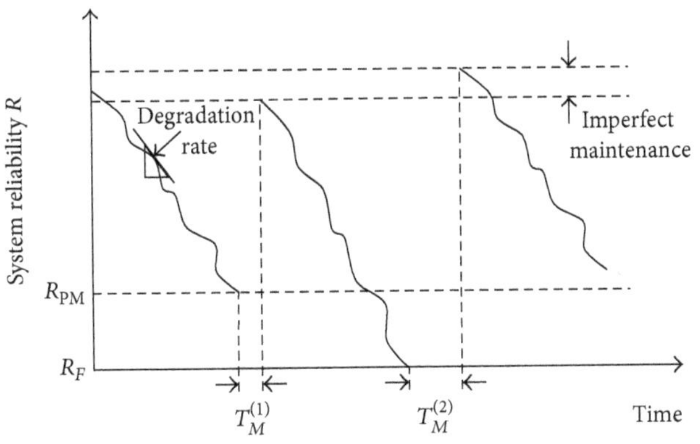

FIGURE 3: Degradation process.

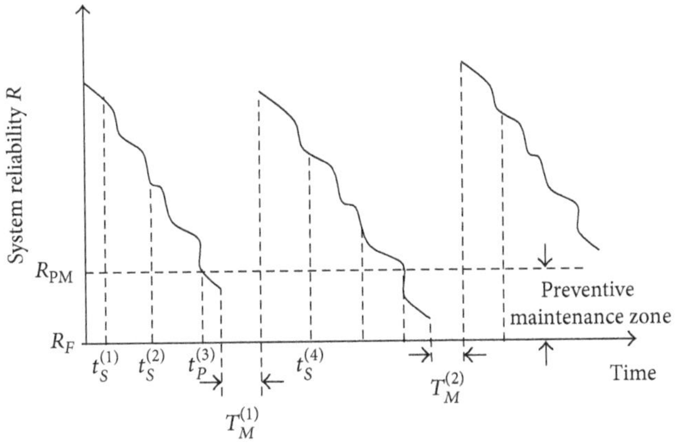

FIGURE 4: Condition based maintenance (CBM) framework.

when there is no ambiguity, the term "maintenance" will also include "repair" operations in this paper. The randomness (being stochastic) is from uncertainties in the degradation rate and maintenance. The latter results from imperfect maintenance.

3.2. Maintenance Model. The purpose of maintenance is to increase the mean time to failure. It is assumed that maintenance will bring about an improvement to the conditions in the previous stage of degradation [31], see Figures 3 and 4. In CBM framework, a maintenance policy relies on continuous condition monitoring which can be carried out by embedded sensors or periodic inspection (i.e., $t_S^{(i)}$ in Figure 4) [11, 14]. This paper assumes that embedded sensors are used to support online continuous condition monitoring (CM). In CBM, instead of traditional fix schedule, maintenance interventions are performed only when the system reliability degrades below a certain preventive maintenance threshold. In this way, unnecessary maintenance actions and unexpected breakdowns can be reduced; a real-time CM system provides an estimate for the reliability level due to degradation. In CBM, when to take maintenance actions, that is, defining the R_{PM} level, is essentially the main design maintenance parameter. This parameter will be based on both the system reliability level at inspection time and the potential evolution of the system's degradation process. R_{PM} must be sufficiently high to allow maintenance actions to be performed before the machine degrades to the failure level R_F.

4. Agent Based Model

DES is widely used in modelling and simulation of manufacturing systems where fixed sequence of operations can be naturally captured. However, in a highly complex process, it is simpler to model using ABS; complex active interactions between entities (e.g., machine, maintenance, and process) can be naturally captured in ABS without having to reformulate a problem into a series of discrete events as required by DES [21, 22]. In our case, CM system, whose self-aware (or active) properties are required in order to trigger maintenance activities, is an example of entities that cannot straightforwardly be modeled using DES; the behaviour of a CM system has to be determined by the system in a DES model (unintuitively being passive). For this reason, ABS is preferred to DES in this paper.

In this paper, for simplicity in the analysis, a CBM enabled ARJ21 assembled system is composed of minimally assembly machines, maintenance and the assembly process itself, and the interdependencies among these components. In ABS, these entities are called agents. Here, we use statechart to model behaviours of an agent and a messaging concept to model interactions between agents [32]. The statechart allows active stochastic behaviours of an agent to be modelled [33]. In ABS, agents independently evolve in parallel using the same universal time tick generated by the simulation environment.

Figure 5 depicts a generic agent based model of the assembly machines described in Section 2, that is, crane, flex tracks and rivet guns. Δt, RUL, ΔR, $\mathcal{X}(\cdot,\cdot)$, μ_{RUL}, σ_{RUL}, $\mu_{\Delta R}$, and $\sigma_{\Delta R}$ are the simulation time step, remaining useful life, degradation rate, random variable, mean remaining useful life after maintenance, uncertainty caused by imperfect maintenance, mean degradation rate threshold, and degradation uncertainty, respectively. A machine can be in either "In Operation" or "Out of Order" states. When in operation, the machine is "Idle" if it is not needed by an assembly process. The changes in state to "Busy" and back to "Idle" are triggered by the messages "In Use" and "Work Completed" sent from the process, respectively. When the machine is in use, the system reliability ~RUL is decreasing at the rate of ΔR. Note that RUL is a function of operating hours not time.

When the RUL falls below the preventive maintenance threshold R_{PM}, if have not yet sent one, a maintenance order, that is, the "Need Maintenance" message, is sent to the related maintenance service. The machine goes into "Out of Order" triggered by either the breakdown (i.e., RUL $\leq$ 0) or "In Maintenance" messages from the related service and is changed back to be operable (either "Idle" or "Busy" depending on which one is the last state) after being maintained or repaired triggered by the message "Maintained". After maintenance, the system reliability is brought back to a certain level randomly defined by the imperfect maintenance

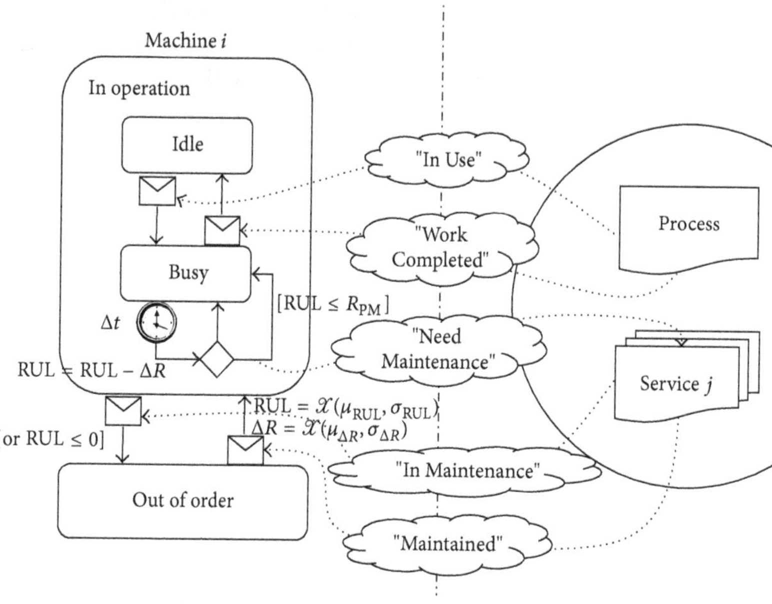

FIGURE 5: Agent based model of machines.

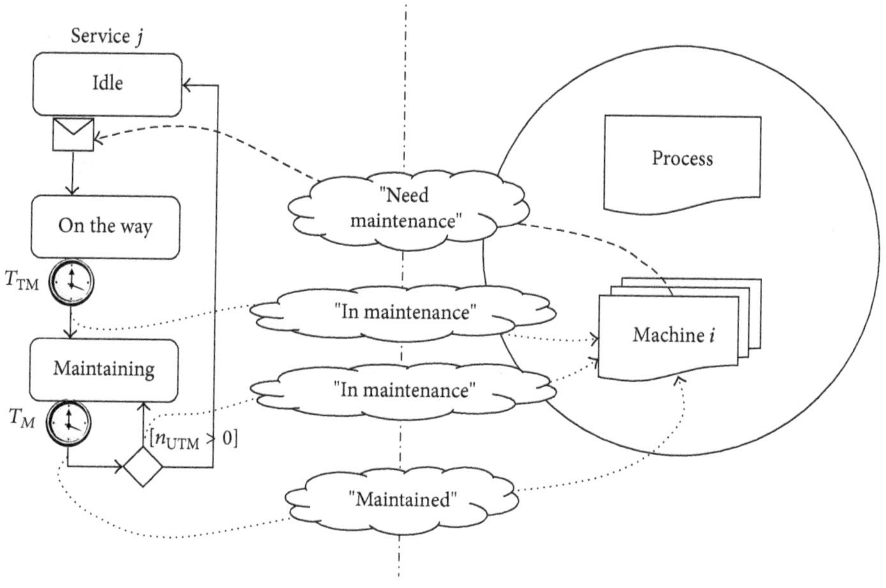

FIGURE 6: Agent based model of services.

parameters μ_{RUL} and σ_{RUL}. Depending on the density function, $\sigma_{...}$ can be either a bound, standard deviation, or no value depending on the probability distribution of the random variable $\mathcal{X}$ for normal. In this model, the stochasticity in the degradation process is simulated by randomly resetting ΔR according to the degradation parameters $\mu_{\Delta R}$ and $\sigma_{\Delta R}$.

An agent based model for maintenance services is shown in Figure 6. T_{TM}, T_M, and n_{UTM} are the time to maintenance caused by delay or travel, maintenance time, and number of machines that require maintenance, respectively. A transition from "Idle" to "On the Way" is triggered by the message "Need Maintenance" sent from one of the related machines. On average, the service needs T_{TM} to respond to the maintenance order. Before the service goes into the "Maintaining" state, a message "In Maintenance" is sent to the machine at the first in the requested list for maintenance. It takes on average T_{TM} for a maintenance action. When the timer expires, the message "Maintenance Complete" is sent to the maintained machine which causes a state transition from "Out of Order" to "In Operation". It is assumed that every triggered maintenance action is completed successfully. The service's new state now depends on the number of machines that required maintenance. If another machine is to be maintained (i.e., $n_{UTM} > 0$), the service chooses the first machine in the requested list and as before maintains the machine. The service's state goes into "Idle" if there are no more machines in the requested list. Whenever the "Need Maintenance" message is received, the requested machine will be added to

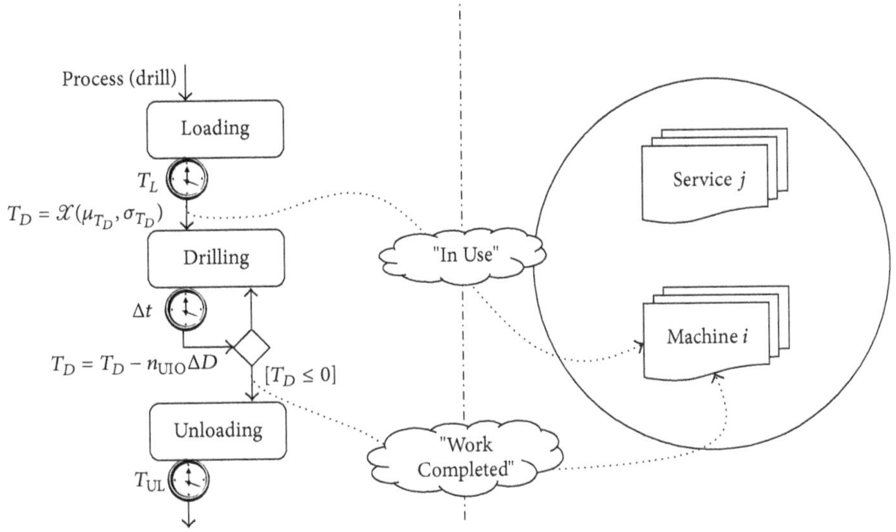

FIGURE 7: Partial agent based model of assembly process.

the requested list for maintenance. This allows the service to respond to the maintenance request even though it is not in the "Idle" state.

Figure 7 depicts an agent based model of the assembly process. For the sake of explanation, it only shows the drill subassembly process. T_L, T_D, μ_{T_D}, σ_{T_D}, ΔD, T_{UL}, and n_{UIO} are the loading time, required drill time, mean required drill time, drilling process uncertainty, drill rate, and number of machines currently in operation, respectively. After the load/locate/align process is completed, the state is transitioned to the "Drill" composite state in which "Loading" is the entry state. On average, it needs T_L to position the two Flex Tracks at the font/middle and middle/aft section joints. When the timer expires, the message "In Use" is sent to the Flex Track machines which causes a state transition from "Idle" to "Busy", and the required drill time is set to a random value determined by the process uncertainty parameters μ_{T_D} and σ_{T_D}. In this paper, it is assumed that maintenance actions do not interrupt the "Loading" and "Unloading" processes. When the machine is in use, the required (or remaining) drill time T_D is decreasing at the rate of $n_{\text{UIO}}\Delta D$. How fast the task can be completed will depend on how many machines are available (i.e., in operation) to perform the tasks. When the drill process is completed (i.e., $T_D \leq 0$), the message "Work Complete" is sent to the related machines, and the process's new state is now "Unloading". It takes on average T_{UL} to remove the machines (i.e., Flex Track in the drill case) from the fuselage. When the timer expires (also means the drill subassembly process completed), the state is then transitioned to the "Rivet" composite state.

5. Optimization Considering Multiple Objectives

5.1. Simulation Optimization. The agent based simulation model described in Section 4 is used to empirically study the impact of CBM on the (semi-) automatic fuselage assembly process. Through simulation, the aim is to find the best CBM configuration that will maximize the production rate and at the same time minimize the maintenance cost of the assembly process. A simulation optimization scheme is shown in Figure 8. $\vec{x}$, Δ, $f(\cdot)$, $g(\cdot)$, and $E[\cdot]$ are the design variable vector, uncertainties, objective function, constraint function, and mean (or expected) value, respectively. In our case, the simulation model (i.e., the agent based simulation model of the fuselage assembly system) takes the controllable maintenance design variables as the input and outputs the simulated time series data which are then used to evaluate the performance objectives of the assembly process. In contrast to the deterministic simulation, two successive simulations of the same input variables return two different simulation results due to the uncertainties in the machine degradation, maintenance, and process lead time. In the simulation optimization, the optimization objective is therefore the mean of the objective evaluations $E[f(\vec{x}, \Delta)]$ from multiple simulation runs (called Monte-Carlo simulation) of the model [35–37]. Here, "Optimal" means on average that this is the best design solution.

5.2. Scatter Search. In this paper, AnyLogic Multimethod Simulation Software is used to implement the agent based fuselage assembly model described in Section 4. OptQuest scatter search package is the only built-in optimization engine/method in AnyLogic [38]. Scatter search is a population-based metaheuristics for optimization and has been successfully applied to many hard optimization problems [28, 39]. In contrast to Evolutionary Algorithms, scatter search does not emphasize randomization. It is instead designed to incorporate strategic responses, both deterministic and probabilistic, that remember which solutions worked well (i.e., both high quality solutions and diverse solutions) and recombined them into new, better solutions [40, 41].

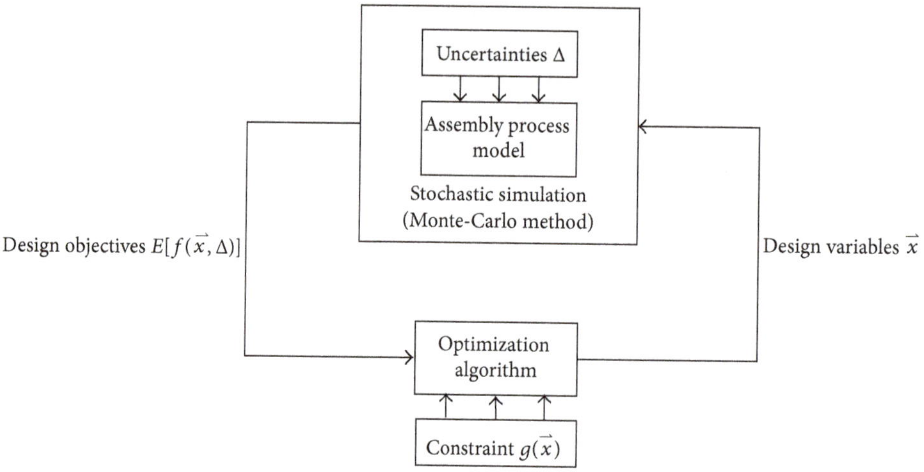

FIGURE 8: Simulation optimization.

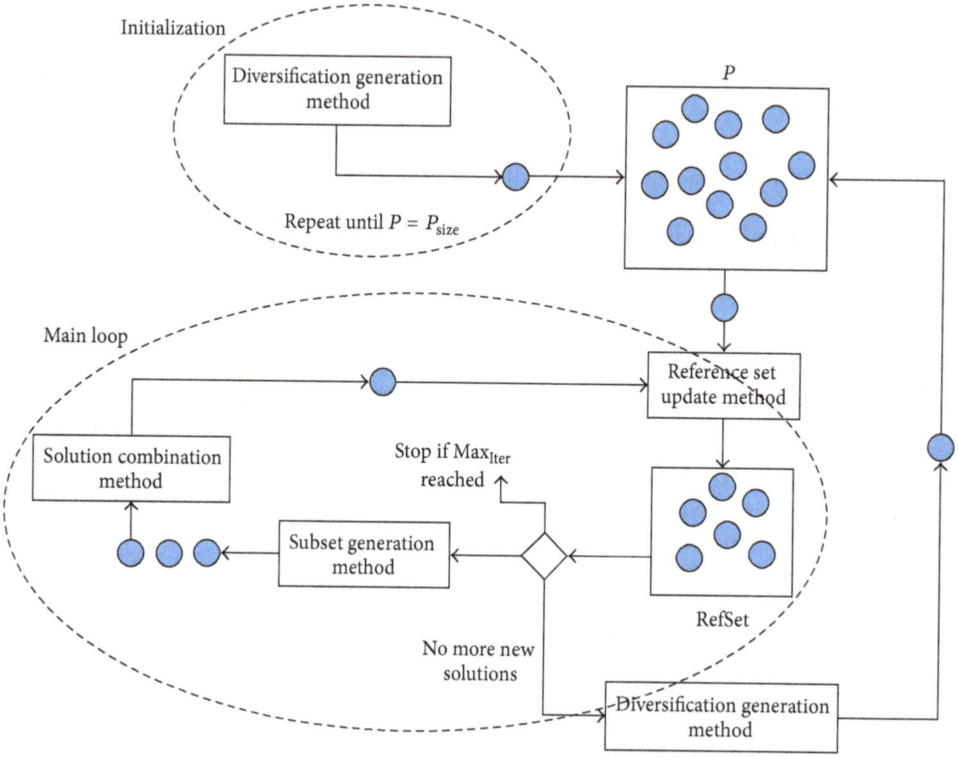

FIGURE 9: Scatter search optimization method [34].

Figure 9 outlines a simplified scatter search optimization method. The search consists of four methods: (1) diversification generation, (2) reference set update, (3) subset generation and (4) solution combination. The search starts with the diversification generation method which is used to generate a large set P of diverse solutions (i.e., scattered across the parameter space) that are the basis for initializing the search. The process repeats until $|P| = P_{Size}$. The initial RefSet is built according to the reference set update method, which can take the b best solutions (as regards their quality or diversity in the problem solving) from P to compose the RefSet. The search is then initiated by applying the subset generation method which produces subsets of reference solutions as the input to the combination method. The solution combination method uses these subsets to create new combined solution vectors. The reference set update method is applied once more to build the new RefSet and the main loop repeats again. The iteration stops when the maximum iteration is reached. However, at the end of each iteration, if no more better solutions are found, the new

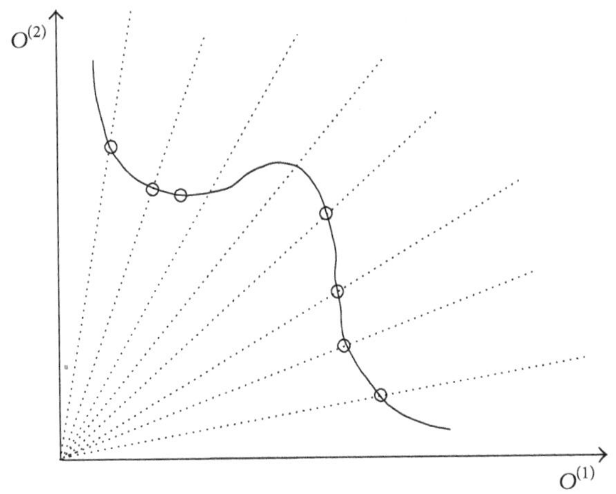

FIGURE 10: Pareto optimal solutions corresponding to a set of target vectors.

RefSet is partially rebuilt by refilling the RefSet with a new diverse solution generated using the diversification method [26, 34, 38, 42].

5.3. Multiple Single Objective Pareto Sampling. In addition to the production rate, the cost of maintenance is another important objective in the integrated maintenance (semi-) automatic aircraft assembly process. In deterministic multi-objective optimization, nondominated ranking methods are generally used in finding the Pareto optimal solutions [43–45]. These methods rely on a large number of populations to find all Pareto solutions and consequently are practically not feasible for simulation optimization primarily due to the add-on computational burden required by the Monte-Carlo simulation. Moreover, the optimization experiment in AnyLogic in any case is limited to single objective optimization and possibly because of the computational reason. This limitation in fact applies to most of the commercial of-the-shelf (COTS) DES or ABS software.

In this paper, a multiple single objective Pareto sampling method described in [46, 47] is used to find the Pareto optimal solutions. It is based on the classical weighed metric method. The method locates some specific solutions on the Pareto front corresponding to a given set of target vectors $V = \{\vec{v}_1, \ldots, \vec{v}_T\}$, where $\vec{v}_i = (1/w_i^{(1)}, \ldots, 1/w_i^{(\mathrm{NOBJ})})$ is the direction corresponding to the given ith weight vector, and $w_i^{(k)}$ is the weight of the kth objective, see Figure 10. For each rerun $\vec{v}_i$, the corresponding optimal solution is found by minimizing the objective function [48–50]

$$L_i(\vec{x}) = E[f_i(\vec{x}, \Delta)], \quad (1)$$

where

$$f_i(\vec{x}, \Delta) = \max\{w_i^{(1)} O_i^{(1)}, \ldots, w_i^{(\mathrm{NOBJ})} O_i^{(\mathrm{NOBJ})}\}. \quad (2)$$

The uncertainties Δ represent a Monte-Carlo simulation replication. In this way, the multiobjective optimization can be carried out using the AnyLogic built-in optimization engine and as well computationally is more plausible.

The primary advantage of this method is that the target vectors can be arbitrarily generated focusing on the regions of interest. Note also that if the optimization process converges to a solution that exactly matches the weight vector, then $w^{(1)}O^{(1)} = \cdots = w^{(\mathrm{NOBJ})}O^{(\mathrm{NOBJ})}$. Thus, the angle between the vectors $\vec{v}$ and $\vec{O}$ indicates whether the solution lies in where it was expected or not. If the vector $\vec{v}$ lies within a discontinuity of the Pareto set or is outside of the entire objective space, then the angle between the two vectors will be significant. By observing the distribution of the final angular errors across the total weight set, the limits of the objective space and discontinuities within the Pareto set can be identified.

6. Results

6.1. Measures of Performance. The agent based simulation model described in Section 4 is by design developed to empirically study the impact of CBM on the (semi-) automatic fuselage assembly process and find a best CBM system configuration for maximizing production throughput while keeping related costs minimum. From Table 1, it can be seen that most parts of the fuselage joint sequence are dependent on the operational availability of the assembly machines, that is, crane, Flex Tracks, and rivet guns. The downtime of these machines will essentially incur loss in the production. For consistency, "Optimize" means "Minimize" in this paper.

For the first performance measure, in order to maximize the production rate, we seek to minimize the production loss objective defined by

$$O^{(1)} = \frac{N_A - n_A}{\epsilon N_A}. \quad (3)$$

n_A, N_A, and ϵ are the number of aircrafts produced per year, maximum number of aircrafts per year without part delays, machine breakdowns and maintenance or other factors, and in percentage maximum allowable production loss from breakdown and maintenance, respectively. The product ϵN_A is the normalization factor of the incurred production loss.

Besides the production loss, at what cost in keeping the machines operable is another performance measure in the integrated CBM aircraft assembly system. In this paper, the maintenance cost consists of the fixed maintenance cost and service level related maintenance cost, and its corresponding performance objective is defined by

$$O^{(2)} = \frac{C_M + C_{\mathrm{MRT}}(T_R)}{C_M + C_{\mathrm{MRT}}(T_{\mathrm{NR}})}. \quad (4)$$

C_M, T_R, and T_{NR} and $C_{\mathrm{MRT}}(\cdot)$ are the fixed maintenance cost, maintenance response time, nominal maintenance response time, and cost function related to required maintenance response time defined by $\rho T_R/T_{\mathrm{NR}}$, respectively. The response time parameters T_{NR} and T_R can be considered as the level of service required for machine maintenance. In simulation, these parameters define T_{TM} the time to maintenance caused by delay or travel shown in Figure 6. For this objective, the nominal maintenance cost $C_M + C_{\mathrm{MRT}}(T_{\mathrm{NR}})$ is the normalization factor.

In this paper, CBM is applied to multiple assembly machines, which will individually be contributing to the overall maintenance cost. The fixed and service level maintenance costs can simply be defined as linearly formulated as

$$C_M = \sum_{i=1}^{N} C_M^{(i)},$$
$$C_{\text{MRT}} = \sum_{i=1}^{N} C_{\text{MRT}}^{(i)} \left(T_R^{(i)}\right). \tag{5}$$

N and ith are the maximum number of assembly machines (i.e., 5 = 1 crane + 2 Flex Tracks + 2 pairs of rivet guns) and the machines' index, respectively. ith indexes the machine specific related costs and required response times.

In CBM, a maintenance order is triggered when the system reliability falls below a predefined preventive maintenance threshold R_{PM}, see Section 3. Moreover, together with R_{PM}, when the maintenance is actually performed after being triggered, T_R will essentially determine the operation availability of the machine, which indirectly determines the production rate of the aircraft assembly process. The cost of maintenance is determined by how frequent maintenance is performed and maintenance service level. These two cost factors are as well determined by the parameters R_{PM} and T_R. In this study, there are 3 types of assembly machines. Hence, R_{PM}^C, T_R^C, $R_{\text{PM}}^{\text{FT}}$, T_R^{FT}, $R_{\text{PM}}^{\text{RG}}$, and T_R^{RG} will be in total the 6 design variables for the optimization of the CBM enabled ARJ21 assembly process. R_{PM}^C and T_R^C are for the crane. $R_{\text{PM}}^{\text{FT}}$ and T_R^{FT} are for the two Flex Tracks. $R_{\text{PM}}^{\text{RG}}$ and T_R^{RG} are for the two pairs of rivet guns. In our ABS, these design variables are corresponding to the machine i's R_{PM} and service j's T_{TM}.

6.2. Trade-Off Analysis. In the paper, AnyLogic and OptQuest scatter search are employed for the simulation optimization experiments of the agent based CBM enabled ARJ21 assembly system described in Section 4. The target vectors were distributedly formed to outline the Pareto front using the weights $w^{(1)} \in [0.5, 0.99]$ and $w^{(2)} \in [0.01, 0.5]$. In this paper, the uncertainties in lead time, maintenance, and machine degradation are modelled using the triangular distribution. The lower and upper uncertainty bounds are assumed at ±25% of the mean values. The design variables are in the intervals $R_{\text{PM}}^{(\cdot)}, T_R^{(\cdot)} \in [0, \text{MTBM}^{(\cdot)}]$, where the subscript $(\cdot)$ is the machine specific type indices C, FT, or RG. Table 3 summarizes the other related simulation parameters required for the evaluation of the performance objectives $O^{(1)}$ and $O^{(2)}$ defined in (3) and (4). It is not possible to disclose related costs in term of money figures due to commercial reasons. In this paper, the ratios and normalization values are alternatively used to quantify the parameters related to the maintenance cost.

Figure 11 shows the Pareto optimal solutions obtained by multiple simulation optimization runs for different weightings (or target vectors). The simulation optimization experiments were based on 20 and 100 Monte-Carlo replications, see also Sections 5.1 and 5.3. Note that the optimization process can be rerun to ensure the consistencies of the Pareto solution (for both 20 and 100 Monte-Carlo replications). The uncertainties in the simulation can be observed from the fluctuations on the Pareto surfaces which is more apparent in the case of 20 Monte-Carlo replications. The resulting Pareto optimal solutions are more likely to represent a true Pareto surface if the optimization objectives are evaluated from a large number of Monte-Carlo simulation runs, and this can be seen in the case of 100 replications shown in Figure 11.

The Pareto surface shows the trade-off between the production loss and maintenance cost conflicting objectives. A reduction in product loss will generally increase in maintenance cost. This is in particular when the production loss $O^{(1)}$ is approximately less than 0.38 (i.e., 1.585 per year); a small decrease in production loss will result in a very significant increment in the maintenance cost. As a mean to increase the machines' operational availability with less number of maintenance performed, the response time T_R will have to be relatively small so that the machines are maintained before

TABLE 3: Parameters for evaluation of performance objectives.

Simulation parameters	Values
Maximum number of aircraft per year N_A	83.429
Allowable production loss ϵ	0.05
Crane	
Fixed maintenance cost C_M^C	1
Ratio for service level maintenance cost ρ^C	1
Nominal response time T_{NR}^C	168 hrs
Flex track	
Fixed maintenance cost C_M^{FT}	2
Ratio for service level maintenance cost ρ^{FT}	1.5
Nominal response time $T_{\text{NR}}^{\text{FT}}$	504 hrs
Rivet gun	
Fixed maintenance cost C_M^{RG}	0.4
Ratio for service level maintenance cost ρ^{RG}	0.6
Nominal response time $T_{\text{NR}}^{\text{RG}}$	336 hrs

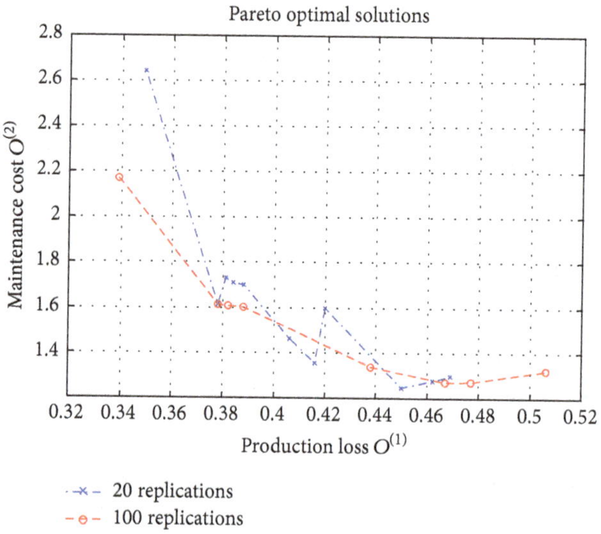

FIGURE 11: Trade-off surface of optimal CBM configurations.

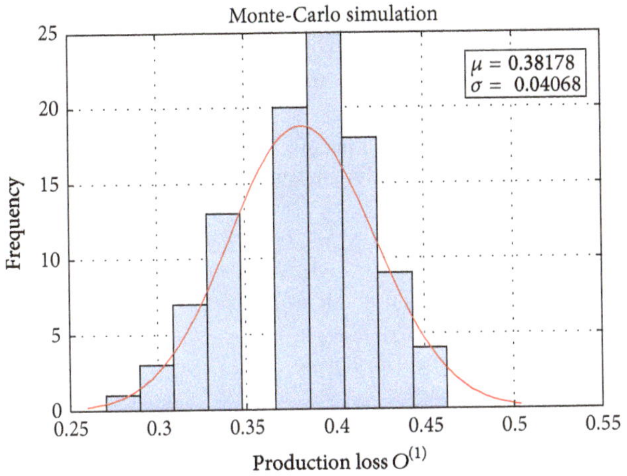

FIGURE 12: Monte-Carlo simulation results for production loss $O^{(1)}$.

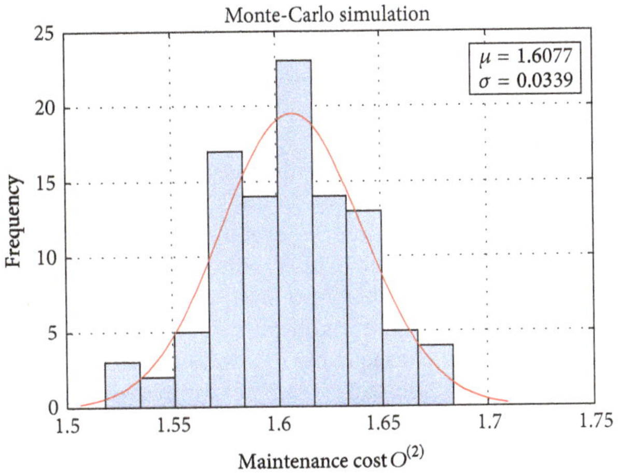

FIGURE 13: Monte-Carlo simulation results for maintenance cost $O^{(2)}$.

breakdowns happen. This will essentially increase the service level related maintenance cost C_{MRT}. This is an important insight as for $O^{(1)} < 0.38$ a decision maker can avoid a situation where a significant investment on the maintenance is made for a very small gain in the production rate and possibly in overall is not worth the investment.

The trade-off surface also outlines the optimal boundary for the maintenance cost. It can be seen that the maintenance cost $O^{(2)}$ approximately cannot be lower than 1.3 even when we deliberately set multiple target vectors to probe in the >0.5 production loss objective space. This result is from the fact that the assembly machines are essentially degrading no matter whether CBM is applied or not. From the ABS model, maintenance will always be performed on the machines regardless before or after the breakdown. Consequently, this will incur the minimum fixed maintenance cost. In addition, there is also a minimum cost for the required maintenance service level. These explain why the maintenance cost cannot be lower than a certain level, and effectively determine the boundary of the objective space.

6.3. Monte-Carlo Experiment. Suppose the optimal solution (0.382, 1.608) in Figure 11 corresponding to the weights $w^{(1)} = 0.85$ and $w^{(2)} = 0.15$ is preferred. The resulting design variables for this optimal solution are $R^{C}_{PM} = 390.54$, $T^{C}_{R} = 139.8$, $R^{FT}_{PM} = 452.7$, $T^{FT}_{R} = 254.76$, $R^{RG}_{PM} = 458.46$, and $T^{RG}_{R} = 337.02$. In paper, a Monte-Carlo simulation experiment is used to evaluate whether the obtained optimal solution is the performance expected from the system and how sensitive the design solution is to the uncertainties.

Figures 12 and 13 show the results of 100 runs of Monte-Carlo simulation. The mean production loss and maintenance cost are 0.3817 and 1.6077, respectively, and these follow what initially expected from the simulation optimization result. From the histograms, 100 Monte-Carlo simulation runs (or samples) is sufficient to represent the true probability distribution of the performance objective values. With ±25% of the mean values for the lower and upper uncertainties, the standard deviations of the production loss and maintenance cost are 0.04068 and 0.0339, respectively, 11% for the production loss and 2% for the maintenance cost in relation to the mean values. This indicates very low sensitivities for this design solution. In terms of number of aircrafts produced, this means in ~70% of the cases, the production loss will be between ±0.17 aircrafts per year from the expected value.

7. Concluding Remarks

In a current competitive environment, increasing production rate and reducing costs are the key drivers in aircraft manufacturing. More (semi-) automatic assembly machines have increasingly been used in the aircraft assembly lines as a mean to deliver high production rate while meeting high quality requirements. However, the production throughput is effectively dependent on the operational availability of these machines. Integration of CBM into the assembly system has potential benefits as a way to minimize the production loss and maintenance related cost. Maintenance are performed as needed, hence avoiding unnecessary downtime and maintenance cost.

In a CBM enabled aircraft assembly system, there are self-active interactions between the subsystems, for example, CM system self-triggers a maintenance order when the system reliability falls below a certain level. This example of active self-aware behavior cannot straightforwardly be modelled using DESs. In this case, where, besides the assembly process, independent entities are in addition parts of the system, ABS is proved effective as it allows complex active interactions between entities to be naturally captured.

Production rate and maintenance cost are the competing objectives in an integrated CBM aircraft assembly system. Finding trade-offs between the production rate and maintenance cost is equivalent to finding a Pareto optimal surface. The conventional nondominated ranking methods will not be practically feasible due to the computational burden required by the Monte-Carlo simulation. This limitation can be addressed by independently sampling the Pareto surface using the weighted Min-Max method. The approach allows less number of populations to be used in the optimization as it does not need to probe the whole Pareto front, and hence effectively a reduction in the computation intensity is required.

In our ARJ21 case study, the preventive maintenance threshold and required service level are the key design parameters that determine the overall performance of the assembly system. Because of uncertainties, increase production rate will require a high required service level (i.e., fast response time) to avoid breakdowns before the maintenance is performed, and consequently this will increase the maintenance cost and sometimes can be significant. However, compromising on the production rate does not always mean a further decrease in maintenance cost. The minimum cost is from the actual cost in maintaining the machines and the minimum service level. Pareto surface is an important piece of information to the system designer. Together with Monte-Carlo simulation, it can be used to support decision making in terms of cost-benefit of different design solutions and also what could be achieved.

In this example, even though in small scale, it can be seen that CBM has potential to be applicable in (semi-) automatic aircraft assembly lines. However, its claim has to be further researched in comparison with other maintenance regimes and with a high fidelity large-scale aircraft assembly example. Moreover, in terms of optimization, other different optimization methods like genetic algorithms (GAs), simulated annealing, and teaching-learning-based optimization (TLBO) should also be used in the optimization to ensure that the true Pareto font is found and consequently their performance in terms of computation and solution can be compared and analyzed.

Conflict of Interests

The authors declare that there is no conflict of interests regarding the publication of this paper.

Acknowledgments

This work is part of a research program at Cranfield University sponsored by Commercial Aircraft Cooperation of China (COMAC) and China Scholarship Council (CSC).

References

[1] MTorres, *Total Solution for Aircraft Automatic Assembly Jigs*, MTorres, Santa Ana, Calif, USA, 2013.

[2] P. Lute, *An Investigation of Airbus A380 Stage 01 Wing Box Assembly Using Discrete Event Simulation*, Cranfield University, Bedfordshire, UK, 2007.

[3] Electroimpact Inc., *A380 Stage 1 GRAWDE Machine*, Electroimpact Inc., Mukilteo, Wash, USA, 2003.

[4] IBM Corporation, *Predictive Maintenance for Manufacturing*, 2011.

[5] Wikipedia, "Preventive maintenance," Wikipedia, http://en.wikipedia.org/wiki/Preventive_maintenance.

[6] ResolveFM, "Preventive/corrective maintenance," ResolveFM, http://www.resolve.com.au/.

[7] A. Kelly, *Maintenance Strategy*, Butterworth-Heinemann, Oxford, UK, 1997.

[8] NACE, "Maintenance strategies," 2013, http://www.nace.org/, http://events.nace.org/library/corrosion/Inspection/Strategies.asp.

[9] I. Grigoryev, *AnyLogic 6 in Three Days: A Quick Course in Simulation Modeling*, Anylogic North America, 2012.

[10] J. B. Leger, E. Neunreuthe, B. Iung, and G. Morel, "Integration of the predictive maintenance in manufacturing system," in *Advanced in Manufacturing*, pp. 133–144, Springer, London, UK, 1999.

[11] Z. Tian, D. Lin, and B. Wu, "Condition based maintenance optimization considering multiple objectives," *Journal of Intelligent Manufacturing*, vol. 23, no. 2, pp. 333–340, 2009.

[12] J. Yulan, J. Zuhua, and H. Wenrui, "Multi-objective integrated optimization research on preventive maintenance planning and production scheduling for a single machine," *International Journal of Advanced Manufacturing Technology*, vol. 39, no. 9-10, pp. 954–964, 2008.

[13] D. Achermann, *Modelling, Simulation and Optimization of Maintenance Strategies Under Consideration of Logistic Processes*, Südwestdeutscher, 2008.

[14] Š. Valčuha, A. Goti, J. Úradníček, and I. Navarro, "Multi-equipment condition based maintenance optimization by multi-objective genetic algorithm," *Journal of Achievements in Materials and Manufacturing Engineering*, vol. 45, no. 2, pp. 188–193, 2011.

[15] L. Tautou and H. Pierreval, "Using evolutionary algorithms and simulation for the optimization of manufacturing systems," *IIE Transactions*, vol. 29, no. 3, pp. 181–189, 1997.

[16] D. Baglee, "Maintenance strategy development in the UK food and drink industry," *International Journal of Strategic Engineering Asset Management*, vol. 1, no. 3, pp. 289–300, 2013.

[17] J. Reimann, G. Kacprzynski, D. Cabral, and R. Marini, "Using condition based maintenance to improve the profitability of performance based logistic contracts," in *Proceedings of the Annual Conference of the Prognostics and Health Management Society*, 2009.

[18] A. Grall, C. Bérenguer, and L. Dieulle, "A condition-based maintenance policy for stochastically deteriorating systems," *Reliability Engineering and System Safety*, vol. 76, no. 2, pp. 167–180, 2002.

[19] M. Rolón and E. Martínez, "Agent-based modeling and simulation of an autonomic manufacturing execution system," *Computers in Industry*, vol. 63, no. 1, pp. 53–78, 2012.

[20] W. Shen, Q. Hao, H. J. Yoon, and D. H. Norrie, "Applications of agent-based systems in intelligent manufacturing: an updated review," *Advanced Engineering Informatics*, vol. 20, no. 4, pp. 415–431, 2006.

[21] M. A. Majid, U. Aickelin, and P. O. Siebers, "Comparing simulation output accuracy of discrete event and agent based models: a quantitave approach," in *Proceedings of the Summer Computer Simulation Conference (SCSC '09)*, Vista, Calif, USA, 2009.

[22] P. O. Siebers, C. M. MacAl, J. Garnett, D. Buxton, and M. Pidd, "Discrete-event simulation is dead, long live agent-based simulation!," *Journal of Simulation*, vol. 4, no. 3, pp. 204–210, 2010.

[23] L. Holst, *Integrating Discrete-Event Simulation into the Manufacturing System Development Process*, Division of Robotics, Lund, Sweden, 2001.

[24] J. A. B. Montevechi, R. d. C. Miranda, and J. D. Friend, "Sensitivity analysis in discrete-event simulation using design of experiments," in *Discrete Event Simulations-Development and Applications*, InTech, Rijeka, Croatia, 2012.

[25] Y. Carson and A. Maria, "Simulation optimization: methods and applications," in *Proceedings of the Winter Simulation Conference*, pp. 118–126, Atlanta, Ga, USA, December 1997.

[26] R. Martí and M. Laguna, "Scatter search: basic design and advanced strategies," *Revista Iberoamericana de Inteligencia Artificial*, vol. 7, no. 19, pp. 123–130, 2003.

[27] A. Ghosh and S. Dehuri, "Evolution algorithms for multi-criterion optimization: a survey," *International Journey of Computing and Information Sciences*, vol. 2, no. 1, 2004.

[28] M. Laguna, R. Martí, M. Gallego, and A. Duarte, "The scatter search methodology," in *Wiley Encyclopedia of Operations Research and Management Science*, Wiley-Blackwell, Hoboken, NJ, USA, 2011.

[29] COMAC, "ARJ21 regional jet program," 2013, http://english.comac.cc/products/rj/pi2/index.shtml.

[30] Electroimpact, *Flex Track*, Electroimpact, Mukilteo, Wash, USA, 2013.

[31] IEEE/PES Task Force on Impact of Maintenance Strategy on Reliability of the Reliability, Risk and Probability Applications Subcommittee, S. Aboresheid, R. N. Allan et al., "The present status of maintenance strategies and the impact of maintenance on reliability," *IEEE Transactions on Power Systems*, vol. 16, no. 4, pp. 638–646, 2001.

[32] M. Kaegi, R. Mock, and W. Kröger, "Analyzing maintenance strategies by agent-based simulations: a feasibility study," *Reliability Engineering and System Safety*, vol. 94, no. 9, pp. 1416–1421, 2009.

[33] A. Borshchev, *Designing State-Based Behavior: Statecharts*, Anylogic, 2013.

[34] Process Engineering Group, *Introduction to SSm*, Instituto de Investigaciones Marinas (C.S.I.C.), Vigo, Spain, 2009.

[35] J. April, F. Glover, J. P. Kelly, and M. Laguna, "Practical introduction to simulation optimization," in *Proceedings of the Winter Simulation Conference*, pp. 71–78, Boulder, Colo, USA, December 2003.

[36] W. Abo-Hamad and A. Arisha, "Simulation optimisation methods in supply chain applications: a review," *Irish Journal of Management*, vol. 30, no. 2, pp. 95–124, 2011.

[37] C.-H. Chen and L. H. Lee, "Introduction to stochastic simulation optimization," in *Stochastic Simulation Optimization: An Optimal Computing Budget Allocation*, System Engineering and Operations Research, World Scientific, Hackensack, NJ, USA, 2010.

[38] M. Laguna, *OptQuest: Optimization of Complex Systems*, OptTek Systems, 2011.

[39] R. Martí, M. Laguna, and F. Glover, "Principles of scatter search," *European Journal of Operational Research*, vol. 169, no. 2, pp. 359–372, 2006.

[40] OptTek, "How the OptQuest engine works," OptTek, http://www.opttek.com.

[41] F. Glover and A. Reinholz, "Metaheuristics in science and industry: new developments," in *Proceedings of the Metaheuristics International Conference*, Montreal, Canada, June 2007.

[42] I. Boussaïd, J. Lepagnot, and P. Siarry, "A survey on optimization metaheuristics," *Information Sciences*, vol. 237, pp. 82–117, 2013.

[43] K. Deb, *Multi-Objective Optimization Using Evolutionary Algorithms*, Wiley-Blackwell, Chichester, UK, 2001.

[44] M. Gen and R. Cheng, *Genetic Algorithms and Engineering Optimization*, John Wiley & Sons, New York, NY, USA, 2000.

[45] A. Ghosh and S. Dehui, "Evolutionary algorithms for multi-criterion optimization: a survey," *International Journal of Computing and Information Sciences*, vol. 2, no. 1, pp. 38–57, 2004.

[46] E. J. Hughes, "Multiple single objective pareto sampling," *Evolutionary Computation*, vol. 4, pp. 2678–2684, 2003.

[47] T. Screenuch, A. Tsourdos, E. J. Hughes, and B. A. White, "Fuzzy gain-scheduled missile autopilot design using evolutionary algorithms," *IEEE Transactions on Aerospace and Electronic Systems*, vol. 42, no. 4, pp. 1323–1339, 2006.

[48] Y. Jin and J. Branke, "Evolutionary optimization in uncertain environments—a survey," *IEEE Transactions on Evolutionary Computation*, vol. 9, no. 3, pp. 303–317, 2005.

[49] J. E. Fieldsend and R. M. Everson, "Multi-objective optimisation in the presence of uncertainty," in *Proceedings of the IEEE Congress on Evolutionary Computation*, vol. 1, pp. 243–250, Edinburgh, UK, September 2005.

[50] OptTek, "Multi-objective optimization," OptTek, http://www.opttek.com.

4

Enhancing the Supersonic Wind Tunnel Performance Based on Plenum Temperature Control

A. Nazarian Shahrbabaki, M. Bazazzadeh, A. Shahriari, and M. Dehghan Manshadi

Department of Mechanical & Aerospace Engineering, Malek-Ashtar University of Technology, Shahin Shahr, Isfahan 83145/115, Iran

Correspondence should be addressed to A. Nazarian shahrbabaki; a.n.shahrbabaki@gmail.com

Academic Editors: E. J. Avital and Y. Shi

The application of fuzzy logic controllers (FLCs) to the control of nonlinear processes, typically controlled by a human operator, is a topic of much study. In this paper, the design and application of a FLC is discussed to control the plenum chamber temperature for a blowdown supersonic wind tunnel (BSWT) with the aim of achieving the accurate and desired results. In this regard, first, a nonlinear mathematical model of special BSWT is developed in Matlab/Simulink software environment. Next, an artificially intelligent controller is designed using fuzzy logic approach. For this purpose, a proportional-derivative FLC (PD-FLC) system is developed in the Simulink toolbox to control the plenum stagnation temperature using a heater upstream of the plenum chamber. Finally, the system simulation results inside of the temperature and pressure controllers in comparison with the experimental run are presented. The results for Mach 2.5 blowdown run show the great performance of the Wind Tunnel Simulator Model and its temperature controller system.

1. Introduction

Blowdown supersonic wind tunnels (BSWTs) deliver flow at constant stagnation temperature and pressure. The stagnation temperature is generally regarded to be equal to the plenum temperature which is controlled by heater upstream of the plenum chamber. During a blowdown run, the storage tank temperature and pressure that supplies plenum chamber flow decrease continuously. Thus, to maintain a constant plenum pressure as close as possible to a setpoint pressure signal, the regulator or control valve must open progressively. Besides, to maintain a constant stagnation temperature in the plenum chamber, a heater must operate continuously during a supersonic run [1–3]. The controller must operate at different stagnation pressures and Mach numbers and has to be robust to accommodate the varying pressure and mass flow requirements safely. New concepts for control are under implementation with the goal of reducing transition phase and overall loads on the models [4, 5].

The block diagram of the control systems in the proposed BSWT is illustrated in Figure 1.

Fuzzy logic has been the area of heated debate and much controversy during the last decades. The first paper in fuzzy set theory, which is now considered to be the seminar paper of the subject, was written by Zadeh et al. [6], who is considered the founding father of the field. In that work, Zadeh was implicitly advancing the concept of approximate human reasoning to make effective decisions on the basis of the available imprecise, linguistic information.

In the 1970s, King and Mamdani [7] studied the application of FLCs to the control of nonlinear industrial processes that typically can only be controlled successfully by a human operator. The idea of FLCs has become a common solution in recent years, with applications ranging from automation of industrial processes to control of electronic devices in consumer products. The design and application of a FLC for the control of plenum chamber temperature in several wind tunnels at NASA's Langley Research Center (LaRC) in Hampton, VA, is described [8].

The intelligent controlling approaches like fuzzy logic (FL) will provide the required scope for wind tunnels to be more efficient, safe, and economic. The approaches will help to enable a level of performance that far exceeds that of today's wind tunnel in terms of reduction of harmful emissions, maximization of run time, and minimization of noise, while improving system affordability and safety.

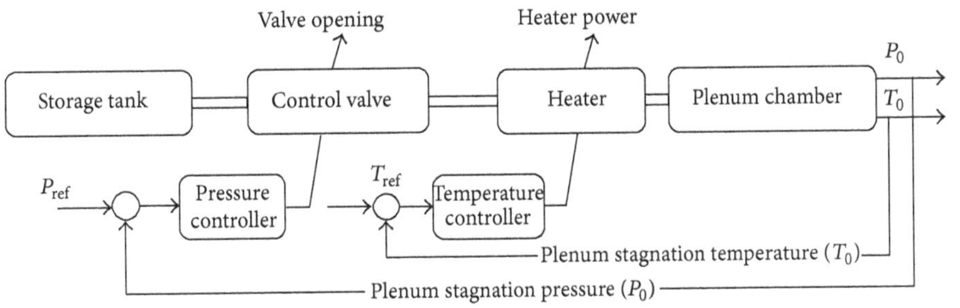

FIGURE 1: Schematic diagram of the pressure and temperature control systems in the supersonic wind tunnel.

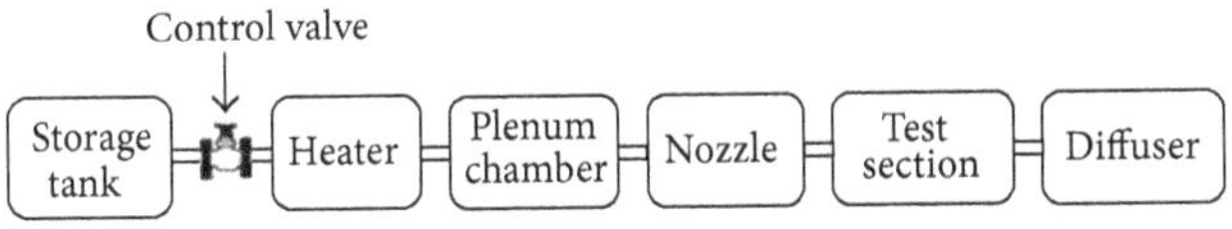

FIGURE 2: Schematic diagram of the blowdown supersonic wind tunnel.

In this paper, a nonlinear mathematical model of a special BSWT is developed in Simulink software simulator. Next, a PD-FLC system is developed in the Simulink toolbox to control the plenum stagnation temperature. Finally, the simulation results for a Mach 2.5 blowdown run in comparison with the experimental results are presented along.

2. Wind Tunnel Mathematical Model

The schematic diagram of the proposed BSWT is shown in Figure 2. In the figure, the major subsystems are a storage tank, a gate valve, a pressure regulator valve or control valve, a heater, a plenum chamber, a supersonic nozzle, a test section, a diffuser, and a silencer.

2.1. Heater. There are two ways in which BSWTs are customary operated: (1) operation with constant pressure and (2) operation with constant-mass flow rate [8].

For constant pressure operation, the only control necessary is a pressure control valve that holds the stagnation pressure constant. By the way, for constant-mass runs, the stagnation temperature must be held constant and either a heater or a thermal mass external to the storage tank is required. Even more important, the constant temperature of the constant-mass run keeps the Reynolds number constant in the test section. This drop in stagnation temperature affects the Mach number in the test section only secondarily through a change in boundary layer thickness, but it can affect balance strain gage readings significantly, and it does change the Reynolds number during a run.

Obviously a heater downstream of the storage tank could be designed to yield essentially zero temperature drops, and such heaters are sometimes employed. In this regard, the controlled flow stagnation temperature in the heater output is presented as [9]

$$T_H = T_{\text{in}} + \left[\frac{W_{\text{th}}}{q_v C_p}\right], \tag{1}$$

where T_H is the flow stagnation temperature in the heater output (K), T_{in} is the air initial stagnation temperature (K), W_{th} is the thermal input power (W), q_v is the mass flow rate through the valve (kg·s^{-1}), and C_p is the gas constant pressure specific heat (J · kg^{-1} · K^{-1}).

2.2. Control Valve. BSWTs are generally operated with a constant stagnation pressure in the plenum chamber, with control usually provided by one or more pressure control valves. In this regard, a main step of the control valve simulation is calculating the mass flow rate through the valve. This parameter is determined using the standard valve sizing relation as [10, 11]

$$q_V = C_v N_8 F_P Y P_T \sqrt{\frac{(XM_a)}{(T_T Z)}}. \tag{2}$$

In (2), C_v is the valve sizing coefficient, N_8 is a constant for units, F_P is a correction factor that accounts for pressure losses due to piping fittings and elbows, Y is expansion factor, P_T is the tank stagnation pressure, X is the valve pressure drop ratio, M_a is the air molecular weight, T_T is the tank stagnation temperature, and Z is the air compressibility factor.

It should be noted that C_v is the functions of control valve opening and should be provided by the manufacturer of the control valve. In this paper, for the valve opening simulation during the tunnel run, some lookup tables were set to the valve manufacturer's data.

3. Wind Tunnel Simulator Model (WTSM)

3.1. Facility and Requirements. The tunnel has a Mach number range of 1.5 to 3, a Reynolds number range of 25~100 × 10^6 per meter, and up to 70 seconds of usable run time. Also the tunnel is operated using a 12-inch diameter, pneumatically driven automatic butterfly valve (Bray control valve). The storage tank has a volume of 220 m^3 and it is filled with

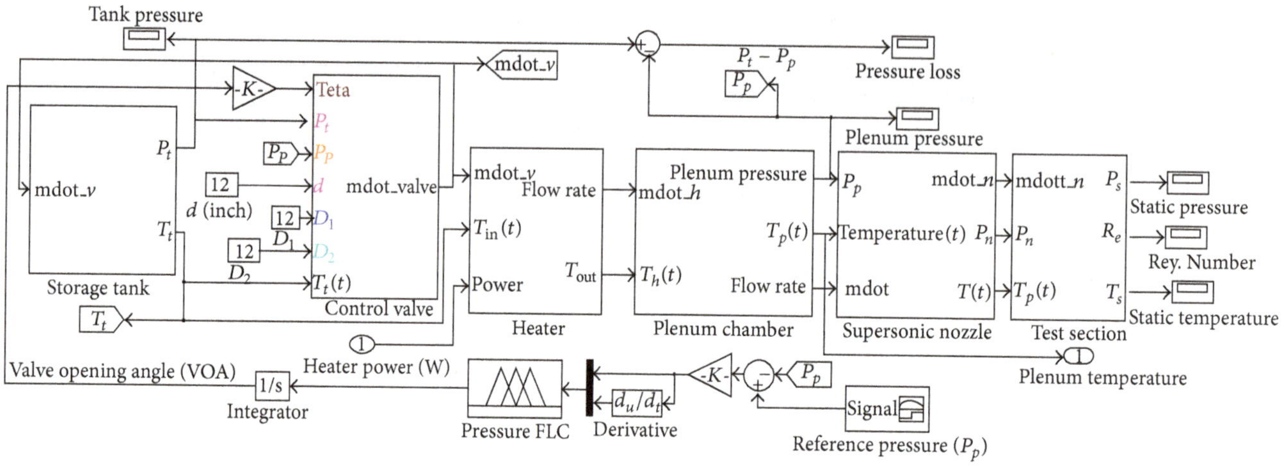

FIGURE 3: Wind tunnel simulator model and pressure controller development in Matlab/Simulink software environment.

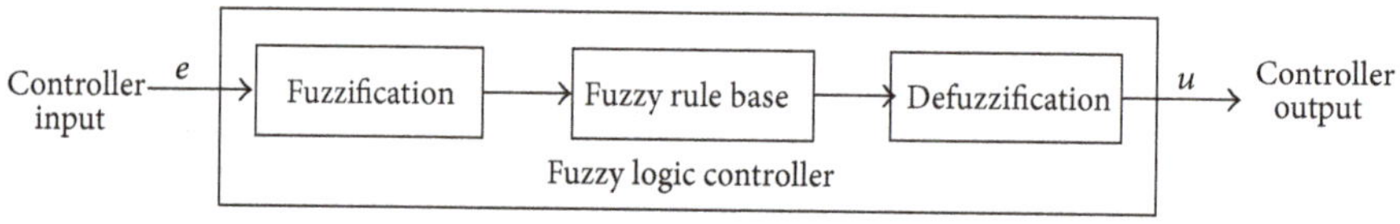

FIGURE 4: The block diagram of the fuzzy logic controller.

compressed, dry air at 1.2 MPa before the run. The heaters for the BSWT system are electric heaters requiring three-phase AC voltage input which is controlled by a silicon controlled rectifier (SCR). The air flowing into the heaters passes through a bundle of electrical resistance heated tubes to maximize air contact with heated metal.

3.2. WTSM Development. Mathematical models of the wind tunnel process are very complicated because they involve viscous effects and distributed characteristics. In this paper, a nonlinear mathematical model of a BSWT based on isentropic relation was developed in Simulink software environment as shown in Figure 3.

The wind tunnel model is constructed with a component approach for ease of modification and replacement with different tunnel components. The WTSM includes some component modules such as storage tank, control valve, heater, plenum pressure, nozzle, and test section.

4. Fuzzy Logic Controller for the Plenum Temperature

The control objective is to provide the plenum chamber temperature desired by the researcher performing tests in the wind tunnel. The application of proportional-integral-derivative (PID) control loops to the temperature processes in the system results in unacceptable control with respect to operation within system temperature constraints and regulation of the controlled process at the desired temperature. A PID control loop is unable to reasonably handle the long delay time in temperature response associated with air flow through long lengths of metal piping which absorb thermal energy. The recent application of a FLC provides appropriate control of the heaters to obtain the desired plenum chamber temperature.

A FLC utilizes FL to convert linguistic information based on expert knowledge into an automatic control strategy. In order to use the FL for control purposes, a front-end "fuzzifier" and a rear-end "defuzzifier" are added to the usual input-output data set. A FLC commonly consists of four sections: rules, fuzzifier, inference engine, and defuzzifier. Once the rule has been established, the controller can be considered as a nonlinear mapping from the input to the output [12]. The block diagram of the generalized indistinct controller is presented in Figure 4.

The designed FLC would determine the amount of heater thermal power over its transient operation. The FLC and the available WTSM constitute a closed loop which is shown in Figure 5.

The first input of the fuzzy controller is the temperature error (Error). The second input is the first input variations over time (Error Rate). The defined membership functions (MFs) for the inputs are shown in Figure 6. The FLC output is heater power over time. The defined fuzzy MFs for the FLC output are shown in Figure 7.

Conventionally, fuzzy rules are established by a combination of knowledge, experience, and observation and may thus not be optimal. Additionally, in spite of efforts to formalize a development standard for fuzzy controllers, fine tuning its performance is still a matter of trial and error [13, 14].

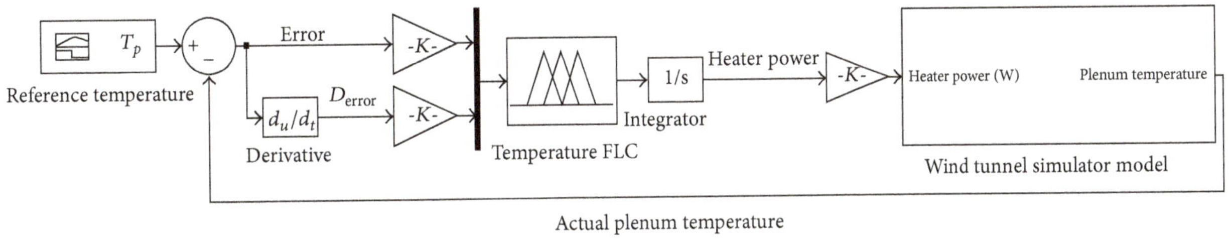

FIGURE 5: Layout of the wind tunnel model and the temperature controller.

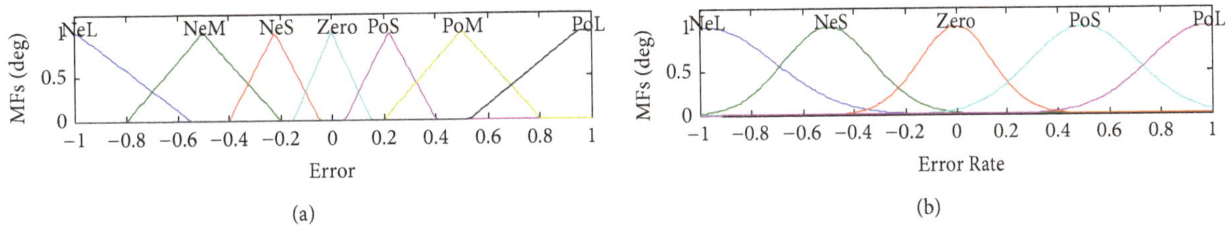

FIGURE 6: The defined membership functions for (a) Error and (b) Error Rate.

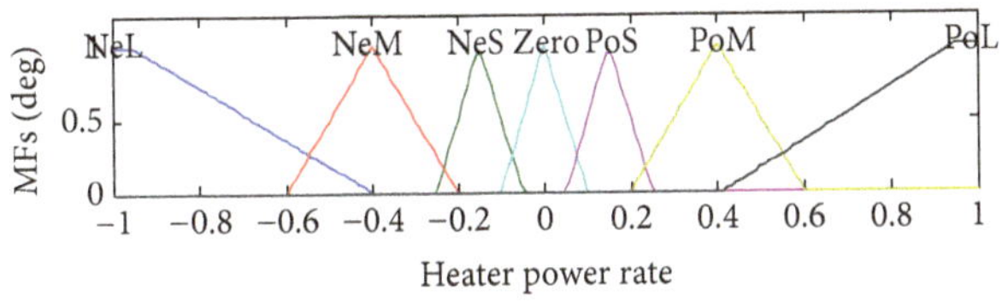

FIGURE 7: Heater power rate membership functions (FLC output).

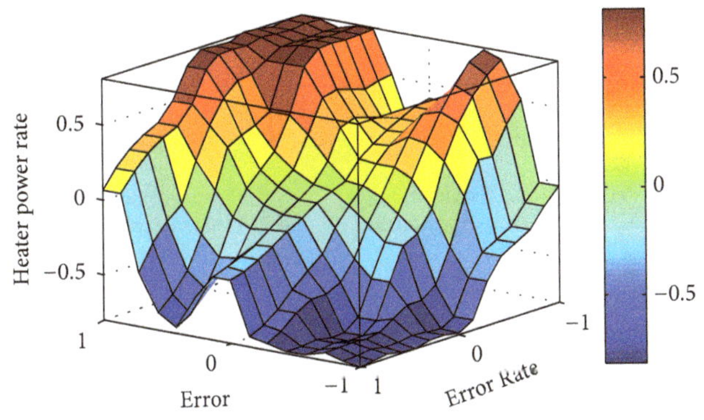

FIGURE 8: Control surface output of the FLC.

TABLE 1: The fuzzy rules list.

Heater power rate	Error Rate NeL	NeS	Zero	PoS	PoL
Error					
NeL	Zero	NeM	NeL	NeL	NeL
NeM	PoL	Zero	NeS	NeL	NeL
NeS	PoM	PoS	Zero	NeM	NeL
Zero	PoM	PoS	Zero	NeS	NeM
PoS	PoL	PoM	Zero	NeS	NeM
PoM	PoL	PoL	PoS	Zero	NeL
PoL	PoL	PoL	PoL	PoM	Zero

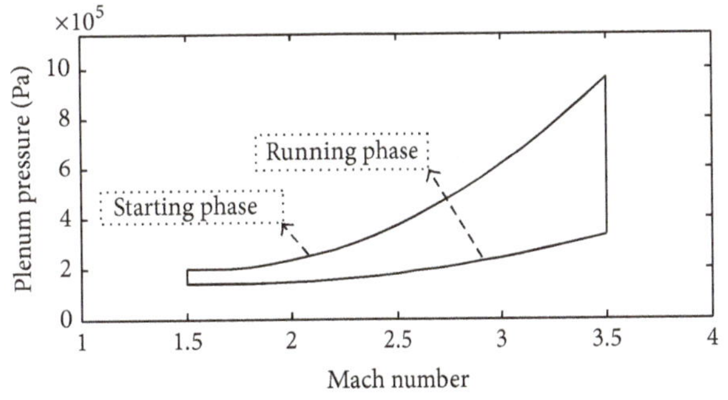

FIGURE 9: The tests range for the proposed BSWT.

The fuzzy rules list is presented in Table 1. For instance, one of the table entries is the equivalent of "if Error is Positive-Large and Error Rate is Zero then Heater Power Rate is Positive Large."

The defuzzification process takes place after the generation of the fuzzy control signals is completed using the inference mechanism. The resulting fuzzy set must be converted to a quantity which would be sent to the process regulating valve as a control signal. In this part, the inference results of all activated logic rules are synthesized into crisp output for making a decision. In this research study, the logic AND has been implemented with the minimum operator, and the defuzzification method is based on bisector area. The variation of heater power rate versus the two controller inputs is depicted in Figure 8.

By designing the controller, the influence of controller parameters on wind tunnel should be examined in more detail. So it would be essential to simulate the wind tunnel and controller simultaneously.

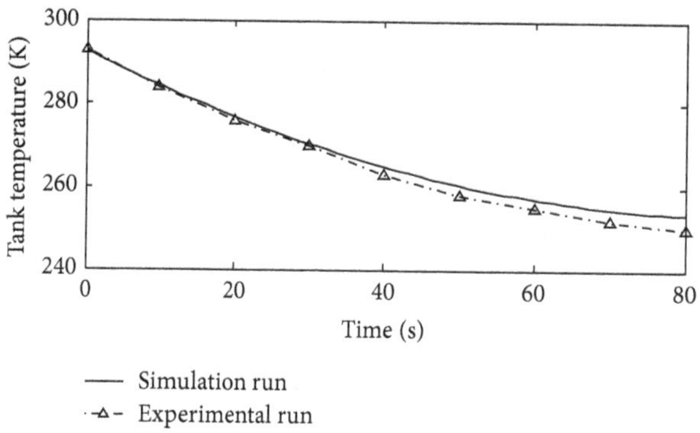

FIGURE 10: Temperature variation for a Mach 2.5 run.

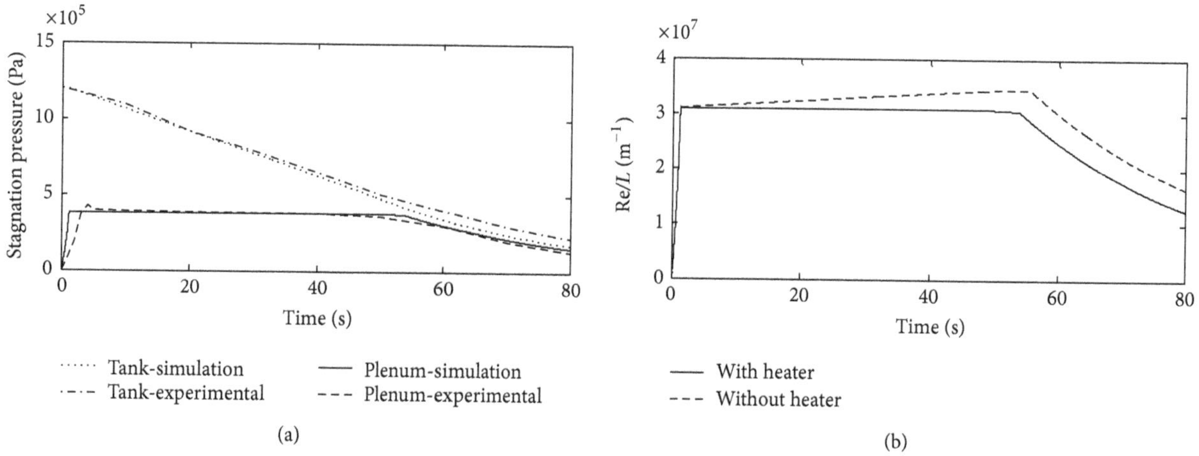

FIGURE 11: Simulation results for a Mach 2.5 run: (a) plenum and storage tank pressures with experimental results and (b) Reynolds number in the test section.

5. Simulation Results

This paper is focused on the tunnel performance for Mach 2.5 blowdown run. The tunnel operational map (Figure 9), which is based on the tests data, is used to select the proper pressure controller setpoint for the proposed Mach number in the test section. It is previously noted that the stagnation temperature drops according to the polytropic process in the storage tank. In this regard, Figure 10 shows the temperature drops during the wind tunnel run without heater for a Mach 2.5 run [1].

The wind tunnel simulation results with temperature controller for Mach 2.5 are presented in Figure 11. It should be noted that the reference temperature for heater controller is considered to be equal to 300 K. Also experimental verification has been carried out for supersonic conditions to show the accuracy of the fuzzy logic applied to the BSWT.

Figure 11(a) shows plenum and tank stagnation pressures for a Mach 2.5 run. Also the figure shows the pressure variation in comparison with experimental run. By the way, Figure 11(b) shows that using heater is caused to maintain the Reynolds number in the test section during the tunnel operation.

Also the valve opening and flow rate variations for a Mach 2.5 blowdown run are presented in Figures 12(a) and 12(b).

According to Figure 12(a), after 54 seconds of tunnel operation, the control valve opening reaches its maximum opening condition. In this situation, the plenum pressure and Reynolds number are dropped suddenly.

Plenum temperature and heater power variations are presented in Figures 12(c) and 12(d). According to the figures, plenum temperature is kept as close as the reference temperature (300 K) with the lowest overshoot, settling time, and steady state error.

6. Conclusion

In this paper, first, a nonlinear mathematical model of special supersonic wind tunnel is developed in Simulink software environment. Then, a fuzzy logic controller system is developed to control the plenum stagnation temperature. Finally, the simulation results for Mach 2.5 blowdown run in comparison with the experimental results are presented. The stagnation temperature affects the Reynolds number and the Mach number of a blowdown supersonic test. For example, in the proposed wind tunnel, the Reynolds number increases 13.16 percent from 30.8×10^6/m at Mach 2.5 when the stagnation temperature drops from 290 to 260 K. In this paper, the temperature drop is suppressed by installing heater

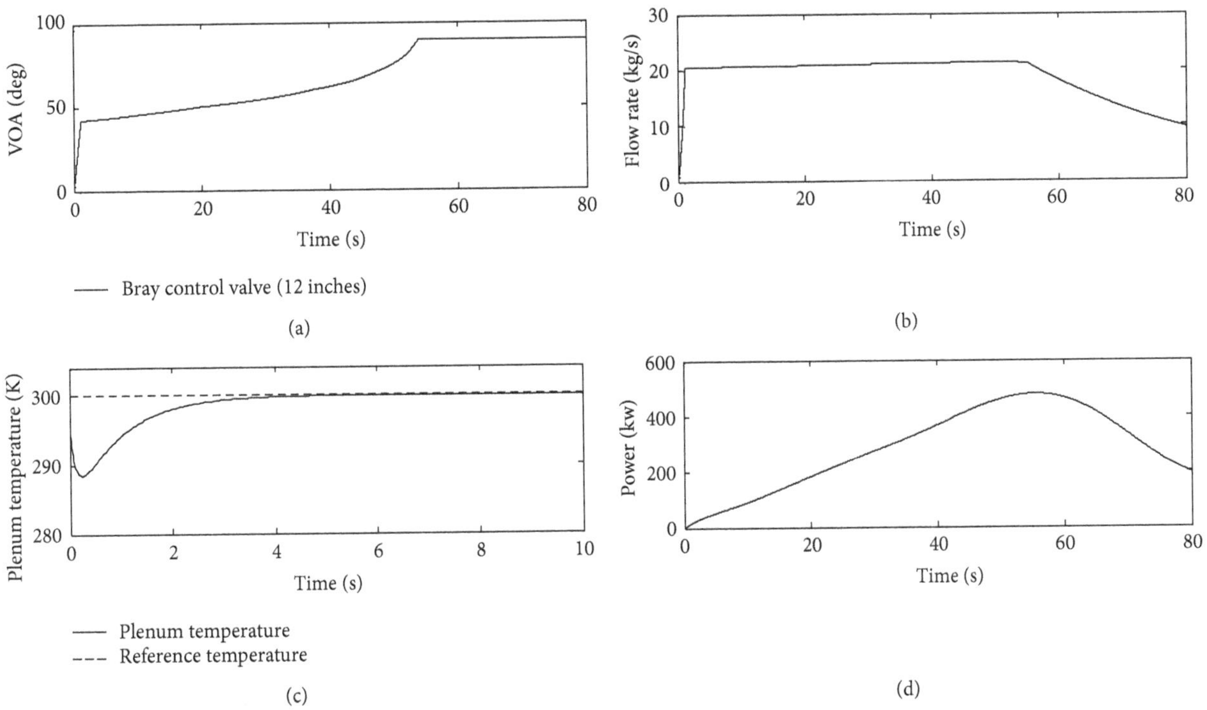

FIGURE 12: Simulation results for a Mach 2.5 run: (a) valve opening angle (VOA) function, (b) valve flow rate, (c) plenum stagnation temperature, and (d) heater power variation during the tunnel run.

and temperature controller upstream of the plenum chamber to maintain a constant Reynolds number with keeping the stagnation pressure constant. In this way, the Error in the heater performance during the run is minimized with the control program for the wind tunnel facility characteristics. The new intelligent control system leads to a control response with the lowest overshoot, settling time, and steady state error.

Conflict of Interests

The authors declare that there is no conflict of interests regarding the publication of this paper.

References

[1] A. Nazarian Shahrbabaki, M. Bazazzadeh, A. Shahriari, and M. D. Manshadi, "Intelligent controller design for a blowdown supersonic wind tunnel," *SERSC International Journal of Control and Automation*, vol. 7, pp. 409–426, 2014.

[2] F. K. Lu, D. R. Wilson, and J. Matsumoto, "Rapid valve opening technique for supersonic blowdown tunnel," *Experimental Thermal and Fluid Science*, vol. 33, no. 3, pp. 551–554, 2009.

[3] C. R. Nott, S. M. Ölçmen, D. R. Lewis, and K. Williams, "Supersonic, variable-throat, blow-down wind tunnel control using genetic algorithms, neural networks, and gain scheduled PID," *Applied Intelligence*, vol. 29, no. 1, pp. 79–89, 2008.

[4] S. H. Rajani, M. K. Bindu, and N. Usha, "Stability analysis and temperature effect on the settling chamber pressure of a hypersonic wind tunnel," in *Proceedings of the IEEE International Conference on Computational Intelligence and Computing Research (ICCIC '12)*, 2012.

[5] D.-S. Hwang and P.-L. Hsu, "A robust controller design for supersonic intermittent blowdown-type windtunnels," *Aeronautical Journal*, vol. 102, no. 1013, pp. 161–169, 1998.

[6] L. A. Zadeh, G. J. Klir, and B. Yuan, *Fuzzy Sets, Fuzzy Logic, Fuzzy Systems*, World Scientific Publishing, Hackensack, NJ, USA, 1996.

[7] P. J. King and E. H. Mamdani, "The application of fuzzy control systems to industrial processes," *Automatica*, vol. 13, no. 3, pp. 235–242, 1977.

[8] A. Pope and K. L. Goin, *High-Speed Wind Tunnel Testing*, John Wiley & Sons, New York, NY, USA, 1965.

[9] L. Biagioni, F. Scortecci, and F. Paganucci, "Development of pulsed arc heater for small hypersonic high-enthalpy wind tunnel," *Journal of Spacecraft and Rockets*, vol. 36, no. 5, pp. 704–710, 1999.

[10] Fisher Controls International, *Control Valve Handbook*, Fisher International Company, Marshalltown, Ia, USA, 2nd edition, 2005.

[11] L. Driskell, *Control-Valve Selection and Sizing*, Instrument Society of America, Research Triangle Park, NC, USA, 1st edition, 1983.

[12] J. H. Lilly, *Fuzzy Control and Identification*, John Wiley & Sons, New York, NY, USA, 2010.

[13] D. Driankov, H. Hellendoorn, and M. Reinfrank, *An Introduction to Fuzzy Control*, Springer, Berlin, Germany, 2nd edition, 1996.

[14] K. M. Passino and S. Yurkovich, *Fuzzy Control*, Addison-Wesley, San Francisco, Calif, USA, 1998.

5

Designing Stipulated Gains of Aircraft Stability and Control Augmentation Systems for Semiglobal Trajectories Tracking

Mohamed Mostafa Y. B. Elshabasy,[1] Yongki Yoon,[2] and Ashraf Omran[3]

[1] *Department of Mechanical Engineering, Faculty of Engineering, Alexandria University, Alexandria 21544, Egypt*
[2] *Whirlpool Corporation, Saint Joseph Technology Center, 303 Upton Drive, Saint Joseph, MI 49085, USA*
[3] *CNH Industrial, 6900 Veterans Ave., Burr Ridge, IL 60527, USA*

Correspondence should be addressed to Mohamed Mostafa Y. B. Elshabasy; mohamed_elshabasy@alexu.edu.eg

Academic Editors: J. Yao and C. Yuan

The main objective of the current investigation is to provide a simple procedure to select the controller gains for an aircraft with a largely wide complex flight envelope with different source of nonlinearities. The stability and control gains are optimally devised using genetic algorithm. Thus, the gains are tuned based on the information of a single designed mission. This mission is assigned to cover a wide range of the aircraft's flight envelope. For more validation, the resultant controller gains were tested for many off-designed missions and different operating conditions such as mass and aerodynamic variations. The results show the capability of the proposed procedure to design a semiglobal robust stability and control augmentation system for a highly maneuverable aircraft such as F-16. Unlike the gain scheduling and other control design methodologies, the proposed technique provides a semi-global single set of gains for both aircraft stability and control augmentation systems. This reduces the implementation efforts. The proposed methodology is superior to the classical control method which rigorously requires the linearization of the nonlinear aircraft model of the investigated highly maneuverable aircraft and eliminating the sources of nonlinearities mentioned above.

1. Introduction

Due to stringent performance and robustness requirements, modern control techniques have been widely used to design the flight control systems (FCS^s). However, researchers have been facing the difficulties of the complex nature and the nonlinearity strength embedded in the aircraft's dynamical model. For example, inertia coupling and attitude representations (Euler angles representation or quaternion representation) of the aircraft rigid body motions require nonlinear mathematical models [1]. Special impact on aircraft model comes from the nonlinear aerodynamic submodel such that aerodynamics coefficients significantly change with operating conditions. This leads to a significant change in the stability and performance of the aircraft dynamics. In addition, many other sources of nonlinearities appear in actuator nonlinear subsystems, sensor nonlinear subsystem, and engine nonlinear subsystems.

In order to address the designing FCS, gain scheduling, one of the popular methodologies to design controllers for nonlinear systems has been adopted to design stability augmentation system (SAS), and control augmentation system (CAS) [2–4]. In the conventional gain scheduling approach, the nonlinear system is linearized at several equilibrium operation conditions. Local linear controllers are designed at each of these points. The linear controller gains are then scheduled between the selected equilibrium points to obtain a semiglobal nonlinear controller. There is, however, still a difficulty as to how to schedule the gain from point to point in the operation regime. Even though the gain scheduling method breaks down the nonlinear model into linear models, operating point might have a significant nonlinearity, which cannot be overlooked in the control design. Besides, the selection of the operating points and the design of interpolation scheme remain a time-consuming procedure. There are several approaches to resolve the issue of highly nonlinear operating conditions such as the dynamic inversion [5], adaptive control [6], and sliding mode control [7]. The more nonlinearity these approaches account for, the more complicated it is to implement the resultant controller on-board. The main goal of this paper is to provide a simple procedure to select the controller gains for an aircraft with a largely wide complex flight envelope such as F-16 with different sources of nonlinearities. Thus, the gains are optimally tuned using genetic algorithm (GA) based on the information for a single designed mission [8–11]. This mission is assigned to cover a wide range of the aircraft's flight envelope. The resultant controller is a semiglobal robust controller and not restricted to any approximation regarding the system's nonlinearity.

Since GA has advantages such as global optimization performance and the ease of distributing its calculations among several processors or computers as it operates on the population of solutions that can be evaluated concurrently, GA is used as an efficient search technique to tune the value of the SAS and CAS controller gains. Thus, there is no need to linearize the model around prescribed patches to design the gains that are only valid for a certain range inside the flight envelope. In other words, the proposed control methodology replaces the classical techniques that depend on scheduling the gains to track the trajectory. Hence, it will be computationally efficient and more realistic because it is based on the nonlinear dynamics of the studied aircraft. To further validate the proposed technique, another mission is tracked using the optimal gains designed by GA to study how close the response of the aircraft is to the off-design mission.

The rest of paper is organized as follows. Section 2 presents the nonlinear dynamic model of F-16 fighter aircraft. Section 3 provides the representation of SAS and CAS. Section 4 reviews the designed flight mission information. Section 5 shows a genetic algorithm to tune the SAS and CAS gains. Section 6 gives the simulation results obtained from the associated nonlinear aircraft model to illustrate the designed-mission/off-designed mission.

2. F-16 Nonlinear Dynamic Model

The F-16 model is considered under many assumptions: the aircraft is a rigid body with six degrees of freedom except for an internal constant spinning engine rotor, the aircraft mass is constant, the aircraft body is symmetric about the XZ plane, the atmosphere is stationary, and the earth is flat with constant gravity. Based on those assumptions, the nonlinear dynamic equations of the F-16 model are first-order ordinary differential equations and can be classified into the following [12, 13] (note that the following variables are listed in the Nomenclature at the end of the paper).

Force equations:

$$\begin{aligned} \dot{u} &= Rv - Qw - g\sin\theta + \frac{1}{m}\left(\overline{X} + F_T\right), \\ \dot{v} &= Pw - Ru + g\sin\varphi\cos\theta + \frac{1}{m}\left(\overline{Y}\right), \\ \dot{w} &= Qu - Pv + g\cos\varphi\cos\theta + \frac{1}{m}\left(\overline{Z}\right). \end{aligned} \tag{1}$$

Motion equations:

$$\dot{P} = \left(c_1 R + c_2 P\right) Q + c_3 \overline{L} + c_4\left(\overline{N} + H_e Q\right), \tag{2}$$

$$\dot{Q} = c_5 PR - c_6\left(P^2 - R^2\right) + c_7\left(\overline{M} + F_T z_T - H_e R\right), \tag{3}$$

$$\dot{R} = \left(c_8 P - c_2 R\right) Q + c_4 \overline{L} + c_9\left(\overline{N} + H_e Q\right). \tag{4}$$

Kinematic equations:

$$\begin{aligned} \dot{\varphi} &= P + \tan\theta\left(Q\sin\varphi + R\cos\varphi\right), \\ \dot{\theta} &= Q\cos\varphi - R\sin\varphi, \\ \dot{\psi} &= \frac{Q\sin\varphi + R\cos\varphi}{\cos\theta}. \end{aligned} \tag{5}$$

Navigational equations:

$$\begin{aligned} \dot{P}_N &= u\cos\psi\cos\theta + v\left(\cos\psi\sin\theta\sin\varphi - \sin\psi\cos\varphi\right) \\ &\quad + w\left(\cos\psi\sin\theta\cos\varphi + \sin\psi\sin\varphi\right), \\ \dot{P}_E &= u\sin\psi\cos\theta + v\left(\sin\psi\sin\theta\sin\varphi + \cos\psi\cos\varphi\right) \\ &\quad - w\left(\sin\psi\sin\theta\cos\varphi + \cos\psi\sin\varphi\right), \\ \dot{h}_E &= u\sin\theta - v\cos\theta\sin\varphi - w\cos\theta\cos\varphi. \end{aligned} \tag{6}$$

In (2) to (4), the constants $c_i^{\prime}$s, where $i = 1, 2, 3, \ldots, 9$, are defined in terms of the moments of inertias; I_X, I_Y, and I_Z and the product of inertia I_{XZ} are defined as

$$\begin{aligned} &\Gamma c_1 = \left(I_Y - I_Z\right) I_Z - I_{XZ}^2 \qquad \Gamma c_4 = I_{XZ} \qquad c_7 = \frac{1}{I_Y}, \\ &\Gamma c_2 = \left(I_X - I_Y + I_Z\right) I_{XZ} - I_{XZ}^2 \qquad c_5 = \frac{I_Z - I_X}{I_Y}, \\ &\Gamma c_8 = I_X\left(I_X - I_Y\right) + I_{XZ}^2 \\ &\Gamma c_3 = I_Z \qquad c_6 = \frac{I_{XZ}}{I_Y} I_{XZ} \qquad \Gamma c_9 = I_X, \end{aligned} \tag{7}$$

where $\Gamma = I_X I_Z - I_{XZ}^2$.

The parameter H_e that exists in the moments equations represents the engine angular momentum which is supposed to be a variable. In the current model, the value of H_e is 160 slug·ft^2/s considering full throttle opening; that is, $\delta_{th} = 100\%$ [1]. The vehicle mass is denoted by m that is assumed to be constant all the time. g represents the gravitational acceleration $g = 32.2\,\text{ft/s}^2$. We assume that the thrust produced by the engine, F_T, acts parallel to aircraft's X-body axis, which makes the thrust vector have only one component acting in the X-body axis. In (3), the constant Z_T, which represents the offset distance of the thrust vector away from the cm, is assumed to be zero.

The aerodynamic forces and moments acting on the aircraft $\overline{X}, \overline{Y}, \overline{Z}, \overline{L}, \overline{M}$, and $\overline{N}$ are well described in Roskam [12] and Stevens and Lewis [13]. Force in X-body axis, for example, can be obtained as follows:

$$\overline{X} = \overline{q} S C_{X_T}\left(\alpha, \beta, P, Q, R, \delta_k, \ldots\right), \tag{8}$$

where C_{X_T} is collected from wind tunnel and flight tests.

The total aerodynamic coefficients C_{X_T}, C_{Y_T}, C_{Z_T}, C_{l_T}, C_{m_T}, and C_{n_T} are computed based on the high fidelity aerodynamic data tables in Nguyen et al. (1979). These coefficients are expressed as a baseline component plus a correction terms that are denoted by the symbol δ. The baseline component is primarily a function of angle of attack α, sideslip angle β, and Mach number M. Mach number dependency can be removed from the baseline component and treated as a correction term in the case of data for subsonic speeds. As the available aerodynamic tables (Nguyen et al., 1979) were conducted at subsonic flow conditions, the effect of Mach number was neglected. In this model, the aerodynamic data shows strong dependency on the horizontal stabilizer deflection δ_h; therefore, δ_h was included as an independent variable for the baseline component. Normally, total coefficient equations have been used to sum the various aerodynamic contributions to given force or moment coefficients as listed in (8). The force coefficient C_{X_T}, for example, is defined as (see [1])

$$C_{X_T} = C_X(\alpha, \beta, \delta_h) + \delta C_{X_{\text{lef}}}\left(1 - \frac{\delta_{\text{lef}}}{25}\right) + \delta C_{X_{\text{sb}}}(\alpha)\left(\frac{\delta_{\text{sb}}}{60}\right) + \frac{Q\bar{c}}{2V_T}\left[C_{X_Q}(\alpha) + \delta C_{X_{Q_{\text{lef}}}}(\alpha)\left(1 - \frac{\delta_{\text{lef}}}{25}\right)\right], \tag{9}$$

where

$$\delta C_{X_{\text{lef}}} = C_{X_{\text{lef}}}(\alpha, \beta) - C_X(\alpha, \beta, \delta_h = 0^\circ). \tag{10}$$

In general, the basic rigid dynamics model will contain 12 state variables that are collected in the state vector $\vec{X}$, where

$$\vec{X} = [P_N P_E h : uvw : PQR : \psi\theta\varphi]. \tag{11}$$

The aircraft model has four inputs represented in the control vector $\vec{U}$:

$$\vec{U} = [\delta_{\text{th}} \;\; \delta_h \;\; \delta_e \;\; \delta_r]. \tag{12}$$

Note that the lower limit of the throttle position δ_{th} is set to 5% to prevent the engine surge. Both of δ_{sb} and δ_{lef} are frozen to their neutral positions as they do not have tangible effect on the aircraft dynamics.

3. Autopilot and Augmented Flight Control System

In this research, inner-loop feedback control systems can be grouped into three broad categories which include SAS, CAS, and Fly-by-Wire (FBW) [14]. The SAS was proposed to give suitable damping and natural frequencies to improve the dynamic stability characteristics as they are referred to as dampers, stabilizers, and stability augmenters. However, SAS itself does not provide efficient control output since there might exist undesired output between the pilot's stick input and aircraft response and nonredundancy of the sensors and control circuits which are not able to make the system reliable. Thus, in order to eliminate these phenomena, the CAS added a pilot command input to the flight control computer using a force sensor on the control stick. With CAS, the aircraft dynamic response is typically welldamped, and control response is scheduled with the control system gains to maintain desirable characteristics throughout the flight envelope. With CAS, both dynamic stability and control response characteristics could be tailored and optimized to the mission of the aircraft.

The F-16 aircraft model under investigation has FBW system as a major part of its flight control system. Therefore, the inner-loop feedback control system of the studied model will be referred to as the digital flight control systems (DFCS) throughout the remaining part of this paper [13, 14].

The overall autopilot system shown in Figure 1 needs three commands in each time step, and the system generates horizontal stabilizer, aileron, and throttle command signals that replace pilot commands. At the same time, the rudder control surface is frozen at the neutral position. The autopilot system receives the three commands from the mission generating logic which provides altitude, velocity, and heading angle. In each time step, these three commands are calculated from the flight trajectory that is generated based on the required vehicle motion of each section of the mission [14].

The controller receives 4 input signals, V_c, H_c, R_c, and that ψ_c. Note R_c is zero in this research. The difference between H_c and H signal is passed to a PI controller, $G_H = K_P^H + (K_I^H/s)$. The controller, then, accounts for pitch angle and rate with the two gains K_θ and K_q. The signal is then passed to the actuator dynamic system, which includes position and rate saturation. Also the actuator includes a lag elevator-actuator with time constant $\tau = 0.136$ sec. The feedback loop in the yaw rate channel provides the wash-out filter so that it operates only transiently and does not contribute to a control law when a high frequency is present. Also, in this loop cross-connection called aileron-rudder interconnection (ARI) is implemented via an alpha-dependent gain to achieve a stability axis roll. In heading angle channel, one static gain is chosen to control the heading angle error signal of the actual and desired heading angle. In the feedback loop of velocity channel, error signals of the command and total velocities are tuned with PI controller; $G_{V_T} = K_P^{V_T} + (K_I^{V_T}/s)$.

Addressing the problem, designer needs to find these gains indicated in Figure 1. Linearization of a large system is applicable, but it needs to overlook the different source of nonlinearities such as saturation. Thus, the mission-based controller design is proposed. Upon these considerations, GA will be employed to find such gains in Section 5.

4. Mission Design

To accomplish the searching process for one semiglobal set of SAS and CAS gains, there is a need to design a special mission called a "designed mission." Thus this designed-mission should be assigned to cover a wide flight range of interest. Generating the design mission is considered under the aircraft performance limitations such as maximum rate of climb, maximum roll rate, minimum/maximum speed, actuator limitations, and maximum g-load. Most of these limitations are nonlinearly defined by the operational flight

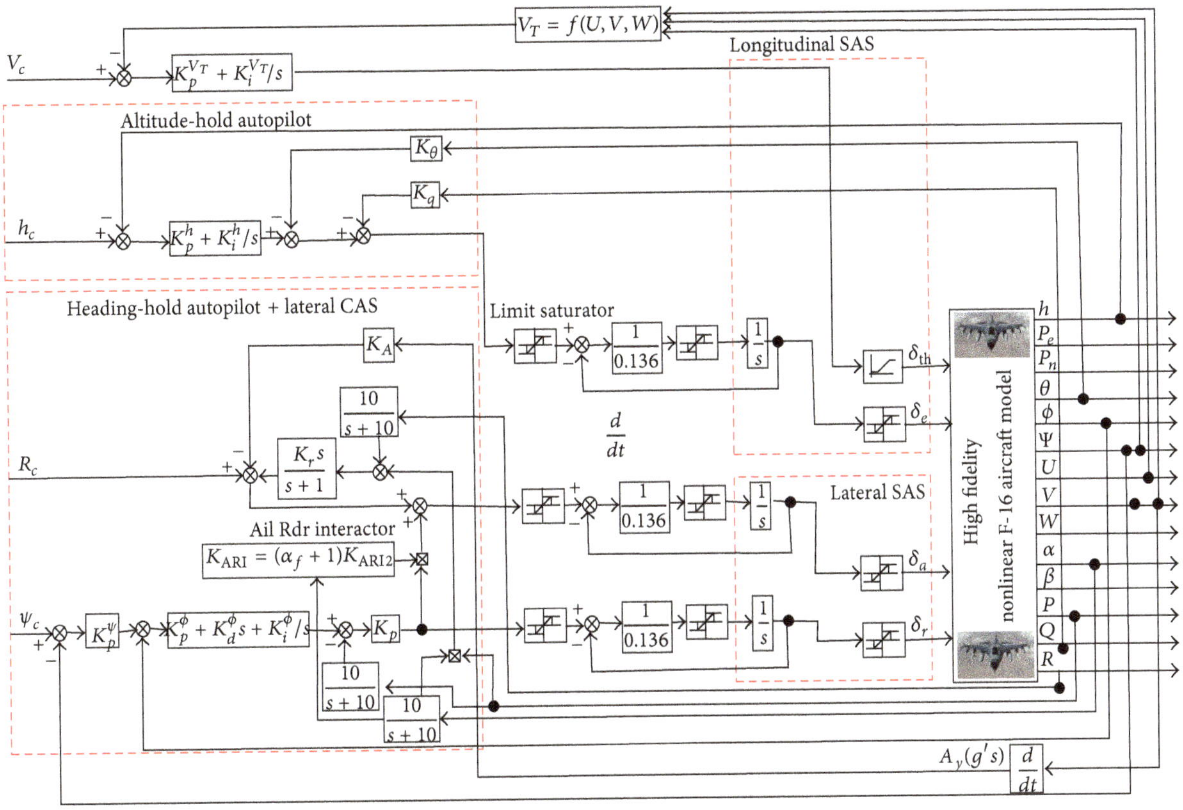

FIGURE 1: Overall system with autopilot, SAS, and CAS.

conditions (total velocity, altitude, etc.) with coupled hypersurfaces. For example, the definition of maximum rate of climb changes with the altitude is defined as the locus of the weight normalized excess power curve over the energy diagram. Counting these types of such nonlinear constrains leads to increase in the complexity of assigning the designed-mission. For that reason, the mission is initially designed without any extreme attitude that leads to violating the above-listed limits. This mission consists of a set of steering points. The steering points or any combination of the velocity and altitude at any instance of the mission is designed to be completely enclosed inside the flight envelope in order not to violate the structural limit, the stall limit, the propulsive limit, and the atmospheric limits of the flight envelope.

Figure 2, considering these characteristics of aircraft, illustrates the overall flight profile for the aircraft with steering points in terms of time, t_n. The trajectory, S, was assigned by four variables, X, Y, Z, and ψ, in terms of four input variables, $\delta_e, \delta_h, \delta_t$, and δ_a, respectively. Since trajectories of the aircraft can be calculated in terms of time step, profile of total velocity and altitude also can be obtained by the following equations:

$$S(t_i) = \lfloor x(t_i) \;\; y(t_i) \;\; z(t_i) \;\; \psi(t_i) \rfloor,$$
$$V(t_i) = \sqrt{\dot{x}^2(t_i) + \dot{y}^2(t_i) + \dot{z}^2(t_i)}, \qquad (13)$$
$$H(t_i) = -z(t_i).$$

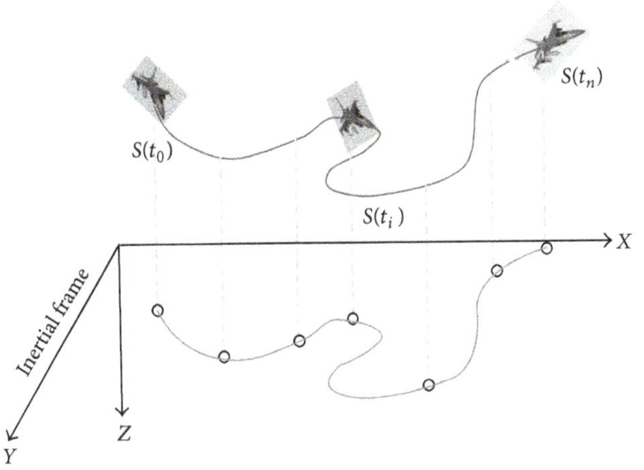

FIGURE 2: Overview of flight trajectory with/without mission design in the inertial frame.

5. Genetic Algorithm

The genetic algorithm (GA) is used as a global constrained and unconstrained optimization technique which is traced back to 1962 when the algorithm was introduced for studying adaptive systems [9, 10]. The other big advantage of this technique is the ease of distributing its calculations among several processors which is the cornerstone of parallel computation. Moreover, the algorithm is more suited to discontinuous

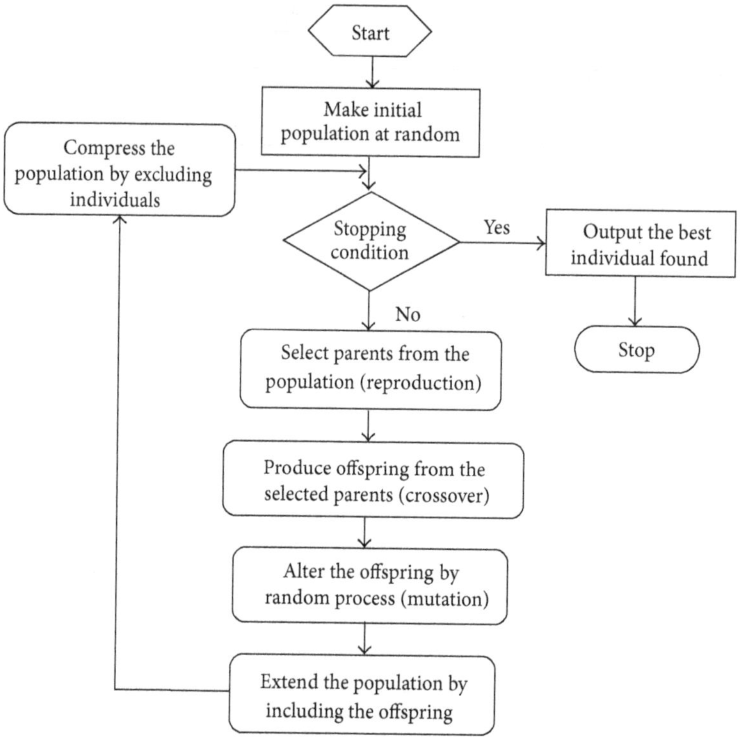

FIGURE 3: Pseudocode of genetic algorithm.

problems unlike the conventional gradient-based searching algorithms.

Fundamentally, the algorithm logic is based on the process that drives biological evolution (natural selection), where the procedure starts by coding the variables to chromosomes. Every chromosome has n genes. The procedure of the GA is controlled by reproduction, crossover, and mutation processes. Through the reproduction process, the parent is selected from a generation, where the selection process is based on survival of the highest performance index individual. The crossover process is then used to swap between two chromosomes using specific probabilistic decisions. The latter process generates offspring carrying mixed information from swapped parents. To prevent the algorithm from centering on local optimal points, the mutation process is implemented by alternation of the gene from zero to one or from one to zero with the mutation point determined uniformly at random. The mutation rate should be selected carefully such that the algorithm convergence will not last long due to high rate and the algorithm will not converge to a local minimum due to selection of low mutation rate. Roughly, a genetic algorithm works as shown in Figure 3.

The objective function to be optimized is a function in terms of the differences between the desired and the actual values of altitudes, total velocities, and heading angles, respectively, at each designated time step. The objective function is represented as

$$J(\chi) = \frac{1}{\sum_{j=1}^{N} \sum_{i=1}^{4} \left\| y_{ij}^{d} - y_{ij}^{a} \right\|}, \quad (14)$$

where N represents the number of points along the whole mission. Note that y can be arbitrary set that fully describes the mission. In this paper, this set is chosen to be $y = [h \ V_T \ \psi \ R]$ where the commanded yaw rate R was frozen to zero during the whole mission. Such set was chosen to avoid further prekinematic calculations which are out of the scope of the current research. However, for the most of flight motion, the trajectory is defined in terms of the navigation variables that include position and orientation. In (14), the vector $\chi = \left[K_P^{V_T} \ K_I^{V_T} \ K_P^{H} \ K_I^{H} \ K_\theta \ K_q \ K_p \ K_P^{\phi} \ K_P^{\psi} \ K_r\right]$ denotes the gains of the SAS and CAS. Searching for optimum χ is bounded between χ_{upper} and χ_{lower} where the system is expected to be stable. Assigning these limits is an iterative process as the system is nonlinear.

6. Results

The results of design mission and off-design missions are discussed in detail in this section.

6.1. Results of Design Mission. An air-to-surface mission for a fixed target is designed to simulate a realistic mission of the studied aircraft. This mission represents a simple striking mission that can be training or real mission. The mission consists of climb, cruise, descent, releasing bomb, another sharp climb to escape from the enemy fires followed by very short cruise at relatively high altitude, descent to the original cruise level, and then final descent at the take-off station. The steering points are listed in Table 1. The continuous flight path is generated based on the mission profile, where, at each

TABLE 1: Velocity, altitude, and heading angle at each steering point.

Steering point	Time (sec)	Altitude (ft)	Velocity (ft/s)	Heading angle (deg)
1	0	1000	540	0
2	100	5000	580	10
3	350	7100	640	80
4	980	7100	700	180
5	1130	1000	810	180
6	1180	1000	820	180
7	1330	20000	790	190
8	1380	20000	760	210
9	1490	17000	750	240
10	1565	17000	750	240
11	1675	14000	690	260
12	1750	14000	620	285
13	2075	1000	540	360

time step, the path consists of sets of three commands—total velocity, altitude, and heading angle. These sets of the three commands are direct input to the autopilot system described before. The autopilot system will follow the prescribed flight path commands through generating the necessary maneuvers for the vehicle. The climb and descent rates and their accelerations are constrained within ±150 ft/s and ±40 ft/s^2, respectively. The heading rate is limited within ±0.027 rad/s, while the angular acceleration of heading angle is limited within ±0.004 rad/s^2. In this research, a nonlinear simulation is developed in order to simulate both systems dynamics and control dynamics using the 4th Runge-Kutta solver. The code of this simulation is executed in the environment of Matlab R2010a through high performance cluster called "Zorka," which has two dual cores of 2.992 GHz Intel Xeon with 8 GB RAM per node. The selected time step is 0.1 second. The time window for simulating the designed-mission is 45 minutes.

The optimal tuning of the controller gains using genetic algorithms starts by generating random initial populations of the controller gains $K_P^{V_T}$, $K_I^{V_T}$, K_P^H, K_I^H, K_θ, K_q, K_p, K_P^ϕ, K_P^ψ, and K_r. The fitness of each individual is evaluated from the inversion of (14). Thus, a genetic algorithm works for maximizing and the control-tuning problem aims to minimize the error. The new generation is selected from the current population and passed to the next iteration of the algorithm. There is a possibility of receiving an unstable behavior for some randomly chosen control gains. In order to avoid such an unstable response, a two-second test simulation is initially conducted to each individual set of gains. If the error during these first 30 seconds violates a specific threshold limit, the simulation will stop and set the fitness value to zero. The parameters of the genetic algorithm are set as follows: (1) the mutation rate is 10%, (2) each generation has a fixed population size of 100 or no generation overlap, and (3) the maximum number of generations is 500. The optimization code was conducted many times with a different starting point (initial solution). In the initial five runs,

TABLE 2: Optimum gain values of the SAS and CAS.

Gains	Values
$K_P^{V_T}$	0.0912
$K_I^{V_T}$	0.0012
K_P^H	0.0661
K_I^H	1.363e − 4
K_θ	−5.1032
K_q	−9.2311
K_p	0.3230
K_P^ϕ	0.2101
K_P^ψ	−193.62
K_r	−1.9453

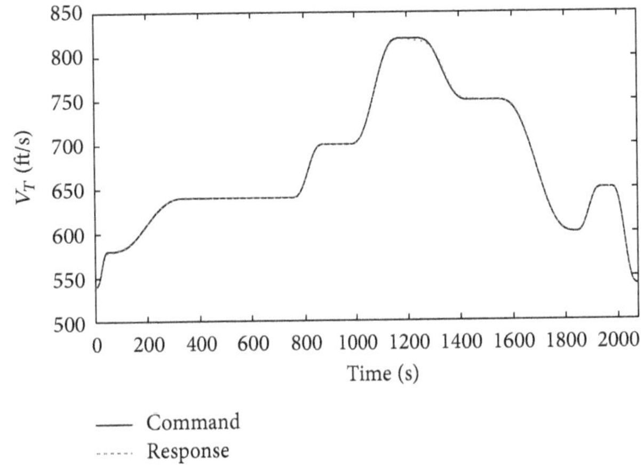

FIGURE 4: The desired and the response velocities of the F-16 model.

GA reaches the maximum number of generations without convergence. Starting from the sixth run, GA converges and this convergence is passed to the next run. After the 8th run, even with changing the mutation rate, the gains keep converging at the same values as listed in Table 2.

The responses of the aircraft in tracking the designed trajectory are listed in Figures 4, 5, and 6. The variations of elevator, aileron, and rudder are, respectively, plotted in Figure 7. In Figure 7, although the frequencies of variations of the aileron or the rudder look high, the zoom-in views show that the cycle time is about 10 sec. Moreover, the maximum range of the cycle is almost 0.15 deg for the aileron and 0.07 deg for the rudder (see Figure 7). For insight analysis of the variations of the aircraft's states during the mission, the time history of each state is plotted in a separate figure. Based on these results, it can be noticed that variation in the control surfaces is close to the deflection values at the trimming conditions. The variations of the roll and pitch angles ϕ, θ during the whole missions are plotted in Figure 8. The angular rates P, Q, and R are plotted versus the time for the whole

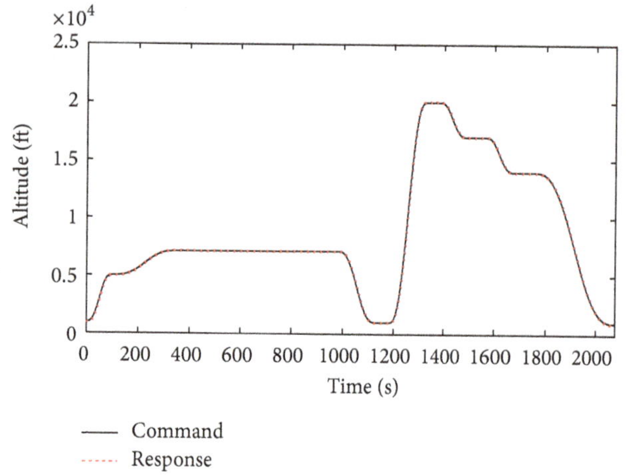

FIGURE 5: The desired and the response altitudes of the F-16 model.

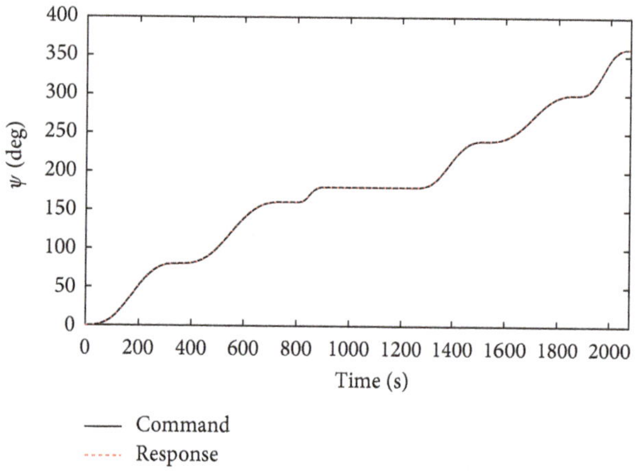

FIGURE 6: The desired and the response heading angles of the F-16 model.

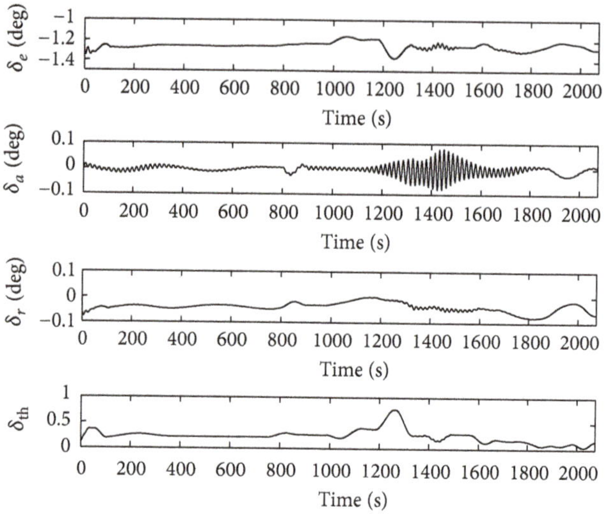

FIGURE 7: The elevator, aileron, rudder, and throttle's variation with time during the on-design mission, respectively.

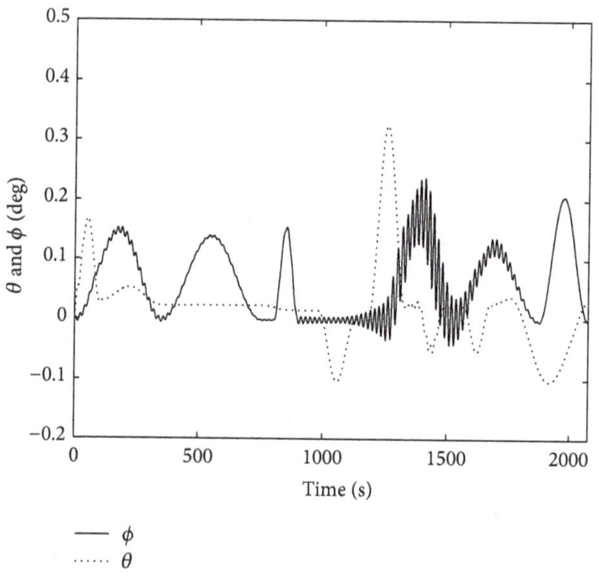

FIGURE 8: The variation of roll and pitch angles with time during the on-design mission.

mission in Figure 9. Notice that none of the control surfaces approaches its mechanical limits [1].

6.2. Results for Off-Design Missions. Now, the resultant mission-based controller using GA is validated for an off-designed mission. Figure 10 shows the difference between the selected off-designed mission and the designed mission in terms of the trajectories of altitudes, total velocities, and the heading angles of the off-design mission. These two mission start from the same position, "O," and end at different ones based on missions.

In order to provide a more realistic simulation of the off-designed mission, system uncertainties and sensor noises are considered [15]. Table 3 shows the standard deviations of sensor noise for the total velocity, altitude, attack angle, and rates of roll, pitch, and yaw. Note that uncertainty was chosen of only 10% increments for the certain parameters which are mass and moment of inertia about X-body axis I_{xx}, while considering 5% decrements of pitching moment coefficient "C_{mT}" and thrust. In Figures 11 and 12, the normalized deviations between actual and desired values of total velocity, altitude, and heading angle were shown with or without noise and uncertainty. Overall, there exists sharp change in certain time frame for velocity and heading angle of the aircraft, while the altitude keeps smooth change over the time range. Figure 13 represents how actual velocity response including noise and uncertainty behaves comparing to command velocity.

7. Conclusion

Using the genetic algorithm to select constant gains during the prescribed missions within a specific patch inside the flight envelope is more efficient than the traditional gain scheduling control scheme. The steps of gain scheduling

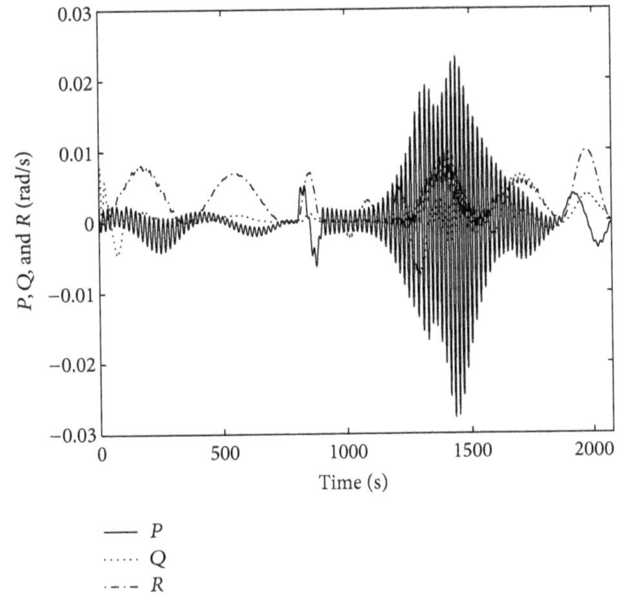

FIGURE 9: The variation of roll, pitch, and yaw rates with time during the on-design mission.

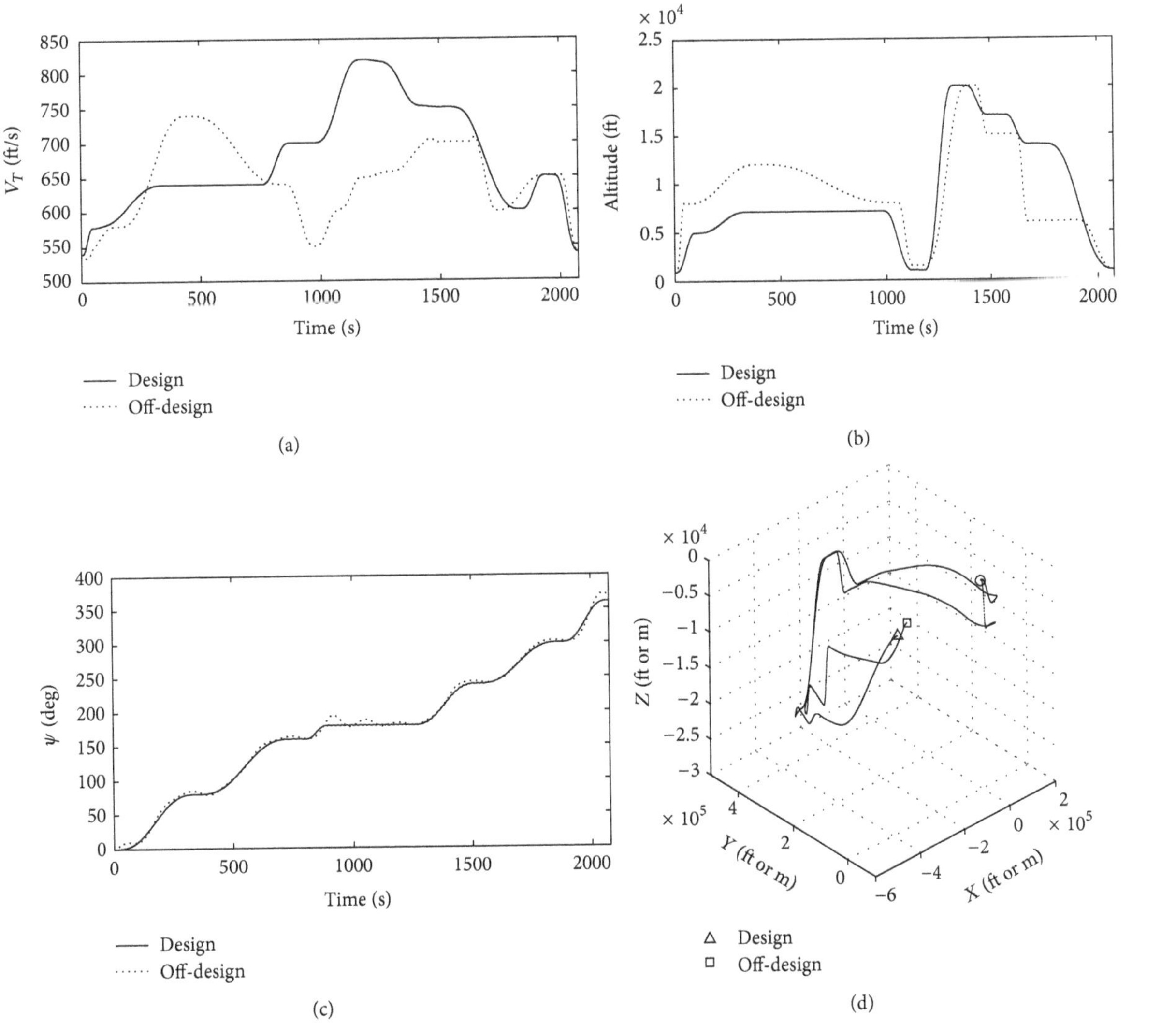

FIGURE 10: Total velocities, altitudes, heading angles, and flight trajectories of design/off-design mission without noise and uncertainty.

TABLE 3: Magnitude of sensor noise standard deviations.

Parameter	Standard deviation
Airspeed indicator	11 ft/sec
Roll rate gyro	0.14 deg/sec
Pitch rate gyro	0.14 deg/sec
Yaw rate gyro	0.14 deg/sec
Longitudinal accelerometer	0.98 ft/sec^2
Lateral accelerometer	0.98 ft/sec^2
Directional accelerometer	0.98 ft/sec^2
Attitude pitch gyro	0.573 deg
Attitude roll gyro	0.573 deg
Attitude yaw gyro	0.573 deg
Altitude rate indicator	0.25 ft/sec
Altitude indicator	10 ft
Angle of attack	0.1 deg
Sideslip angle	0.1 deg

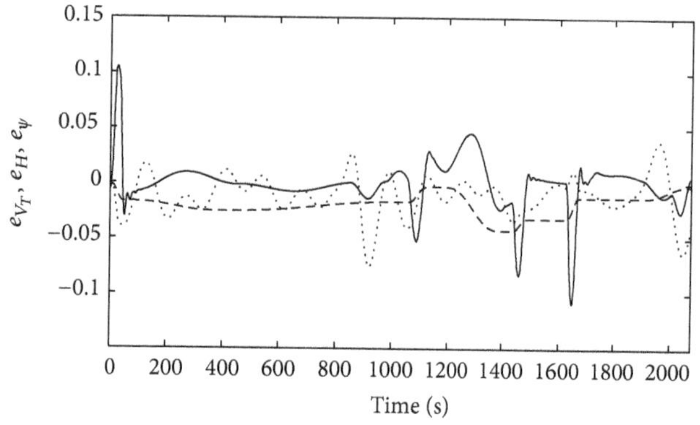

FIGURE 11: Normalized errors of total velocity, altitude, and heading angle of off-design mission without noise and uncertainty.

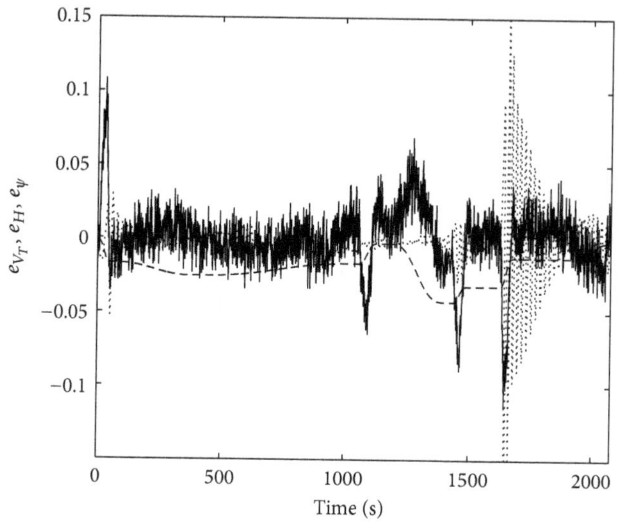

FIGURE 12: Normalized errors of total velocity, altitude, and heading angle of off-design mission with noise and uncertainty.

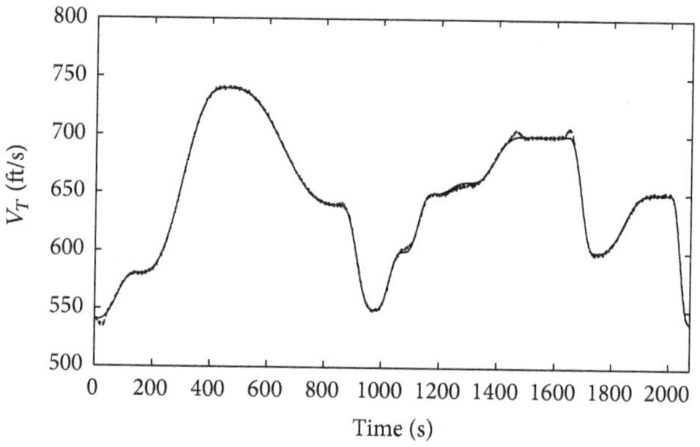

FIGURE 13: The desired and command velocities with noise and uncertainty.

process make it numerically costly as the control system is based upon linearization of the aircraft nonlinear motions around various points along the mission. Moreover, more insight analysis of the dynamics model is mandatory for the acceptable results. On the contrary, the genetic-optimized cost function depends on the errors between the desired and the actual trajectories of the prescribed mission without the time-consuming profound analysis of the aircraft dynamics at various points along certain mission. The genetic-select gains can be used with different off-design missions inside the envelope of the design mission.

Nomenclature

u, v, w: Velocity components in x, y, and z body axes (ft/s)
P, Q, R: Airplane roll, pitch, and yaw rates (rad/s or deg/s)
M: Mach number (dimensionless)
F_T: Total instantaneous engine thrust, horizon time (Ib)
h_E: Altitude (ft)
P_N, P_E: Aircraft position in north and east (ft)
H_e: Engine angular momentum (slug·ft^2/s)
$\overline{X}, \overline{Y}, \overline{Z}$: Aerodynamic force in XYZ body axis (Ib)
$\overline{L}, \overline{M}, \overline{N}$: Aerodynamic moment in XYZ body axis (Ib.ft)
$\overline{q}$: Free stream aerodynamic pressure (Ib/ft^2)
α: Angle of attack (deg)
β: Sideslip angle (deg)
ψ: Heading angle (deg)

φ, θ: Roll and pitch angles (deg)
I_X, I_Y, I_Z: Moment of inertia about X, Y, and Z body axes (slug·ft^2)
I_{XZ}: Product of inertia with respect to X and Z body axes (slug·ft^2)
V_T: Aircraft total velocity (ft/s)
δ_h: Elevator deflection (deg)
δ_{lef}: Leading edge flap deflection (deg)
δ_{sb}: Speed breaker deflection (deg)
θ_{th}: Throttle percentage (dimensionless)
$C_{X_T}, C_{X_Y}, C_{X_Z}$: Total X, Y, and Z axis force coefficients (dimensionless)
$C_{l_T}, C_{m_T}, C_{n_T}$: Total rolling, pitching, and yawing moment coefficients (dimensionless).

Disclosure

The second and third authors would like to acknowledge that the presented work in this paper does not belong or represent their current affiliations at any level.

Conflict of Interests

The authors declare and certify that there is no conflict of interests regarding the publication of this research paper.

References

[1] L. T. Nguyen, M. E. Ogburn, W. P. Gilbert, K. S. Kibler, P. W. Brown, and P. L. Deal, "Simulator study of stall/ post stall characteristics of a fighter airplane with relaxed longitudinal static stability," NASA TP-1538, NASA, Washington, DC, USA, 1979.

[2] T. Richardson, P. Davison, M. Lowenberg, and M. di Bernardo, "Control of nonlinear aircraft models using dynamic state-feedback gain scheduling," in *Proceedings of the AIAA Guidance, Navigation, and Control Conference and Exhibit*, AIAA-2003-5503, Austin, Tex, USA, August 2003.

[3] M. N. Hammoudi and M. H. Lowenberg, "Dynamic gain scheduled control of an F16 model," in *Proceedings of the AIAA Guidance, Navigation and Control Conference and Exhibit*, AIAA-2008-6487, Honolulu, Hawaii, USA, August 2008.

[4] W. J. Rugh and J. S. Shamma, "Research on gain scheduling," *Automatica*, vol. 36, no. 10, pp. 1401–1425, 2000.

[5] S. A. Snell, D. F. Enns, and W. L. Garrard Jr., "Nonlinear inversion flight control for a supermaneuverable aircraft," *Journal of Guidance, Control, and Dynamics*, vol. 15, no. 4, pp. 976–984, 1992.

[6] G. J. Balas, "Flight control law design: an industry perspective," *European Journal of Control*, vol. 9, no. 2-3, pp. 207–226, 2003.

[7] E. Promtun and S. Seshagiri, "Sliding mode control of pitch-rate of an F-16 aircraft," *International Journal of Engineering and Applied Sciences*, vol. 5, article 1, 2009.

[8] J. Holland, *Adaptation in Natural and Artificial Systems: An Introductory Analysis with Applications to Biology, Control, and Artificial Intelligence*, University of Michigan Press, Ann Arbor, Mich, USA, 1992.

[9] E. Goldberg, *The Design of Innovation: Lessons from and for Competent Genetic Algorithms*, Kluwer Academic, Norwell, Mass, USA, 2002.

[10] C.-D. Yang, C.-C. Luo, S.-J. Liu, and Y.-H. Chang, "Applications of Genetic-Taguchi algorithm in flight control designs," *Journal of Aerospace Engineering*, vol. 18, no. 4, pp. 232–241, 2005.

[11] A. Omran and A. Kassem, "Optimal task space control design of a Stewart manipulator for aircraft stall recovery," *Aerospace Science and Technology*, vol. 15, no. 5, pp. 353–365, 2011.

[12] J. Roskam, *Airplane Flight Dynamics and Automatic Controls*, University of Kansas, Lawrence, Kan, USA, 6th edition, 2001.

[13] B. L. Stevens and F. L. Lewis, *Aircraft Control and Simulation*, Wiley-Interscience, New York, NY, USA, 2nd edition, 2003.

[14] V. R. Schmitt Morris, J. W. Morris, and G. D. Jenny, *Fly-by-Wire: A Historical and Design Perspective*, Society of Automotive Engineers, 1998.

[15] P. Motyka, W. Bonnice, S. Hall, E. Wagner et al., "The evaluation of failure detection and isolation algorithms for restructurable control," NASA Contractor Report 177983, National Aeronautics and Space Administration, Langley Research Center, Hampton, Va, USA, 1985.

6

Combination of Two Nonlinear Techniques Applied to a 3-DOF Helicopter

P. Ahmadi,[1] M. Golestani,[1] S. Nasrollahi,[2] and A. R. Vali[1]

[1] *Department of Electrical and Electronic Engineering, Malek Ashtar University of Technology, Tehran, Iran*
[2] *Department of Aerospace Engineering, Sharif University of Technology, Tehran, Iran*

Correspondence should be addressed to P. Ahmadi; peyman.ahmadi.1366@gmail.com

Academic Editors: S. Aus Der Wiesche, P.-C. Chen, and R. K. Sharma

A combination of two nonlinear control techniques, fractional order sliding mode and feedback linearization control methods, is applied to 3-DOF helicopter model. Increasing of the convergence rate is obtained by using proposed controller without increasing control effort. Because the proposed control law is robust against disturbance, so we only use the upper bound information of disturbance and estimation or measurement of the disturbance is not required. The performance of the proposed control scheme is compared with integer order sliding mode controller and results are justified by the simulation.

1. Introduction

Helicopters are versatile flight vehicles that can perform aggressive maneuvers because of their unique thrust generation and operation principle. They can perform many missions that are dangerous or impossible for human to perform them. Helicopter is a multi-input multioutput (MIMO) highly nonlinear dynamical system, so most of the existing results to date have been based on the linearization model or through several linearization techniques [1, 2]. The linearization method provides local stability. In presence of disturbance or uncertainties the linearization can lead state variables of system to instability.

In recent years, many papers have been published about control design of helicopter. The sliding mode approach has been employed for helicopter's altitude regulation at hovering [3, 4]. H-infinity approach has been also used to design a robust control scheme for helicopter [5, 6]. In [5] a robust H-infinity controller has been presented using augmented plant and the performance and robustness of the proposed controller have been investigated in both time and frequency domain. The proposed controller in [6] is based on H-infinity loop shaping approach and it has been shown that the proposed controller is more efficient than classical controller such as PI and PID controller. A robust linear time-invariant controller based on signal compensation has been presented in [7]. By suitably combining feedforward control actions and high-gain and nested saturation feedback laws, a new control scheme has been presented in [8]. Intelligent methods such as fuzzy [9] and neural network theory [10] have been used to design controller. Furthermore, in [11] a new intelligent control approach based on emotional model of human brain has been presented.

The history of fractional calculus goes back 300 years ago. For many years, it has remained with no applications. Recently, this branch of science has become an attractive discussion among control scholars [12]. Good references on fractional calculus have been presented in [13, 14]. The sliding mode control (SMC) has been also extended in [12, 15–19]. In [15] a PID controller based on sliding mode strategy is designed for linear fractional order systems. In [17] a single input fractional order model, described by a chain of integrators, is considered for nonlinear systems. In [16–18] sliding mode method has been applied to synchronize fractional order nonlinear chaotic systems. In [12, 19] sliding surface has been defined as an expressed manifold with fractional order integral.

In this paper, we propose a combination of two nonlinear control techniques, fractional order (FO) sliding mode and feedback linearization control methods. The performance of the proposed control scheme is compared with integer order (IO) sliding mode controller and results are justified by the simulation.

This paper is organized as follows: in Section 2, the integer order SMC and feedback linearization techniques are briefly reviewed, and, in Section 3, some preliminaries and definitions of fractional order calculus are introduced. In Section 4, the 3-DOF helicopter model description is presented. In Section 5, the proposed controller scheme is given, and, in Section 6, the proposed approach is applied to 3-DOF helicopter model and experimental results are provided. Finally, conclusion is addressed in Section 7.

2. Review of IO SMC and Feedback Linearization

In this section, definitions of sliding mode and feedback linearization control methods are briefly reviewed for essential preparation of the combination of two nonlinear control techniques, FO SMC and feedback linearization.

2.1. Integer Order Sliding Mode Control. Consider a second-order nonlinear dynamical system as

$$\dot{x}_1 = x_2 \qquad \dot{x}_2 = f(x) + g(x)u, \tag{1}$$

where $x = [x_1\ x_2]^T$ is the system state vector, $f(x)$ and $g(x)$ are nonlinear functions of x, and $u \neq 0$ is the scalar input. Sliding mode control approach consists of two parts, correctivecontrol law (u_c) that compensates the deviations from the sliding surface to reach the sliding surfaceand equivalent control law (u_{eq}) that makes the derivative of the sliding surface equal zero to stay on the sliding surface. This control law is represented as

$$u = u_{eq} + u_c. \tag{2}$$

In general, the sliding surface is $S = x_2 + \beta x_1$, where $\beta > 0$.

To guarantee the existence of sliding mode, the control law must satisfy the condition

$$\frac{1}{2}\frac{d}{dt}s^2 < 0. \tag{3}$$

To have a fast convergence, it is sufficient to modify the sliding surface.

2.2. Input-Output Linearization Control. Consider the following nonlinear dynamical system:

$$\dot{x} = f(x) + g(x)u \qquad y = h(x). \tag{4}$$

The control method consists of the following steps.

(a) Differentiate y until u appears in one of the equations for the derivatives of y. Consider

$$y^{(r)} = \alpha(x) + \beta(x)u. \tag{5}$$

(b) Choose u to give $y(r) = v$, where v is the synthetic input. Consider

$$u = \frac{1}{\beta(x)}[-\alpha(x) + v]. \tag{6}$$

(c) Then the system has the following form:

$$\frac{Y(s)}{V(s)} = \frac{1}{s^r}. \tag{7}$$

Design a linear control law for this r-integrator linear system.

(d) Check internal dynamics.

3. Basic Description of Fractional Calculus

There exist many definitions of fractional derivative. Two of the most commonly used definitions are the Riemann-Liouville (RL) and Caputo (C) definitions [14].

Definition 1 (see [20]). The left RL fractional derivative is described by

$${}^{\mathrm{RL}}_{a}D^{\alpha}_{t}f(t) = \frac{1}{\Gamma(n-\alpha)}\frac{d^n}{dt^n}\int_a^t \frac{f(\tau)}{(t-\tau)^{\alpha-n+1}}d\tau, \tag{8}$$

where n is the first integer which is not less than α, that is, $n-1 \leq \alpha < n$ and $\alpha > 0$, and $\Gamma(\cdot)$ is the gamma function.

Definition 2 (see [21]). The Laplace transform of the RL definition is described as

$$\int_0^{\infty} e^{-st}\,{}_aD^{\alpha}_{t}f(t)\,dt = s^{\alpha}F(s) - \sum_{k=0}^{n-1} s^k\,{}_aD^{\alpha-k-1}_{t}f(t)\Big|_{t=0}. \tag{9}$$

Lemma 3 (see [20]). *The following equality is satisfied for all $\alpha > 0$ and n is a natural number. Consider*

$$\begin{gathered}\frac{d^n}{dt^n}\left\{{}^{RL}D^{\alpha}f(t)\right\} = {}^{RL}D^{n+\alpha}f(t),\\ {}^{C}D^{\alpha}\left\{\frac{d^n}{dt^n}f(t)\right\} = {}^{C}D^{n+\alpha}f(t).\end{gathered} \tag{10}$$

Lemma 4 (see [21]). *Fractional differentiation and fractional integration are linear operations. Consider*

$${}_0D^{\alpha}_{t}(af(t) + bg(t)) = a\,{}_0D^{\alpha}_{t}f(t) + b\,{}_0D^{\alpha}_{t}g(t). \tag{11}$$

Lemma 5 (see [20]). *The following equality for left RL definition is satisfied for all $\alpha \in (0,1)$:*

$${}^{RL}D^{\alpha}\{I^{\alpha}f(t)\} = f(t), \quad \alpha \in (0,1). \tag{12}$$

However the opposite is not true, since

$$I^{\alpha}\left\{{}^{RL}D^{\alpha}f(t)\right\} = f(t) - \frac{f_{1-\alpha}(0)}{\Gamma(\alpha)}t^{\alpha-1}, \tag{13}$$

where $f_{1-\alpha}(0) = \lim_{t\to 0}I^{1-\alpha}f(t)$.

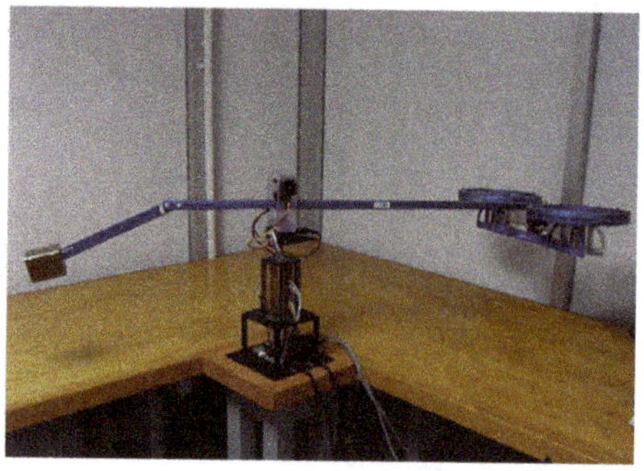

FIGURE 1: The 3-DOF helicopter system.

4. System Description

In this work, a 3-DOF helicopter model is considered (see Figure 1).

The following notations are used to describe the 3-DOF helicopter system dynamics:

θ: is angular position of the pitch axis, rad;

φ: is angular position of the roll axis, rad;

ψ: is angular position of the yaw axis, rad;

J_θ: is moment of inertia of the system around the pitch axis, kg m^2;

J_φ: is moment of inertia of the helicopter body about the roll axis, kg m^2;

J_ψ: is moment of inertia of the helicopter body about the yaw axis, kg m^2;

V_l: is voltage applied to the left motor, V;

V_r: is voltage applied to the right motor, V;

K_f: is force constant of the motor combination, N;

l_a: is distance between the base and the helicopter body, m;

l_h: is distance from the pitch axis to either motor, m;

T_g: is effective gravitational torque, N m;

K_p: is constant of proportionality of the gravitational force, N.

The 3-DOF helicopter control system consists of two DC motors at the end of the arm as shown in Figure 1. Figure 2 shows a physical model of the 3-DOF helicopter. The following equations (see Figure 2) describe the 3-DOF helicopter dynamics:

$$\ddot{\psi} = -\frac{K_p l_a}{J_\psi} \sin(\varphi), \quad (14)$$

$$\ddot{\theta} = -\frac{T_g}{J_\theta} + \frac{K_f l_a}{J_\theta}\left(V_l + V_r + \xi_\varphi\right), \quad (15)$$

$$\ddot{\varphi} = \frac{K_f l_h}{J_\varphi}\left(V_l - V_r\right), \quad (16)$$

$$y = [\psi \; \theta \; \varphi]^T. \quad (17)$$

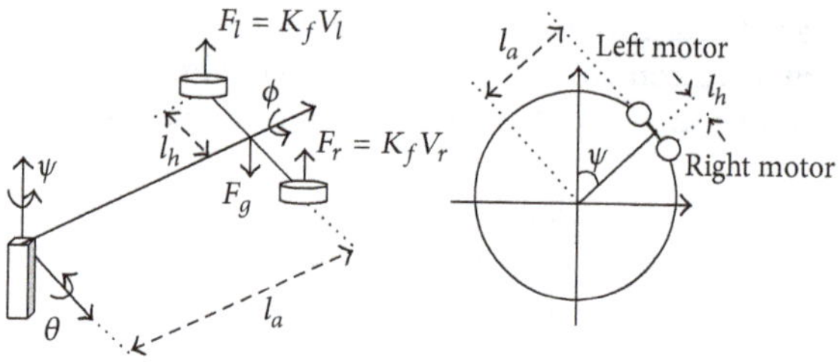

FIGURE 2: Schematic diagram of the 3-DOF helicopter.

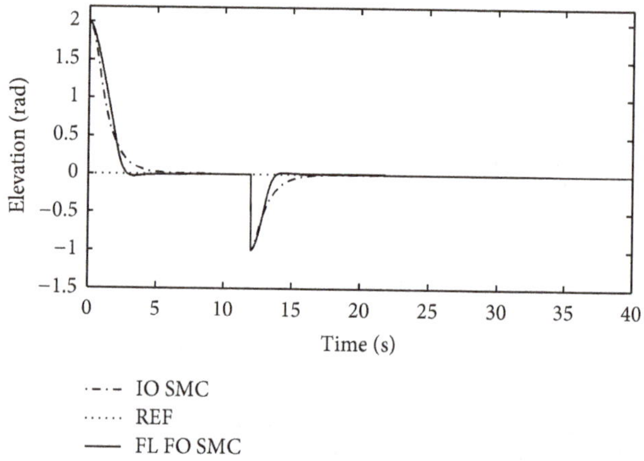

FIGURE 3: Stability curve of elevation.

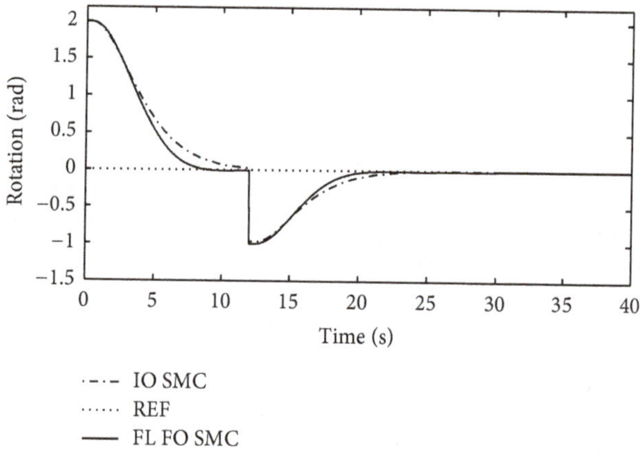

FIGURE 4: Stability curve of rotation.

Let define $u_1 = V_l + V_r$ and $u_2 = V_l - V_r$; for the sake of simplicity we define state vector as $x = [\psi \; \dot{\psi} \; \theta \; \dot{\theta} \; \varphi \; \dot{\varphi}]^T$ and the model (14)–(16) can be rewritten as

$$\begin{aligned} \dot{x}_1 &= x_2 \qquad \dot{x}_2 = -\frac{T_g}{J_\theta} + \frac{K_f l_a}{J_\theta}\left(u_1 + \xi_\varphi\right), \\ \dot{x}_3 &= x_4 \qquad \dot{x}_4 = -\frac{K_p l_a}{J_\psi}\sin(x_5), \\ \dot{x}_5 &= x_6 \qquad \dot{x}_6 = \frac{K_f l_h}{J_\varphi} u_2. \end{aligned} \quad (18)$$

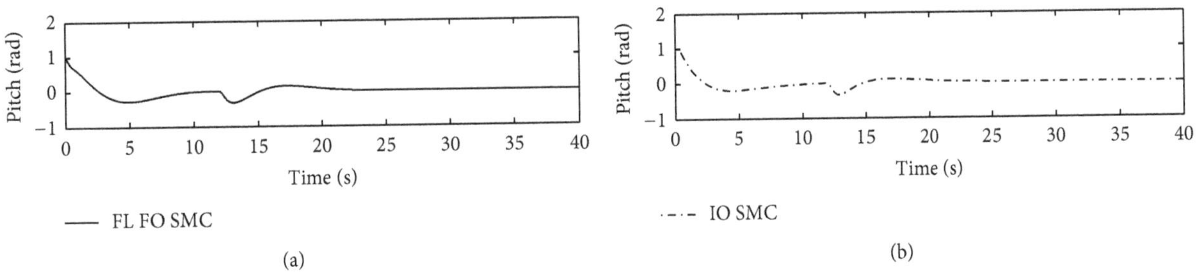

FIGURE 5: Stability curve of pitch.

Furthermore, let vectors X_1 and X_2 represent the partitions of the state vector, respectively. Thus, $x = [X_1{}^T, X_2{}^T]^T$, where $X_1 \in R^2$ and $X_2 \in R^4$. Hence, $\dot{X}_1$-subsystem and $\dot{X}_2$-subsystem are presented as follows:

$$\dot{X}_1\text{-subsystem: } \begin{cases} \dot{x}_1 = x_2 \\ \dot{x}_2 = -\dfrac{T_g}{J_\theta} + \dfrac{K_f l_a}{J_\theta}\left(u_1 + \xi_\varphi\right), \end{cases}$$

$$\dot{X}_2\text{-subsystem: } \begin{cases} \dot{x}_3 = x_4 \\ \dot{x}_4 = -\dfrac{K_p l_a}{J_\psi}\sin(x_5) \\ \dot{x}_5 = x_6 \\ \dot{x}_6 = \dfrac{K_f l_h}{J_\varphi}u_2. \end{cases} \quad (19)$$

5. Design of Controller

It will be seen that $\dot{X}_1$-subsystem and $\dot{X}_2$-subsystem have no coupling. We know feedback linearization scheme is suitable approach for multivariable systems but is not robust inherently. Sliding mode control is well known as a robust nonlinear control scheme by disturbance rejection capability. Because feedback linearization is a simple method and the FO SMC method is robust against disturbance, so we combine these two approaches. Firstly, feedback linearization method is applied; secondly, FO SMC method is used. Our control approach is formed in two steps. In the first step, a fractional order sliding mode controller for $\dot{X}_1$-subsystem is designed; then, since the $\dot{X}_2$-subsystem is multivariable and nonlinear, feedback linearization control is applied. For $\dot{X}_2$-subsystem, controller must be robust against disturbance. After linearization, a fractional order sliding mode controller is proposed for linearized subsystem. This helicopter model is an underactuated system by three outputs and two inputs, so we can control two outputs and third output only is kept limited.

Consider $\dot{X}_1$-subsystem as

$$\ddot{x}_1 = f + gu_1. \quad (20)$$

Let define sliding surface as $S_1 = {}^{\mathrm{RL}}D^{\alpha}\ddot{e}_1 + \lambda_1 e_1$ and the error signal is defined as $e_1 = x_{1d} - x_1$. Also, disturbance is imposed to roll and yaw channel.

The fractional order sliding mode control input signal can be defined as follows for subsystem (20), [12]:

$$u_1(t) = g^{-1}\left[\ddot{x}_d(t) - f + {}^{\mathrm{RL}}D^{-1-\alpha}(\lambda_1\dot{e}_1) + {}^{\mathrm{RL}}D^{-1-\alpha}(k_1 S_1 + k_1\,\mathrm{sign}(S_1))\right], \quad (21)$$

where k_1 and λ_1 are positive constants.

In [12], it has been shown that this control input signal guarantees the stability of the closed-loop $\dot{X}_1$-subsystem and the tracking error converges to zero in finite time. Then, in order to control $\dot{X}_2$-subsystem, at first we use input-output feedback linearization method. Feedback linearization control law is defined as

$$u_2(t) = \left(-\frac{K_p l_a}{J_\psi}\cos(x_5)\right)^{-1}\left[V - \frac{K_p l_a}{J_\psi}x_6\sin(x_5)\right]. \quad (22)$$

V is new control law that will be designed for linearized system. So we want to obtain overall control law that it has robustness behavior and good performance. We use fractional order control instead of integer order to design control input V.

The sliding mode control law, V, is designed as follows:

$$V = x_{3d}^{(4)} - \lambda_2^3\,{}^{\mathrm{RL}}D^{\beta}\dot{e}_2 - 3\lambda_2^2\,{}^{\mathrm{RL}}D^{\beta}\ddot{e}_2 - 3\lambda_2\,{}^{\mathrm{RL}}D^{\beta}e_2^{(3)} - k_2\mathrm{Sgn}(S_2)\,. \quad (23)$$

Let the sliding surface equation be proposed as follows:

$$S_2 = e_2^{(3)} + \lambda_2^3\,{}^{\mathrm{RL}}D^{\beta}e_2 + 3\lambda_2^2\,{}^{\mathrm{RL}}D^{\beta}\dot{e}_2 + 3\lambda_2\,{}^{\mathrm{RL}}D^{\beta}\ddot{e}_2\,. \quad (24)$$

The error signal is defined as $e_2 = x_3 - x_{3d}$ and the Lyapunov function is to be defined as

$$\widetilde{V} = \frac{1}{2}S_2^2. \quad (25)$$

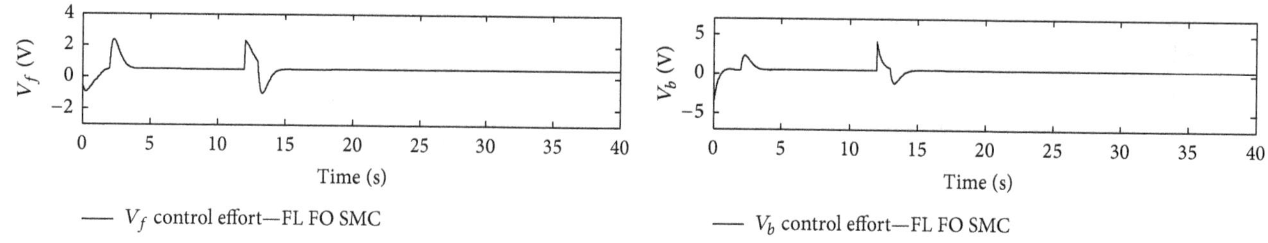

(a) (b)

FIGURE 6: Force control inputs of proposed controller (stability).

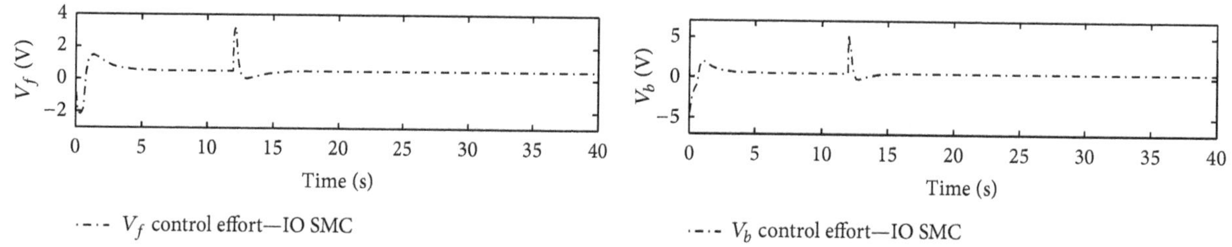

(a) (b)

FIGURE 7: Force control inputs of IO SMC (stability).

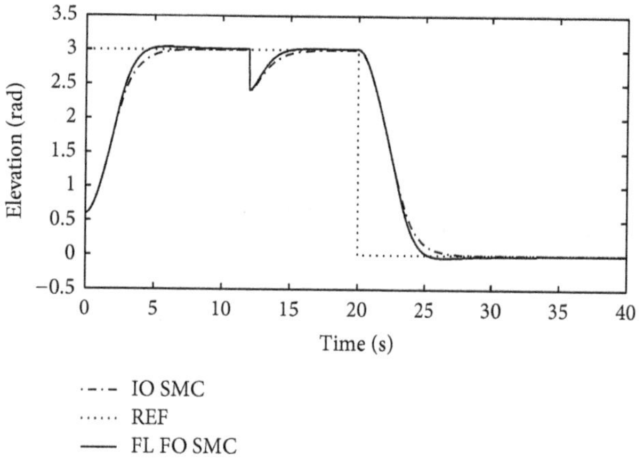

FIGURE 8: Tracking curve of elevation.

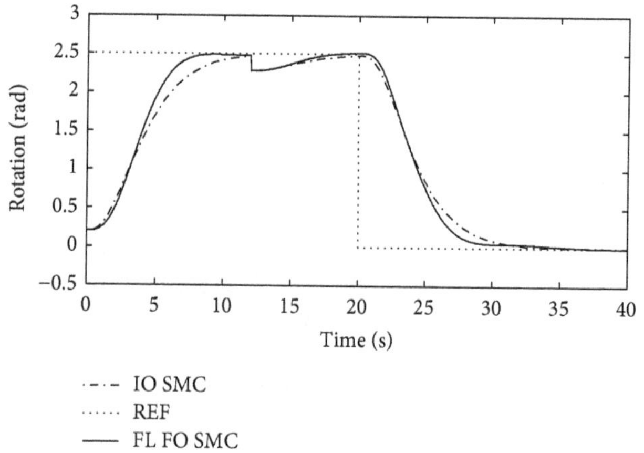

FIGURE 9: Tracking curve of rotation.

Consider the sufficient condition for the existence finite time convergence is

$$\frac{1}{2}\frac{d}{dt}S_2^2(t) \le -k_2\left|S_2(t)\right|, \tag{26}$$

where k_2 is a positive constant. Taking time derivative of both sides of (25), according to Lemmas 3 and 5, then we have

$$\begin{aligned}\dot{\tilde{V}} &= S_2\dot{S}_2 = S_2\left(e_2^{(4)} + \lambda_2^3\,{}^{\mathrm{RL}}D^{\beta+1}e_2 + 3\lambda_2^2\,{}^{\mathrm{RL}}D^{\beta+1}\dot{e}_2 + 3\lambda_2\,{}^{\mathrm{RL}}D^{\beta+1}\ddot{e}_2\right)\\ &= S_2\left(V - x_{3d}^{(4)} + \lambda_2^3\,{}^{\mathrm{RL}}D^{\beta+1}e_2 + 3\lambda_2^2\,{}^{\mathrm{RL}}D^{\beta+1}\dot{e}_2 + 3\lambda_2\,{}^{\mathrm{RL}}D^{\beta+1}\ddot{e}_2\right).\end{aligned} \tag{27}$$

Replacing (23) into (27) gives

$$\dot{\tilde{V}} = S_2\dot{S}_2 = S_2\left(-k_2\operatorname{sgn}(S_2)\right) \le -\left(k_2 + F\right)\left|S_2\right|, \tag{28}$$

where F is maximum of disturbance. So, total feedback linearization fractional order sliding mode control (FL FO

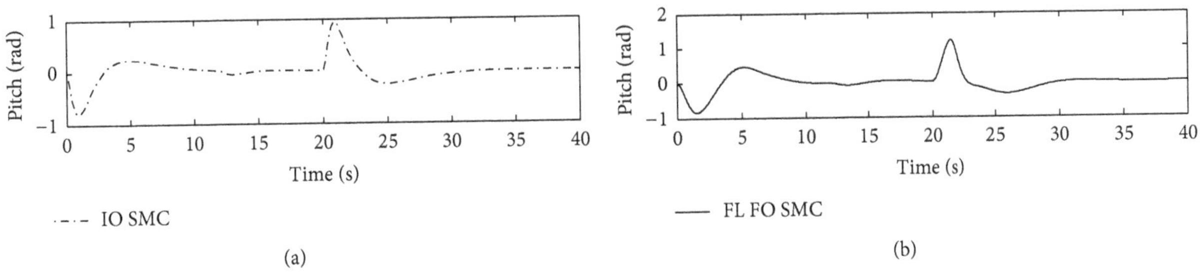

FIGURE 10: Tracking curve of pitch.

SMC) law is achieved by combination of (22) and (23) that is represented as follows:

$$u_2(t) = \left(-\frac{K_p l_a}{J_\psi}\cos(x_5)\right)^{-1} \times \left[x_{3d}^{(4)} - \lambda_2^3\,{}^{RL}D^{\beta}\dot{e}_2 - 3\lambda_2^2\,{}^{RL}D^{\beta}\ddot{e}_2 - 3\lambda_2\,{}^{RL}D^{\beta}e_2^{(3)} - k_2\mathrm{Sgn}(S_2) - \frac{K_p l_a}{J_\psi}x_6\sin(x_5)\right]. \quad (29)$$

These two control laws, u_1 and u_2, are applied to 3-DOF helicopter model and simulation results are shown in the next section.

TABLE 1: Parameters' values.

Parameter	Value	Unit
J_θ	0.91	(kgm^2)
J_φ	0.0364	(kgm^2)
J_ψ	0.91	(kgm^2)
K_f	0.5	(N)
K_p	0.5	(N)
l_a	0.66	(m)
l_h	0.177	(m)
T_g	0.33	(Nm)

6. Simulation Results

To see the performance of the proposed controller, we have simulated the controlled system. The results have been compared with integer order sliding mode controller. The helicopter nominal parameters are shown in Table 1. We choose $\alpha = 0.1$, $\beta = 0.15$ in this simulation. Furthermore, a step disturbance is applied in $t = 12$ sec.

To investigate the stability of the closed-loop system, the initial conditions are applied and the reference input is considered to be zero. The time response of the states of the system (18) in the presence of the control laws (21) and (29) is illustrated in Figures 3, 4, and 5. It can be seen that under the proposed controller the output converges to zero and when a disturbance applies the proposed controller is able to nullify the output again. It is clear that the convergence rate in the proposed controller is more than the integer order controller.

The force control inputs are given in Figures 6, and 7. Although the proposed controller is faster than integer order controller in states convergence rate, it requires less control signals energy compared with the integer order SMC controller.

The simulation results of the reference tracking under the disturbance are illustrated in Figures 8, 9, and 10. These figures show that the helicopter can follow the reference trajectory under the proposed control method, and the output response under the proposed control scheme is faster than another one in spite of disturbance. Also the pitch angle remained limited.

The control efforts have been shown in Figures 11 and 12. According to these figures it can be seen that although the proposed controller is faster than integer order controller, it needs less control effort. To implement the controller it is essential that the control effort has an acceptable value.

According to these simulations, it can be seen that the tracking performance of proposed controller is more efficient, with less control effort compared with the integer order sliding mode controller in spite of disturbances.

7. Conclusion

In this paper, we presented a combination of two nonlinear control techniques applied to 3-DOF helicopter. Also we considered disturbance in roll and yaw channel. In fact we have used feedback linearization fractional order sliding mode theory to control roll and yaw channel. It is desirable that the outputs track the reference input in less time. In practical application, a high-gain controller may be undesirable. So in this paper a fractional order controller was developed to increase the convergence rate in less control signal energy. To verify the performance of the proposed controller, it was compared with integer order sliding mode theory. The results show that the proposed method also simplifies the design process due to the use of feedback linearization control mechanism for multivariable systems, the sliding mode method advantages, such as robustness is retained. The

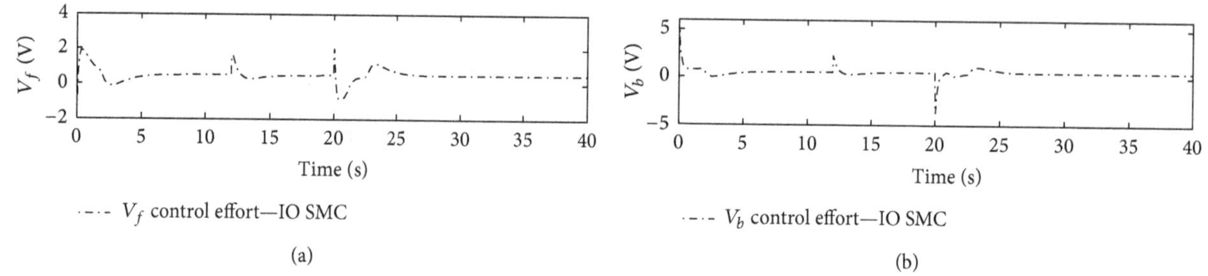

FIGURE 11: Force control inputs of proposed controller (tracking).

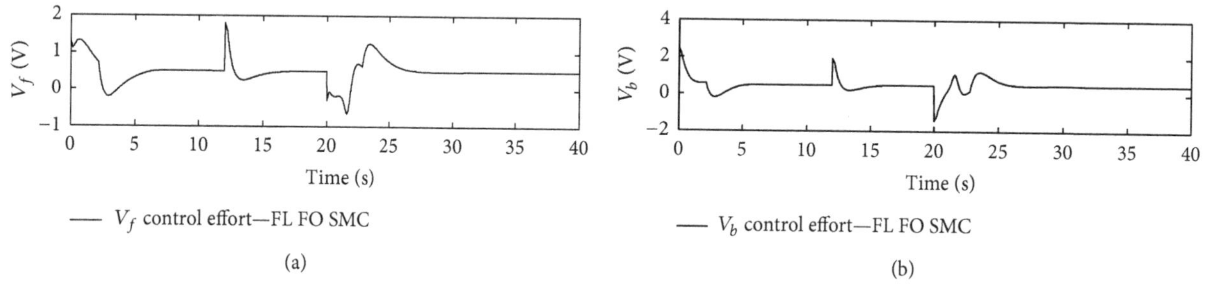

FIGURE 12: Force control inputs of pitch (tracking).

reason is that after linearization, a fractional order sliding mode controller is designed for linearized subsystem.

Conflict of Interests

The authors declare that there is no conflict of interests regarding the publication of this paper.

References

[1] K. H. Kienitz, Q.-H. Wu, and M. Mansour, "Robust stabilization of a helicopter model," in *Proceedings of the 29th IEEE Conference on Decision and Control*, pp. 2607–2612, December 1990.

[2] W. Garrad and E. Low, "Eigenspace design of helicopter flight control systems," Tech. Rep., Department of Aerospace Engineering and Mechanics, University of Minnesota, 1990.

[3] K. M. Im and W. Ham, "Sliding mode control for helicopter attitude regulation at hovering," *Journal of Control, Automation and Systems Engineering*, vol. 3, no. 6, pp. 563–568, 1997.

[4] A. Ferreira de Loza, H. Ríos, and A. Rosales, "Robust regulation for a 3-DOF helicopter via sliding-mode observation and identification," *Journal of the Franklin Institute*, vol. 349, no. 2, pp. 700–718, 2012.

[5] D. Jeong, T. Kang, H. Dharmayanda, and A. Budiyono, "*H*-infinity attitude control system design for a small-scale autonomous helicopter with nonlinear dynamics and uncertainties," *Journal of Aerospace Engineering*, vol. 25, no. 4, pp. 501–518, 2011.

[6] M. Boukhnifer, A. Chaibet, and C. Larouci, "*H*-infinity robust control of 3-DOF helicopter," in *Proceedings of the 9th International Multi-Conference on Systems, Signals and Devices (SSD '12)*, 2012.

[7] Y. Yu and Y. Zhong, "Robust attitude control of a 3DOF helicopter with multi-operation points," *Journal of Systems Science and Complexity*, vol. 22, no. 2, pp. 207–219, 2009.

[8] L. Marconi and R. Naldi, "Robust full degree-of-freedom tracking control of a helicopter," *Automatica*, vol. 43, no. 11, pp. 1909–1920, 2007.

[9] H. T. Nguyen and N. R. Prasad, "Development of an intelligent unmanned helicopter," in *Fuzzy Modeling and Control: Selected Works of M. Sugeno*, pp. 13–43, CRC Press, NewYork, NY, USA, 1999.

[10] A. J. Calise and R. T. Rysdyk, "Nonlinear adaptive flight control using neural networks," *IEEE Control Systems Magazine*, vol. 18, no. 6, pp. 14–24, 1998.

[11] S. Jafarzadeh, R. Mirheidari, and M. Jahed Motlagh, "Intelligent autopilot control design for a 2-DOF helicopter model," *International Journal of Computers, Communications & Control*, vol. 3, pp. 337–342, 2008.

[12] S. Dadras and H. R. Momeni, "Fractional terminal sliding mode control design for a class of dynamical systems with uncertainty," *Communications in Nonlinear Science and Numerical Simulation*, vol. 17, no. 1, pp. 367–377, 2012.

[13] I. Podlubny, *Fractional Differential Equations*, Academic Press, New York, NY, USA, 1999.

[14] A. Kilbas, H. Srivastava, and J. Trujillo, *Theory and Application of Fractional Differential Equations, North Holland Mathematics Studies*, Edited by J. van Mill, Elsevier, 2006.

[15] A. J. Calderón, B. M. Vinagre, and V. Feliu, "Fractional order control strategies for power electronic buck converters," *Signal Processing*, vol. 86, no. 10, pp. 2803–2819, 2006.

[16] R. El-Khazali, W. Ahmad, and Y. Al-Assaf, "Sliding mode control of generalized fractional chaotic systems," *International Journal of Bifurcation and Chaos*, vol. 16, no. 10, pp. 3113–3125, 2006.

[17] M. O. Efe and C. A. Kasnakoğlu, "A fractional adaptation law for sliding mode control," *International Journal of Adaptive Control and Signal Processing*, vol. 22, no. 10, pp. 968–986, 2008.

[18] M. S. Tavazoei and M. Haeri, "Synchronization of chaotic fractional-order systems via active sliding mode controller," *Physica A*, vol. 387, no. 1, pp. 57–70, 2008.

[19] A. Si-Ammour, S. Djennoune, and M. Bettayeb, "A sliding mode control for linear fractional systems with input and state delays," *Communications in Nonlinear Science and Numerical Simulation*, vol. 14, no. 5, pp. 2310–2318, 2009.

[20] A. Pisano, M. Rapaic, Z. Jelicic, and E. Usai, "Nonlinear fractional PI control of a class of fractional order systems," in *IFAC Conference on Advances in PID Control*, pp. 637–642, 2012.

[21] C. A. Monje, Y. Q. Chen, B. M. Vinagre, D. Xue, and V. Feliu, *Fractional Order Systems and Control: Fundamentals and Applications*, Springer, 2010.

7

Nonlinear Control of a Satellite Electrical Power System Based on the Sliding Mode Control

Mohammad Rasool Mojallizadeh[1] and Bahram Karimi[2]

[1] *Department of Electrical Engineering, Najafabad Branch, Islamic Azad University, University Blvd., Najafabad, Isfahan, Iran*
[2] *Department of Electrical Engineering, Malek Ashtar University of Technology, Shahinshahr, Isfahan, Iran*

Correspondence should be addressed to Mohammad Rasool Mojallizadeh; r.mojalli@gmail.com

Academic Editors: R. V. Kruzelecky, Z. Mazur, S. Simani, and I. Taymaz

The power electronic interface between a satellite electrical power system (EPS) with a photovoltaic main source and battery storage as the secondary power source is modelled based on the state space averaging method. Subsequently, sliding mode controller is designed for maximum power point tracking of the PV array and load voltage regulation. Asymptotic stability is ensured as well. Simulation of the EPS is accomplished using MATLAB. The results show that the outputs of the EPS have good tracking response, low overshoot, short settling time, and zero steady-state error. The proposed controller is robust to environment changes and load variations. Afterwards, passivity based controller is provided to compare the results with those of sliding mode controller responses. This comparison demonstrates that the proposed system has better transient response, and unlike passivity based controller, the proposed controller does not require reference PV current for control law synthesis.

1. Introduction

As space missions are getting more involved, satellites systems are getting more complex in parallel. Even the size of satellites are getting smaller due to budgetary constraints, the amount of power required to run the complete system is getting bigger resulting in larger PV arrays, higher battery capacity, and a much more sophisticated electrical power system (EPS). The primary function of EPS is to supply and manage uninterrupted power to its subsystems and payloads. These subsystems include power generation subsystems such as PV arrays, power storage subsystems which are batteries with different chemical structures, power control and distribution subsystems like power converters, power distribution units, power conditioning units, and battery charging units [1].

In the present space power domain, most of the satellite power systems use PV arrays as their power core. Despite all the advantages presented by the generation of energy through PV cells, the efficiency of energy conversion is currently low; thus, it becomes necessary to use techniques to extract the maximum power from these panels, in order to achieve maximum efficiency in operation. The requirement for maximum power point tracking (MPPT) is raised by the fact that the MPP of the PV array continuously varies with temperature and illumination changes. Due to the nonlinear characteristic of the PV array and drastic changes in irradiance and temperature, design of the MPPT unit is important. Several studies have been carried out, such as sliding mode control [2], adaptive control [3], neural networks [4], PSO [5, 6], fuzzy logic [7], chaos search [8], and GA-PI [9]. There are also some works about comparing different MPPT algorithms, such as [10–12]. This study uses a sliding mode controller for the MPPT. Unlike passivity based controller which is introduced in [13], sliding mode control approach which is proposed in this paper does not require reference current for control law synthesis.

Sliding mode control (SMC) is popular to converters [14]. The application of SMC to DC/DC converters can be traced back to 1983 [15]. The SMC design theory and application examples are available in [16]. SMC offers several benefits, namely, large signal stability, robustness, good dynamic response, system order reduction, and simple implementation [17]. A typical sliding mode control has two modes of operation. One is called the approaching mode, where

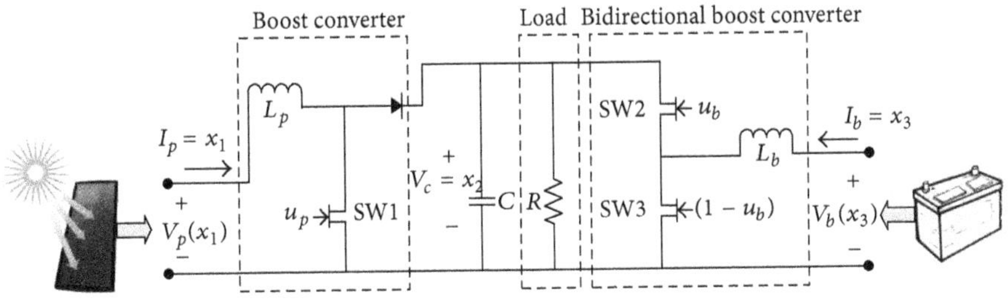

FIGURE 1: Proposed system diagram.

the system state converges to a predefined manifold named sliding function in finite time. The other mode is called the sliding mode, where the system state is confined on the sliding surface and is driven to the origin [16].

Passivity based control (PBC) was introduced by Ortega et al. [18], as a controller design methodology that achieves stabilization by passivation. Two theories for PBC have been developed: Euler Lagrange (EL)-PBC and interconnection and damping assignment (IDA)-PBC [13]. In this paper, EL-PBC [13] is provided to compare the results with those of the proposed sliding mode controller responses.

In this study, SMC approach is used for maximum power point tracking of the PV array, load voltage regulating, and charging or discharging the battery. Asymptotic stability of the proposed system is confirmed by using Lyapunov theory.

The paper is organized as follows. System modeling is introduced in Section 2. In Section 3, design and analysis of the SMC are presented. PBC is provided in Section 4 for comparison with the proposed control approach. Simulation results in MATLAB environment are then used in Section 5 to demonstrate the effectiveness of the proposed controller. Finally, Section 6 concludes the paper.

2. System Modeling

The proposed satellite EPS is depicted in Figure 1. This EPS comprises a PV array, a battery storage, DC/DC converters, and load. The battery storage has been considered for conditions in which the load power exceeds the generating power of the PV array.

2.1. Photovoltaic Energy System. A solar cell is the fundamental component of a PV system, which converts the solar energy into electrical energy. A PV cell consists of a p-n junction semiconductor material. A PV array consists of a certain number of PV cells connected in series/parallel to provide the desired voltage and current. The equivalent circuit of a PV cell is depicted in Figure 2.

The equivalent circuit mainly consists of a current source (i_{ph}), in which its amplitude depends on irradiance and temperature, diode, and internal resistance (r'_p). The P-I characteristic of a PV cell is highly nonlinear and is given by the following equations [3]:

$$V_p = \left(\frac{k_b T A}{q}\right) \ln\left(\frac{\left(i_{\text{ph}} + i_0 - I_p\right)}{i_0}\right) - I_p\, r_p, \quad (1)$$

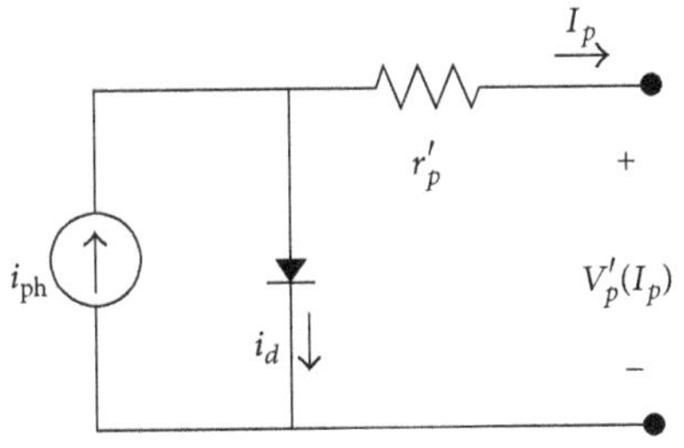

FIGURE 2: Equivalent circuit of the PV cell.

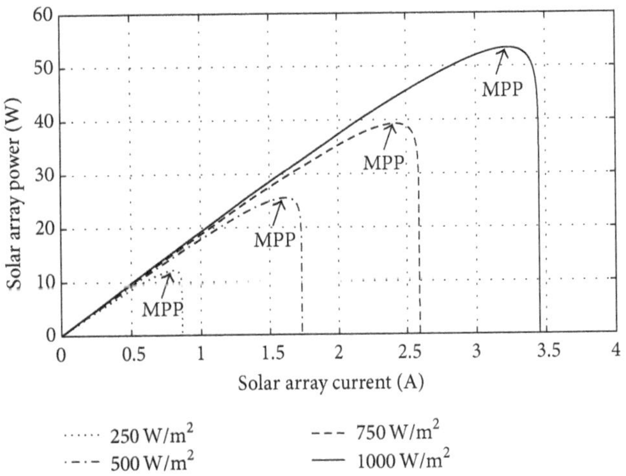

FIGURE 3: PV array characteristic under different irradiance levels.

$$i_{\text{ph}} = \lambda\left[i_{\text{sc}} + k_i\left(T - T_r\right)\right],$$
$$i_0 = i_r\left[\frac{T}{T_r}\right]^3 \exp\left(\left(\frac{qE_g}{(k_b A)}\right)\left(\frac{1}{T_r} - \frac{1}{T}\right)\right), \quad (2)$$

where V_p and I_p are the PV cell voltage and current, respectively, i_0 is the diode reverse saturation current, q is the electron charge, A is the ideality factor of the p-n junction, k_b is the Boltzmann constant, k_i is the temperature coefficient, T is the cell temperature, T_r is the reference temperature, E_g is the bandgap energy, i_r is the saturation current at T_r, and i_{sc} is the short circuit current.

Figure 3 shows the P-I curve of the SM-55 PV array under different irradiance conditions ($T = 300^k$). The power delivered by the PV module depends on the PV irradiance and cell temperature. Thus, maximum power must

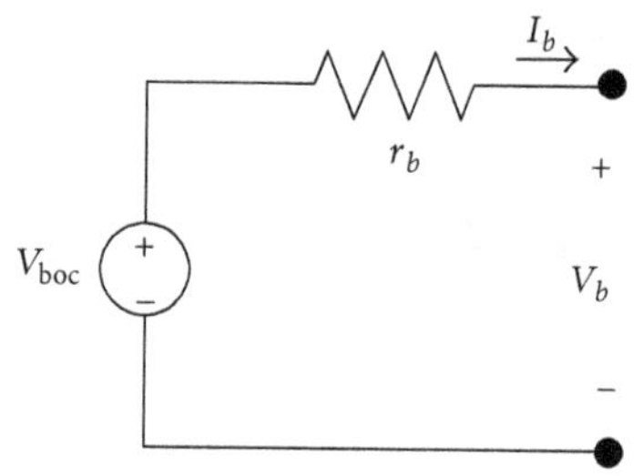

FIGURE 4: Equivalent circuit of the battery storage.

be available under different conditions. The typical approach used to maximize the power drawn from PV arrays under varying atmospheric conditions is the MPPT algorithm, which provides a reference voltage or current for the DC/DC converter. Unlike common approaches [13], in this study, the maximum power is drawn from PV array by implementing the sliding mode controller, and no desired PV reference is required.

2.2. Battery Storage. Storage devices are utilized for energy storage in EPS. The batteries store energy in the electrochemical form. In this study, the battery is modelled based on the generic Thevenin model [19]. Figure 4 shows equivalent circuit of the battery, where V_{boc} is the open circuit voltage and r_b is the equivalent resistance.

2.3. DC/DC Converters. The amplitude of the DC output current (voltage) of the PV array depending on the solar irradiance delivered to the PV arrays. Therefore, a boost DC/DC converter is utilized to adjust the output current of the PV system. Also, a bidirectional DC/DC converter is used for load voltage regulation and charge or discharge the battery storage. The DC/DC converters have been depicted in Figure 1.

From Figure 5, the system can be written in four sets of state equation depends on the position of switches SW1, SW2, and SW3. By utilizing state space averaging method [20], dynamic equations of the system can be expressed as:

$$\dot{x}_1 = \frac{\left(V_p(x_1) - x_2 + x_2 u_p\right)}{L_p},$$
$$\dot{x}_2 = \frac{\left(-(1/R)\, x_2 + x_1 - x_1 u_p + x_3 u_b\right)}{C}, \quad (3)$$
$$\dot{x}_3 = \frac{\left(V_b - x_2 u_b\right)}{L_b},$$

where $\mathbf{x} = [x_1, x_2, x_3]^T = [i_{L_p}, v_C, i_{L_b}]^T$ is the state vector, $0 < u_p < 1$ is the duty cycle of SW1 which is also a control input for MPPT, and $0 < u_b < 1$ is the duty cycle of SW2 which is also a control input for regulating load voltage and charging or discharging battery. Equation (3) can be written in general form of the nonlinear time invariant system:

$$\dot{\mathbf{x}} = f(\mathbf{x}) + g(\mathbf{x})\,\mathbf{u}. \quad (4)$$

3. Sliding Mode Controller Design

In the proposed EPS, the main objectives are maximum power point tracking of the PV array and load voltage regulation. In this study, an MIMO sliding mode controller is designed for these purposes. The proposed controller produces two control signals. The first control signal (u_p) is applied to the boost converter for the MPPT of the PV array, and the second control signal (u_b) is applied to the bidirectional boost converter for load voltage regulation. Unlike other approaches [13], in this study, no desired PV reference required for control law synthesis. Thus, the proposed system is robust to operation conditions and PV array's parameter changes.

Figure 3 shows P-I curve of the PV array under uniform insolation conditions. By selecting the PV sliding surface as (5), it is guaranteed that the system state will hit the surface and produce maximum power persistently:

$$\frac{\partial P_P}{\partial I_P} = \frac{\partial\left(I_P^2 R_P\right)}{\partial I_P} = I_P\left(2R_P + I_P\frac{\partial R_P}{\partial I_P}\right) = 0, \quad (5)$$

where $R_p = V_p/I_p$ is the equivalent load. The nontrivial solution of (5) selected as PV sliding surface (s_p):

$$s_p \triangleq 2R_P + I_P\frac{\partial R_P}{\partial I_P}. \quad (6)$$

With $R_p = V_p/I_p$, (6) can also be written as:

$$s_p \triangleq 2\frac{V_p(x_1)}{x_1} + x_1\frac{\partial\left(V_p(x_1)/x_1\right)}{\partial x_1}. \quad (7)$$

Voltage regulation sliding surface (s_b) is also selected as:

$$s_b = x_3 - x_{3d}, \quad (8)$$

where x_{3d} is the desired battery current and can be described by:

$$P_b = P_L - P_p \longrightarrow x_{3d} = \frac{1}{V_b}\left(\frac{1}{R}x_{2d}^2 - V_p\, x_1\right), \quad (9)$$

where P_b, P_L, and P_p are battery, load, and PV powers respectively. By considering the sliding surface vector as (10) it is guaranteed that the system will reach its desired states:

$$\mathbf{s} = \begin{bmatrix} s_p \\ s_b \end{bmatrix} = \begin{bmatrix} 2\frac{v_p(x_1)}{x_1} + x_1\frac{\partial\left(v_p(x_1)/x_1\right)}{\partial x_1} \\ x_3 - x_{3d} \end{bmatrix}. \quad (10)$$

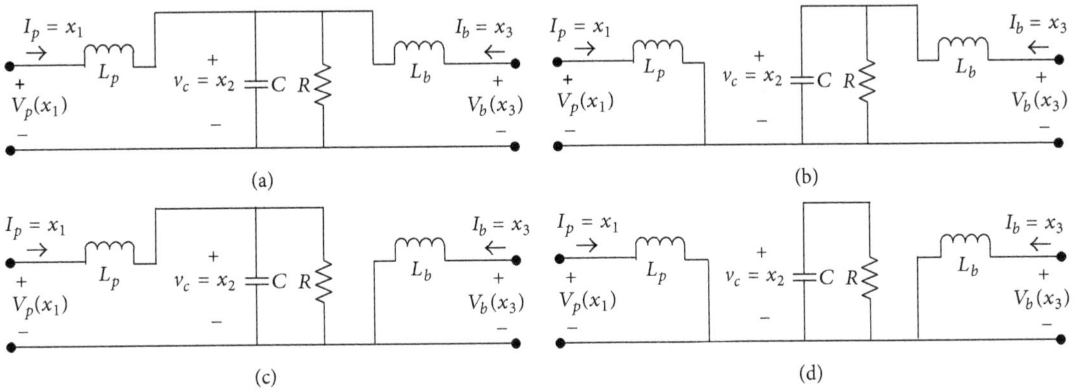

FIGURE 5: Different operating conditions. (a): SW1 = O, SW2 = C, SW3 = O. (b): SW1 = C, SW2 = C, SW3 = O. (c): SW1 = O, SW2 = O, SW3 = C. (d): SW1 = C, SW2 = O, SW3 = C. (open = O and close = C).

In order to get the equivalent control ($\mathbf{u}_{\text{eq}}$) suggested by [16], the equivalent control is determined from the following condition:

$$\dot{\mathbf{s}} = \begin{bmatrix} \dfrac{\partial s_p}{\partial x_1}\dot{x}_1 + \dfrac{\partial s_p}{\partial x_2}\dot{x}_2 + \dfrac{\partial s_p}{\partial x_3}\dot{x}_3 \\ \dfrac{\partial s_b}{\partial x_1}\dot{x}_1 + \dfrac{\partial s_b}{\partial x_2}\dot{x}_2 + \dfrac{\partial s_b}{\partial x_3}\dot{x}_3 \end{bmatrix}_{\mathbf{u}=\mathbf{u}_{\text{eq}}} = \begin{bmatrix} \dfrac{\partial s_p}{\partial x_1}\dot{x}_1 \\ \dfrac{\partial s_b}{\partial x_3}\dot{x}_3 \end{bmatrix}_{\mathbf{u}=\mathbf{u}_{\text{eq}}} = \begin{bmatrix} 0 \\ 0 \end{bmatrix}. \tag{11}$$

The equivalent control is then derived:

$$\mathbf{u}_{\text{eq}} = \begin{bmatrix} u_{\text{peq}} \\ u_{\text{beq}} \end{bmatrix} = \begin{bmatrix} 1 - \dfrac{V_p(x_1)}{x_2} \\ \dfrac{V_b}{x_2} \end{bmatrix}. \tag{12}$$

Since the range of duty cycle must lie in $0 < (u_p, u_b) < 1$, the real control signal is proposed as

$$\mathbf{u} = \begin{bmatrix} u_p \\ u_b \end{bmatrix}, \tag{13}$$

where u_p and u_b can be written as

$$u_p = \begin{cases} 0 & u_{\text{peq}} + k_p s_p < 0 \\ u_{\text{peq}} + k_p\, s_p & 0 < u_{\text{peq}} + k_p s_p < 1 \\ 1 & 1 < u_{\text{peq}} + k_p s_p, \end{cases}$$

$$u_b = \begin{cases} 0 & u_{\text{beq}} + k_b \operatorname{Sat}\left(\dfrac{s_b}{\phi}\right) < 0 \\ u_{\text{beq}} + k_b \operatorname{Sat}\left(\dfrac{s_b}{\phi}\right) & 0 < u_{\text{beq}} + k_b \operatorname{Sat}\left(\dfrac{s_b}{\phi}\right) < 1 \\ 1 & 1 < u_{\text{beq}} + k_b \operatorname{Sat}\left(\dfrac{s_b}{\phi}\right), \end{cases} \tag{14}$$

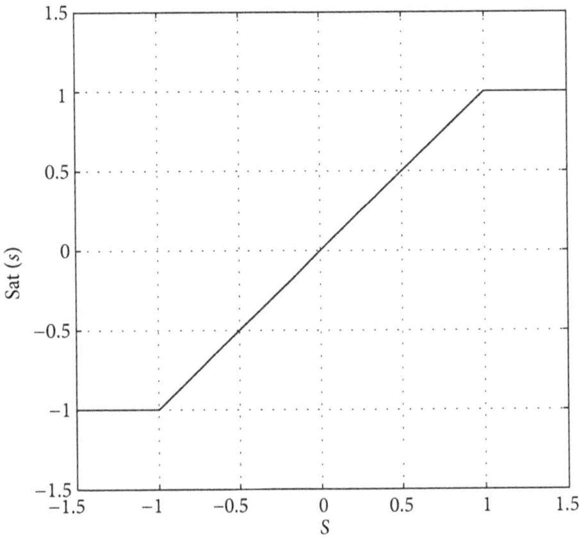

FIGURE 6: Saturation function.

where k_p and k_b are constant coefficients and are determined by trial and error method by using computer simulations, Sat(s) is the saturation function which is shown in Figure 6, and ϕ is a small constant and is selected for chattering avoidance [16].

The existence of the approaching mode of the proposed sliding function **s** is provided. A Lyapunov function is defined as

$$v = v_M + v_R, \tag{15}$$

where v_M and v_R are positive definite terms and defined as

$$v_M = \frac{1}{2}s_p^2, \qquad v_R = \frac{L_b}{2}s_b^2. \tag{16}$$

The time derivative of v can be written as

$$\dot{v} = \dot{v}_M + \dot{v}_R. \tag{17}$$

The achievability of $\mathbf{s} = \mathbf{0}$ will be obtained by $\dot{v} < 0$. It can be shown that both $\dot{v}_M$ and $\dot{v}_R$ are negative definite. $\dot{v}_R$ can be written as follows:

$$\dot{v}_R = s_b \left(V_b - x_2 u_b\right). \tag{18}$$

Three cases should be examined for the fulfillment of $\dot{v}_R < 0$.

For $\mathbf{0 < u_b < 1}$

$$\dot{v}_R = -x_2\ k_b \left(x_3 - x_{3d}\right) \operatorname{sat}\left(\frac{x_3 - x_{3d}}{\phi}\right) < 0. \tag{19}$$

Since $(x_3 - x_{3d})$ always has the same sign of $\operatorname{sat}((x_3 - x_{3d})/\phi)$ and $x_2\ k_b > 0$, the sign of (19) is negative definite.

For $\mathbf{u_b = 1}$

$$\dot{v}_R = s_b \left(V_b - x_2\right) < 0. \tag{20}$$

In this case, the load voltage (x_2) is higher than the battery voltage (V_b) and $s_b > 0$. From (20) it results $\dot{v}_R < 0$.

For $\mathbf{u_b = 0}$

$$\dot{v}_R = s_b V_b < 0. \tag{21}$$

In this case, $s_b < 0$ is obtained and $\dot{v}_R < 0$.

From the discussion above, $\dot{v}_R < 0$ is obtained.

$\dot{v}_p$ can also be written as

$$\dot{v}_M = s_p \dot{s}_p. \tag{22}$$

The time derivative of s_p can be written as

$$\dot{s}_p = \left(3\frac{\partial R_p}{\partial x_1} + x_1 \frac{\partial^2 R_p}{\partial x_1^2}\right) \frac{\left(v_p\left(x_1\right) - x_2 + x_2 u_p\right)}{L_p}. \tag{23}$$

Replacing R_p by the definition of $R_p = V_p / I_p$,

$$\begin{aligned}\frac{\partial R_p}{\partial x_1} &= \frac{\partial}{\partial x_1}\left[\frac{V_p}{x_1}\right] = \frac{1}{x_1}\frac{\partial V_p}{\partial x_1} - \frac{V_p}{x_1^2}, \\ \frac{\partial^2 R_p}{\partial x_1^2} &= \frac{1}{x_1}\frac{\partial^2 V_p}{\partial x_1^2} - \frac{2}{x_1^2}\frac{\partial V_p}{\partial x_1} + \frac{2V_p}{x_1^3}.\end{aligned} \tag{24}$$

By (1), the following equations will be obtained:

$$\begin{aligned}\frac{\partial V_p\left(x_1\right)}{\partial x_1} &= -\frac{k_b T A}{q}\frac{i_0}{i_{\text{ph}} + i_0 - x_1} - r_p < 0, \\ \frac{\partial^2 V_p\left(x_1\right)}{\partial x_1^2} &= -\frac{k_b T A}{q}\frac{i_0}{\left(i_{\text{ph}} + i_0 - x_1\right)^2} < 0.\end{aligned} \tag{25}$$

Substituting (25) into (23) yields

$$\begin{aligned}\frac{\partial s_p}{\partial x_1} &= 3\frac{\partial R_p}{\partial I_p} + I_p \frac{\partial^2 R_p}{\partial I_p^2} \\ &= \frac{\left(\partial v_p\left(x_1\right)/\partial x_1\right)}{x_1} + \frac{\partial^2 v_p\left(x_1\right)}{\partial x_1^2} - \frac{v_p}{x_1^2} < 0.\end{aligned} \tag{26}$$

According to the result of (25) and $(V_p, x_1) > 0$, the sign of (26) is negative definite. The achievability of $\dot{v}_p < 0$ will be obtained by $s_p \dot{s}_p < 0$ for all u_p discussed as follows.

For $\mathbf{0 < u_p < 1}$

$$\begin{aligned}\dot{x}_1 &= \frac{\left(v_p\left(x_1\right) - x_2\left(1 - u_{\text{peq}} - k_p s_p\right)\right)}{L_p} \\ &= \frac{x_2 k_p s_p}{L_p}.\end{aligned} \tag{27}$$

Based on the result of (26) and (27), $\dot{s}_p$ always has inverse sign of s_p. Therefore, $s_p \dot{s}_p < 0$ is obtained for $0 < u_p < 1$.

For $\mathbf{u_p = 1}$

$$\dot{x}_1 = \frac{V_p\left(x_1\right)}{L_p} > 0. \tag{28}$$

By (26) and (28), $\dot{s}_p < 0$ with $u_p = 1$, two cases should be examined for the fulfillment of $s_p \dot{s}_p < 0$.

(a) $u_{\text{peq}} = 1$.

If $u_{\text{peq}} = 1$, it implies that $V_p(x_1) = 0$ which means s_p is negative for this case. Therefore, $u_{\text{peq}} + k_p s_p$ will be less than 1, which contradicts to the assumption of $u_p = 1$.

(b) $u_{\text{peq}} < 1$ and $u_{\text{peq}} + k_p s_p \geq 1$.

If $u_{\text{peq}} < 1$, but $u_{\text{peq}} + k_p s_p \geq 1$, it implies that $s_p > 0$ and $s_p \dot{s}_p < 0$.

It conclude that $s_p \dot{s}_p < 0$ for $u_p = 1$.

For $\mathbf{u_p = 0}$

$$\dot{x}_1 = \frac{\left(V_p\left(x_1\right) - x_2\right)}{L_p}. \tag{29}$$

In this case load voltage (x_2) is higher than the PV voltage $(V_p(x_1))$. From (26) and (29), it results that $\dot{s}_p > 0$. Two cases for $u_p = 0$ are examined as follows.

(a) $u_{\text{peq}} = 0$.

$u_{\text{peq}} = 0$ implies $V_p(x_1) = x_2$, which corresponding to the situation that the PV array is directly connected to the load and operates in the region $s_p > 0$. As the results $u_p > 0$ and it contradicts to the assumption of $u_p = 0$.

(b) $u_{\text{peq}} > 0$ and $u_{\text{peq}} + k_p s_p \leq 0$.

In this case, $s_p < 0$ is obtained and $s_p \dot{s}_p < 0$.

It concludes that $s_p \dot{s}_p < 0$ for $u_p = 0$.

From the discussion above, the stability of the system can be guaranteed using the proposed control law (14).

4. Passivity Based Control

In this study, Euler Lagrange—Passivity based control (EL-PBC) approach [13] is provided to compare the results with those of the proposed sliding mode controller responses. The design of the passivity based (PB) controller is based on the

TABLE 1: System specification.

PV	Model SM-55	Equivalent resistance 0.03 (Ω)	Maximum power 55 (W)	Open circuit voltage 21.7 (V)	Short circuit current 3.8 (A)
Battery		Open circuit voltage 9 (V)		Equivalent resistance 0.08 (Ω)	
DC/DC		L_p 5 (mH)	C 500 (μf)		L_b 10 (mH)

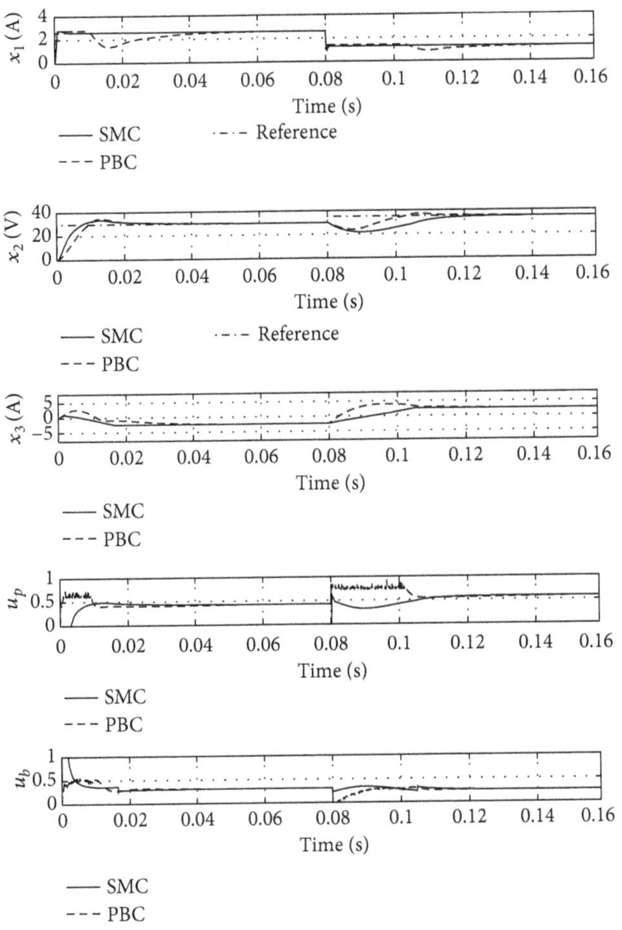

FIGURE 7: State variables and control signals.

Euler Lagrange model of the converters. The PBC control signals which proposed in [13] are shown in

$$u_p = 1 - \frac{1}{x_{2d}}\left(V_p\left(x_1\right) + r_{a1}\left(x_1 - x_{1d}\right)\right),$$
$$u_b = \frac{1}{x_{2d}}\left(V_b\left(x_3\right) + r_{a2}\left(x_3 - x_{3d}\right)\right), \quad (30)$$

where x_{1d} is the reference current and needs to be determined by MPPT algorithms, x_{3d} is the desired battery current and can be described by (9), x_{2d} is the reference of the load voltage. $(r_{a1}, r_{a2}) > 0$ are design parameters (see [13] for more details). Passivity based control approach which is introduced in [13], required reference current x_{1d} for control law synthesis and may lead to a lack of robustness to operation conditions.

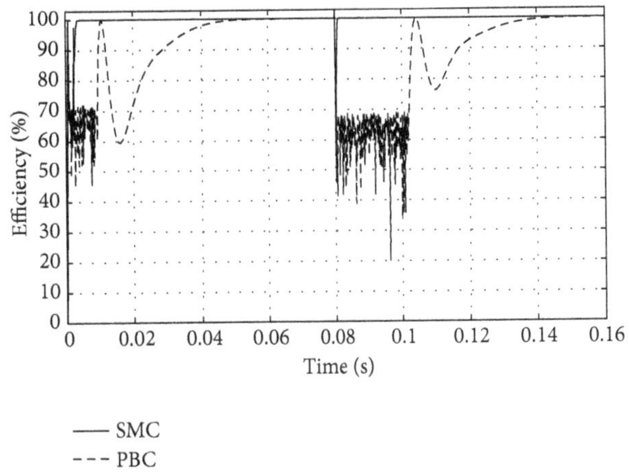

FIGURE 8: Efficiency.

TABLE 2: Parameters variations.

Parameters	0 to 79 (ms)	80 to 160 (ms)
Load resistance (R)	40 (ohms)	30 (ohms)
Reference dc voltage (x_{2d})	30 (v)	35 (v)
PV temperature	290 (k)	310 (k)
PV irradiance level	800 (w/m^2)	400 (w/m^2)

5. Simulation

MATLAB environment is used to investigate the performance of the SMC on the proposed satellite EPS. PBC is provided to compare the results with those of proposed SMC responses. The simulation investigates four system characteristics: robustness against irradiance, temperature, load resistance, and load voltage reference changes. Unlike SMC, PBC approach required reference current (x_{1d}) for control law synthesis which comes from P&O algorithm [13]. It is assumed that optimal reference current (x_{1d}) is available to PBC. The parameters of the components are chosen to deliver maximum 55 W of power generated by SM-55 and battery. The specification of the system is tabulated in Table 1.

According to Table 2, two step changes applied to Load resistance, load voltage reference, irradiance level, and PV array temperature at $t = 0$ and $t = 80$ (ms).

Figure 7 presents the state variables and control signals. PV array current (x_1) based on the proposed controllers

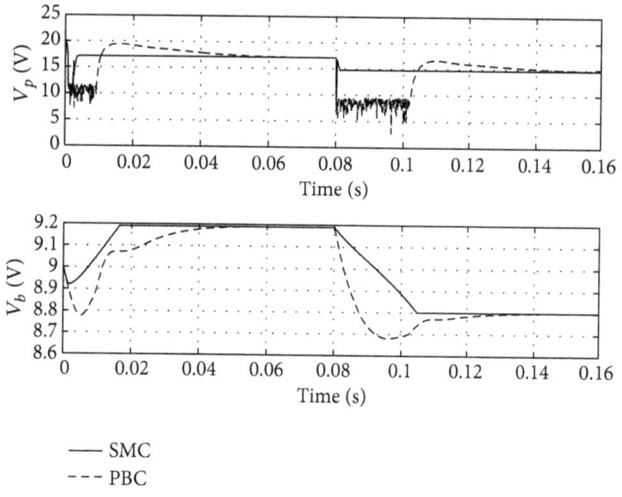

FIGURE 9: PV and battery voltage.

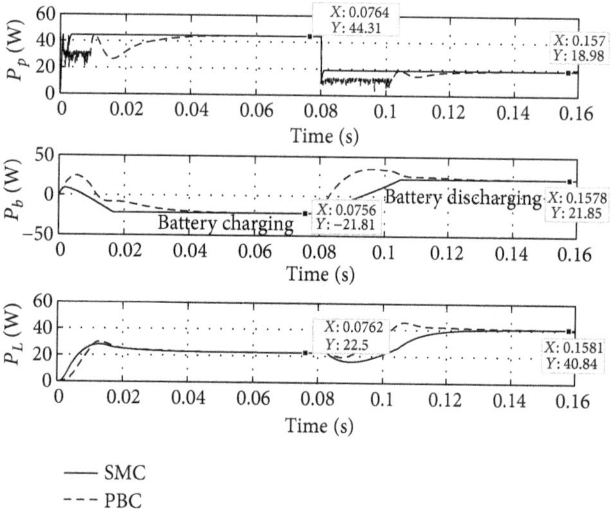

FIGURE 10: PV power, battery power and load power.

has faster response with zero steady state error. The load voltage (x_2) tracks the voltage reference (x_{2d}) well, with low overshoot, short settling time and zero steady-state error. Figure 8 shows the efficiency of the both control approaches. As illustrated in the figure, both approaches have 100% steady state efficiency, but SMC has better transient efficiency. Unlike the SMC, the PBC approach needs optimal reference current which comes from external MPPT algorithms.

Figure 9 shows the PV array and battery voltage. Figure 10 shows PV array power, battery power, and load power. As illustrated, the proposed SMC has better transient response and charges and discharges the battery efficiently.

6. Results

In this paper, a state space averaging model of a satellite electrical power system (EPS) with a PV array as the main source, a battery storage as the secondary source and interfacing DC/DC converters has been presented. Subsequently, sliding mode controller has been designed to control the interfacing DC/DC converters. To investigate the validity of the proposed system, a passivity based controller was provided. Unlike PBC, the proposed system does not require reference current for control law synthesis. Simulation results show that both approaches have zero steady state error. But the proposed control approach has better transient response and does not require external MPPT reference current. Moreover, the aforementioned results demonstrate the robustness of the proposed control approach during the load resistance, battery voltage, solar irradiance, PV array temperature, and load voltage reference changes.

References

[1] M. A. Kaya and M. K. Bayrakceken, "Complete electrical model and simulation of a medium size satellite," in *Proceedings of the 5th International Conference on Recent Advances in Space Technologies (RAST '11)*, pp. 522–525, Istanbul, Turkey, June 2011.

[2] C. C. Chu and C. L. Chen, "Robust maximum power point tracking method for photovoltaic cells: a sliding mode control approach," *Solar Energy*, vol. 83, no. 8, pp. 1370–1378, 2009.

[3] H. El Fadil and F. Giri, "Climatic sensorless maximum power point tracking in PV generation systems," *Control Engineering Practice*, vol. 19, no. 5, pp. 513–521, 2011.

[4] Syafaruddin, E. Karatepe, and T. Hiyama, "Performance enhancement of photovoltaic array through string and central based MPPT system under non-uniform irradiance conditions," *Energy Conversion and Management*, vol. 62, pp. 131–140, 2012.

[5] K. Ishaque, Z. Salam, A. Shamsudin, and M. Amjad, "A direct control based maximum power point tracking metystem under partial shading conditions using particle swarm optimization algorithm," *Applied Energy*, vol. 99, pp. 414–422, 2012.

[6] K. Ishaque, Z. Salam, M. Amjad, and S. Mekhilef, "An improved particle swarm optimization (PSO)-based MPPT for PV with reduced steady-state oscillation," *IEEE Transactions on Power Electronics*, vol. 27, no. 8, pp. 3627–3638, 2012.

[7] M. M. Algazar, H. Al-Monier, H. A. El-Halim, and M. E. E. K. Salem, "Maximum power point tracking using fuzzy logic control," *International Journal of Electrical Power and Energy Systems*, vol. 39, no. 1, pp. 21–28, 2012.

[8] L. Zhou, Y. Chen, Q. Liu, and J. Wu, "Maximum power point tracking (MPPT) control of a photovoltaic system based on dual carrier chaotic search," *Journal of Control Theory and Applications*, vol. 10, no. 2, pp. 244–250, 2012.

[9] H. T. Yau, Q. C. Liang, and C. T. Hsieh, "Maximum power point tracking and optimal Li-ion battery charging control for photovoltaic charging system," *Computers and Mathematics with Applications*, vol. 64, no. 5, pp. 822–832, 2012.

[10] M. A. G. de Brito, L. Galotto, L. P. Sampaio, G. de Azevedo e Melo, and C. A. Canesin, "Evaluation of the main MPPT techniques for photovoltaic applications," *IEEE Transactions on Industrial Electronics*, vol. 60, no. 3, pp. 1156–1167, 2013.

[11] N. Onat, "Recent developments in maximum power point tracking technologies for photovoltaic systems," *International Journal of Photoenergy*, vol. 2010, Article ID 245316, 11 pages, 2010.

[12] W. Xiao, A. Elnosh, V. Khadkikar, and H. Zeineldin, "Overview of maximum power point tracking technologies for photovoltaic power systems," in *Proceedings of the 37th Annual*

Conference of the IEEE Industrial Electronics Society (IECON '11), pp. 3900–3905, Melbourne, Australia, November 2011.

[13] A. Tofighi and M. Kalantar, "Power management of PV/battery hybrid power source via passivity-based control," *Renewable Energy*, vol. 36, no. 9, pp. 2440–2450, 2011.

[14] V. I. Utkin, J. Guldner, and J. X. Shi, *Sliding Mode Control in ElectroMechanical Systems*, Taylor & Francis, London, UK, 2008.

[15] F. Bilalovic, O. Music, and A. Sabanovic, "Buck converter regulator operating in the sliding mode," in *Proceedings of the 7th Power Conversion International Conference*, pp. 331–340, 1983.

[16] J. J. E. Slotine and W. Li, *Applied Nonlinear Control*, Prentice Hall, New York, NY, USA, 1991.

[17] P. Mattavelli, L. Rossetto, and G. Spiazzi, "Small-signal analysis of DC-DC converters with sliding mode control," *IEEE Transactions on Power Electronics*, vol. 12, no. 1, pp. 96–102, 1997.

[18] R. Ortega, J. A. Loría Perez, P. J. Nicklasson, and H. J. Sira-Ramirez, *Passivity-Based Control of Euler-Lagrange Systems: Mechanical, Electrical and Electromechanical Applications*, Springer, New York, NY, USA, 1998.

[19] B. Lin, "Conceptual design and modeling of a fuel cell scooter for urban Asia," in *Mechanical and Aerospace Engineering*, Princeton University, Princeton, NJ, USA, 1999.

[20] P. T. Krein, J. Bentsman, R. M. Bass, and B. L. Lesieutre, "On the use of averaging for the analysis of power electronic systems," *IEEE Transactions on Power Electronics*, vol. 5, no. 2, pp. 182–190, 1989.

8

Wake Measurements behind an Oscillating Airfoil in Dynamic Stall Conditions

A. Zanotti, G. Gibertini, D. Grassi, and D. Spreafico

Politecnico di Milano, Dipartimento di Scienze e Tecnologie Aerospaziali, Campus Bovisa, Via La Masa 34, 20156 Milano, Italy

Correspondence should be addressed to A. Zanotti; alex.zanotti@polimi.it

Academic Editors: J. Meseguer, R. K. Sharma, and W. Zhang

The unsteady flow field in the wake of an NACA 23012 pitching airfoil was investigated by means of triple hot-wire probe measurements. Wind tunnel tests were carried out both in the light and deep dynamic stall regimes. The analysis of the wake velocity fields was supported by the measurements of unsteady flow fields and airloads. In particular, particle image velocimetry surveys were carried out on the airfoil upper surface, while the lift and pitching moments were evaluated integrating surface pressure measurements. In the light dynamic stall condition, the wake velocity profiles show a similar behaviour in upstroke and in downstroke motions as, in this condition, the flow on the airfoil upper surface is attached for almost the whole pitching cycle and the airloads show a small amount of hysteresis. The deep dynamic stall measurements in downstroke show a large extent of the wake and a high value of the turbulent kinetic energy due to the passage of strong vortical structures, typical of this dynamic stall regime. The comprehensive experimental database can be considered a reference for the development and validation of numerical tools for such peculiar flow conditions.

1. Introduction

The aerodynamics of oscillating airfoils is widely investigated as it represents a good model for the study of the dynamic stall of the retreating blade sections [1, 2]. The dynamic stall phenomenon occurs on the retreating side of the helicopter rotor at a high forward flight speed or during maneuvers at high load factors, and it produces several adverse effects on the helicopter performance. The main detrimental effects due to dynamic stall are the limitation of the forward speed and thrust, high control system loads, the introduction of a high level of vibrations affecting the helicopter dynamic performance in terms of maneuver capability and handling qualities, and the occurrence of the aeroelastic problem known as stall flutter [3] causing blade structural damage and excessive cabin vibration.

Therefore, in order to overcome the limitations on helicopter performance introduced by this phenomenon and to expand the helicopter flight envelope and vehicle utility, dynamic stall has become in the recent years one of the most challenging research topics in rotorcraft aerodynamics field. In fact, several research activities both in experimental and numerical fields investigated the effectiveness of different dynamic stall control systems integrated into the blade section (e.g., [4–6]). Moreover, the study of the fine details involved in the physics of this phenomenon was widely investigated in the recent years, as, for instance, in [7–9]. In this research area, the study of the unsteady wake of an oscillating airfoil features a relatively low attention by the literature with respect to the evaluation of the characteristic flow field and the measurement of the airloads on oscillating airfoils in dynamic stall conditions, as demonstrated by the large number of research activities focused on this topic (e.g., [9–12]).

In the recent literature, Jung and Park [13] investigated the unsteady characteristics of the vortex shedding in the near wake of an oscillating airfoil by means of hot-wire anemometry surveys and smoke visualizations, finding that the vortex-shedding characteristics in dynamic conditions are quite different from the steady airfoil case and depends on the state of the boundary layer. Hot-wire measurements were also carried out by Chang [14] to study the Reynolds number effects and the turbulence intensity in the near wake of a NACA 4412 oscillating airfoil. Moreover, Sadeghi et al. [15]

investigated the effects of the operating conditions of the pitching airfoil, as mean angle of attack, reduced frequency, and instantaneous angle of attack on the wake velocity field. In a more recent work by Sadeghi et al. [16], wake measurements behind an oscillating airfoil compared to smoke visualizations were considered to predict the aerodynamic features characteristic of the dynamic stall phenomenon.

The present paper presents the results of an experimental activity that investigated the flow conditions evolution around an NACA 23012 airfoil during its pitching oscillation. The wind tunnel tests were carried out at the Aerodynamics Laboratory of Politecnico di Milano, using an experimental rig designed for testing full-scale helicopter blade sections oscillating in the pitch. In particular, the wake of the pitching blade section model was measured by means of a triple hot-wire probe. The test campaign investigated the different regimes of the dynamic stall occurring on the rotor retreating blade in a forward flight, namely, the light and deep dynamic stall [1]. The analysis of the wake velocity fields was supported by the measurements of unsteady flow fields and airloads [17]. In particular, the flow fields over the upper surface of the oscillating airfoil were measured by means of particle image velocimetry (PIV). Lift and pitching moments were evaluated integrating surface pressure measurements carried out on the midspan contour of the model by means of fast unsteady pressure transducers. The experimental investigation was completed by unsteady pressure measurements carried out on the wind tunnel floor and ceiling.

2. Experimental Setup

The experimental activity was conducted at Politecnico di Milano in the low-speed closed-return wind tunnel of the Aerodynamics Laboratory of DAER. The wind tunnel has a rectangular test section with 1.5 m height and 1 m width. The maximum wind velocity is 55 m/s, and the freestream turbulence level is less than 0.1%.

An NACA 23012 aluminium machined model, with chord $c = 0.3$ m and span $b = 0.93$ m, was used in this activity. The airfoil model has an interchangeable midspan section for the different measurements' techniques employed, one for PIV flow surveys and another for unsteady pressure measurements equipped with pressure taps positioned along the midspan contour. At the maximum angle of the attack tested ($\alpha = 20^\circ$), the wind tunnel blockage due to the oscillating airfoil model is 6%. The model is pivoted around two external steel shafts with the axis at 25% of the airfoil chord. The pitching drive system is composed of a brushless servomotor with a 12:1 gear drive jointed mechanically to one of the model external steel shafts. The end plates were mounted at model extremity to reduce the interference of the wind tunnel walls' boundary layer and to reduce the extremity effects. Two encoders are directly mounted on the shaft on the other side of the model respective to the motor: a 2048 imp/rev absolute digital encoder with EnDat 2.2 protocol is used for feedback control, and a 4096 imp/rev incremental analog encoder is used to get the instantaneous position of the model. Further details about the pitching airfoil experimental rig can be found in [18, 19]. The layout of the experimental rig designed for testing pitching airfoils is illustrated in Figure 1.

2.1. Hot-Wire Measurement Setup. The velocity surveys in the wake of the oscillating airfoil were carried out by means of a constant temperature anemometry (CTA) system Streamline 90N10 by Dantec Dynamics. The system was composed of a one frame with three CTA modules. Every basic anemometer module contains three CTA bridges, a servo loop with programmable gain, filters and cable compensation, and a programmable signal conditioner. The programmable servo loop allows to optimize the dynamic response and the bandwidth of the system, while the signal conditioner provides amplification of the CTA signal before digitizing.

A tri-axial fiber-film probe, Dantec 55R91, was used for the velocity surveys. The sensor probe has three mutually perpendicular sensors, consisting of 70 μm diameter nickel-plated quartz fibres. The probe was calibrated in the laboratory under monitored conditions with respect to Reynolds number and velocity direction. The calibration method took into account the effects of temperature, pressure, and humidity in order to extend the calibration itself to the wind tunnel ambient conditions [20]. The probe was mounted on a supporting strut, and it was moved in the model midspan plane in a vertical direction by means of a single axis traversing system. The velocity profiles were measured 2 chords past the airfoil trailing edge. The velocity time history was acquired for a time corresponding to 150 complete pitching cycles with a sampling frequency of 20 kHz.

Figure 2 illustrates the blade section model inside the wind tunnel; behind the model, the supporting strut for the triple hot-wire probe can be observed.

2.2. PIV Setup. The PIV system used a double shutter CCD camera with a 12-bit, 1280 × 1024-pixel array and a 55 mm lens. The measurement area covers the airfoil upper surface and it was composed by different measurement windows spanning the chord direction to obtain image pairs with a better resolution. The CCD camera was mounted on a dual axis traversing system driven by two stepper motors that allowed to move the measurement window along two orthogonal guides.

An Nd:Yag double pulsed laser, with 200 mJ output energy and wavelength 532 nm, was positioned on the test section roof. The laser was mounted on a single axis traversing system to move the sheet along the wind tunnel flow direction. This solution enabled to use a laser sheet with a smaller width and a higher energy level centered on each measurement window. A particle generator with Laskin nozzles was used for the flow insemination. The tracer particles, consisting of small oil droplets with a diameter in the range of 1-2 μm, were injected in correspondence to a section just after the fans and fulfill the wind tunnel volume with a homogeneous density.

The velocity flow fields phase which averaged over 40 image pairs are presented in the following.

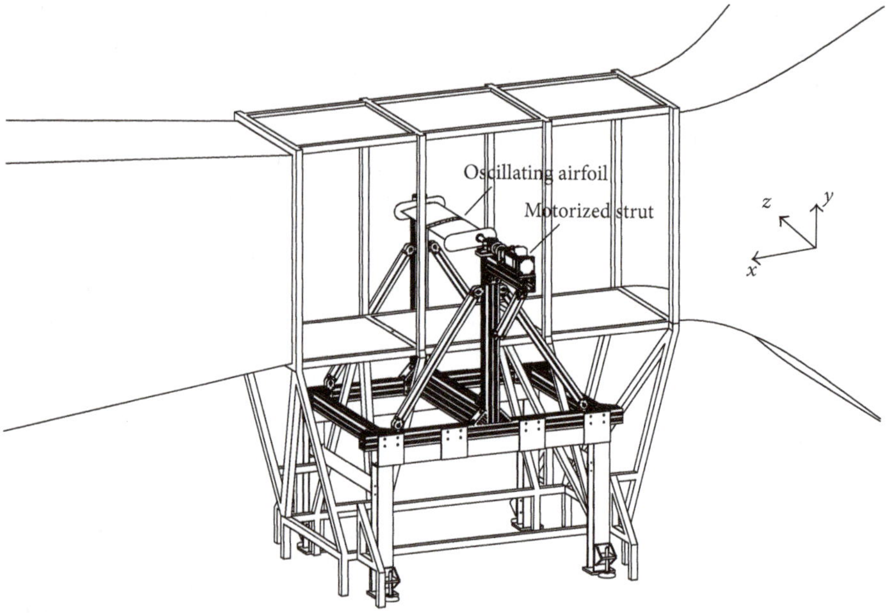

FIGURE 1: Layout of the experimental rig for pitching airfoils tests.

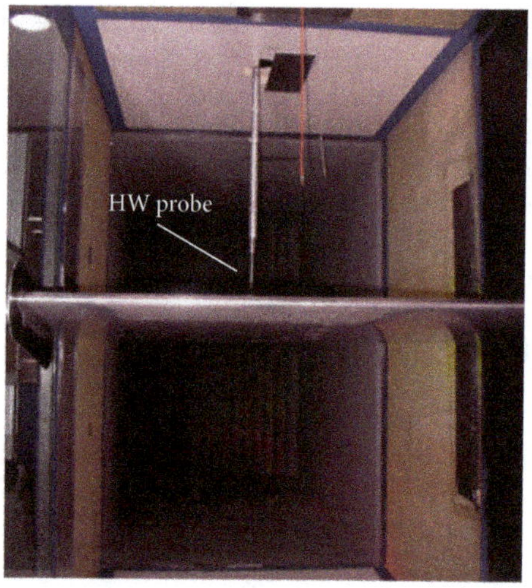

FIGURE 2: NACA 23012 blade section model inside the wind tunnel test section.

2.3. Pressure Measurement Setup

2.3.1. Airfoil Model Surface. The time history of the lift and pitching moments during a pitching cycle was evaluated by the integration of the phase-averaged pressures collected over 30 complete pitching cycles. Pressure was measured on the airfoil midspan contour using 21 Kulite fast-response pressure transducers located inside the model interchangeable midspan section. The locations of the pressure taps around the model midspan section are listed in Table 1 following a closed loop from the upper to the lower surface of the airfoil, starting from the leading edge.

The pressure transducers signals were acquired with a sampling rate of 50 kHz by means of a National Instruments compact data acquisition system equipped with six 24-bit A/D simultaneous bridge modules with 4 channels each.

2.3.2. Wind Tunnel Ceiling and Floor. Pressure measurements were carried out also on the wind tunnel ceiling and floor by means of a pressure rod (see Figure 3(a)). The rod was instrumented with two Kulite fast-response transducers (2 psi F.S.), and it was mounted on the test section ceiling and the floor in two different tests for each investigated dynamic stall condition. The pressure taps were located on the longitudinal midspan plane of the rod, while the transducers were installed in a threaded housing on the lateral side of the rod as it can be observed from the particular of the layout in Figure 3(b). Pressure was measured in correspondence to two pressure taps located 2 chords downstream the airfoil trailing edge (longitudinal position of the HW velocity surveys) and 3 chords upstream the airfoil leading edge.

3. Phase Averaging Method

The phase averaging is the most widely used method to point out a time-varying signal measured in the case of periodic unsteady flows. The measured time-varying signal $s(t)$ can be decomposed as follows:

$$s(t) = \langle s(t) \rangle + s'(t) \tag{1}$$

into a phase average term $\langle s(t)\rangle$ and a fluctuating term $s'(t)$. The phase averaging operator is defined as follows:

$$\langle s(t) \rangle = \lim_{N\to\infty} \frac{1}{N} \sum_{i=1}^{N} s(t + (i-1)T), \tag{2}$$

TABLE 1: Pressure taps location on the NACA 23012 model midspan section.

Tap number	Location x/c
1	0
2	0.01
3	0.044
4	0.096
5	0.164
6	0.28
7	0.358
8	0.453
9	0.618
10	0.76
11	0.9
12	0.9
13	0.767
14	0.628
15	0.459
16	0.373
17	0.285
18	0.185
19	0.118
20	0.06
21	0.02

where T is the period of the cyclic flow and N is the number of cycles, while the fluctuations term is defined as

$$s'(t) = s(t) - \langle s(t) \rangle . \quad (3)$$

In practice, the phase average term depends also on the number of cycles N as the phase averaging method is carried out over a finite number of cycles. Then, the definition of the phase average term in (2) becomes

$$\langle s(t, N) \rangle = \frac{1}{N} \sum_{i=1}^{N} s(t + (i-1)T) . \quad (4)$$

The larger is the number of cycles N, the more converging is the $\langle s(t, N) \rangle$ value towards the theoretical phase average term $\langle s(t) \rangle$. The criterion to determine the number of cycles to be used in the phase averaging method was discussed by Wernert and Favier [21] for different measurement techniques.

4. Experimental Results

The two conditions considered for wake velocity surveys are sinusoidal pitching cycles with reduced frequency $k = 0.1$, oscillation amplitude $\alpha_a = 10^\circ$, and a mean angle of attack of $\alpha_m = 5^\circ$ and $\alpha_m = 10^\circ$ to reproduce, respectively, the light and the deep dynamic stall regimes. The tests were carried out at $U_\infty = 30$ m/s corresponding to a Reynolds number Re $= 6 \times 10^5$ and a Mach number Ma $= 0.09$. Such a low testing speed, which is far beyond the test rig capabilities, was chosen because the long run time required by the HW wake surveys might have raised fatigue issues on the model strut.

Figures 4 and 5 present the phase averaged freestream velocity component profiles measured in the wake of the oscillating airfoil at some selected angles of attack. The wake freestream velocity component was phase averaged over 150 pitching cycles.

For the test condition with $\alpha_m = 5^\circ$, the measured velocity profiles show a small defect of the freestream velocity component, both in upstroke and in downstroke (see Figure 4). The similar behavior shown by the velocity profiles with respect to the direction of the airfoil oscillating motion is in agreement with the behavior of the measured airloads. In fact, as it can be observed in Figures 6(a) and 6(b), for this light dynamic stall condition, both the measured $C_L - \alpha$ and $C_M - \alpha$ curves show a small amount of hysteresis.

Moreover, the vertical extent of the wake velocity defect region is in the order of the airfoil thickness, both in upstroke and in downstroke. Indeed, in this dynamic stall regime the flow on the airfoil upper surface is attached for almost all the pitching cycle, as it can be observed by PIV surveys carried out on the model midspan plane. The PIV results in the following are illustrated by means of the in-plane streamlines patterns. For this light dynamic stall condition, the flow over the upper surface of the airfoil is fully attached at $\alpha = 10^\circ$ both in upstroke and downstroke (see Figures 7(a) and 7(c)), as well as at the top of the upstroke motion $\alpha = 15^\circ$ (see Figure 7(b)).

For the test condition with $\alpha_m = 10^\circ$, the velocity profiles behavior in the upstroke motion is similar to the one measured for the light dynamic stall condition (see Figure 5). The wake profiles measured in upstroke are coherent with the predominant aerodynamic feature due to the dynamic stall phenomenon. Indeed, the stall is delayed with respect to the static case due to a kinematic-induced camber increase produced by the rapid positive pitching rate [2]. As it can be observed from Figure 6(a), the lift grows linearly to a maximum value higher than the static stall value. In this test condition, the flow is attached to the airfoil upper surface for almost all the upstroke motion, as it is illustrated by the PIV survey carried out at $\alpha = 18^\circ$ in upstroke (see Figure 8(a)).

During the downstroke motion, the measured velocity profiles show a higher velocity defect and a higher thickness of the wake with respect to the upstroke. The wake thickness can be considered a parameter to estimate the size of the vortical flow in the wake [13]. Indeed, in this phase of the motion, the flow field on the airfoil upper surface shows a large separation, and it is characterised by the rapid formation, migration, and shedding of strong vortices. These flow phenomena are typical of the deep dynamic stall regime [2], and they are responsible of the conspicuous and rapid variations of the airloads measured during this phase of the pitching motion (see Figures 6(a) and 6(b)). Moreover, the observed asymmetric behavior of the flow field with respect to the direction of the airfoil motion results in the large airloads' hysteresis observed in Figures 6(a) and 6(b).

In particular, at $\alpha = 15^\circ$, the wake thickness reaches about half of the test section height, as it can be observed

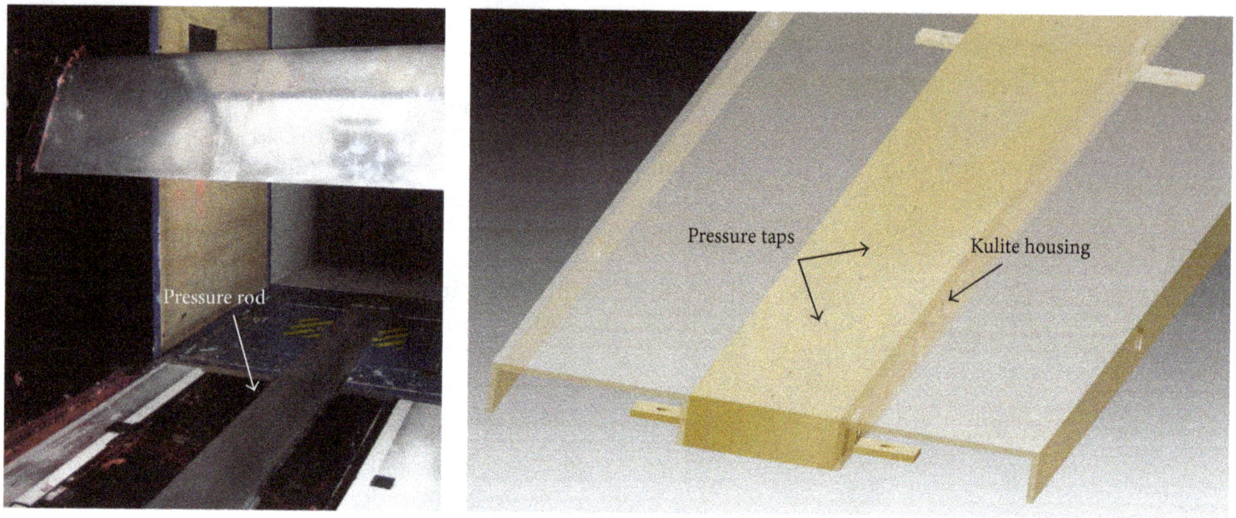

(a) (b)

FIGURE 3: (a) Aluminium rod for pressure measurements installed in the test section; (b) particular of the pressure rod.

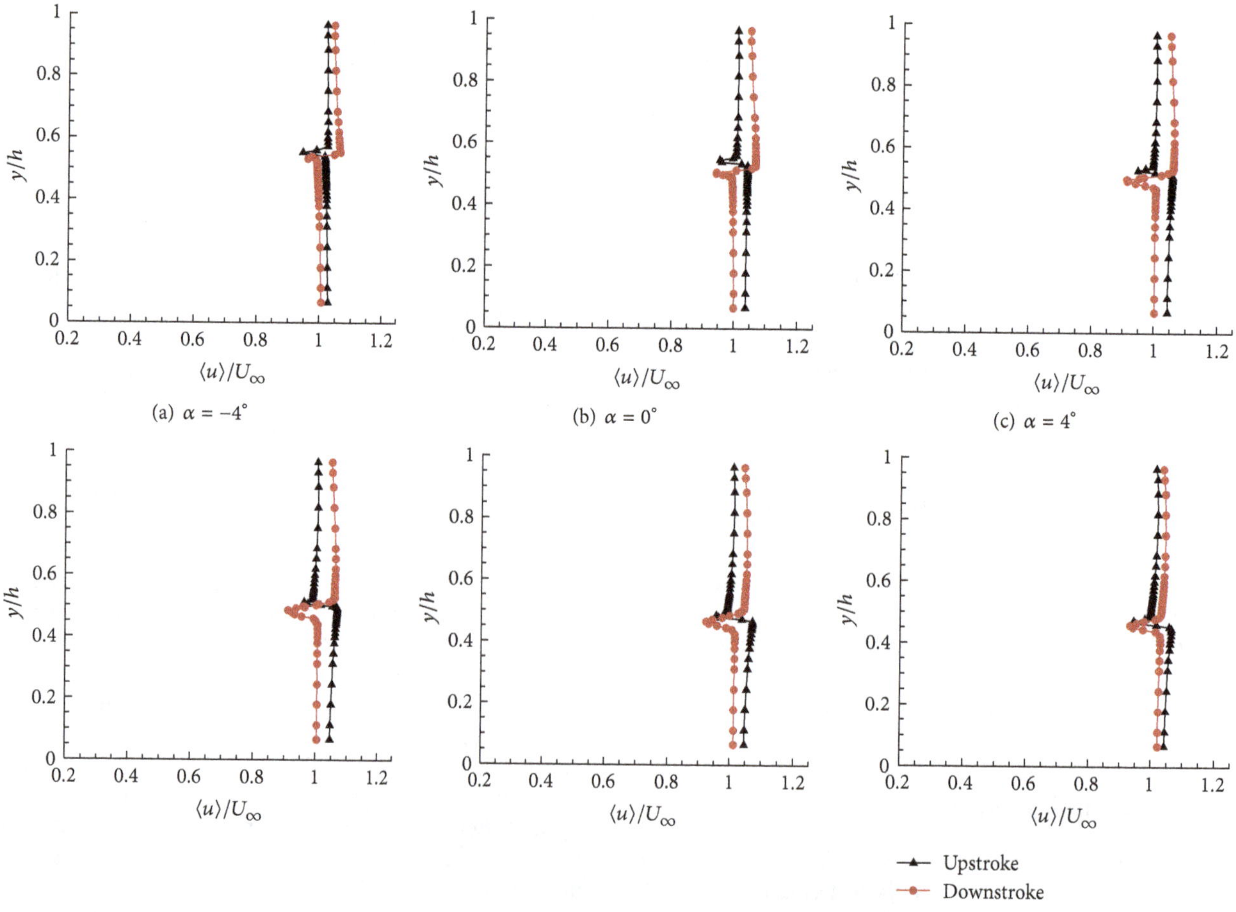

(a) $\alpha = -4^\circ$ (b) $\alpha = 0^\circ$ (c) $\alpha = 4^\circ$

(d) $\alpha = 8^\circ$ (e) $\alpha = 12^\circ$ (f) $\alpha = 14^\circ$

FIGURE 4: Streamwise velocity profiles for $\alpha = 5^\circ + 10^\circ \sin(\omega t)$ and $k = 0.1$ ($\mathrm{Re} = 6 \times 10^5$ and $\mathrm{Ma} = 0.09$).

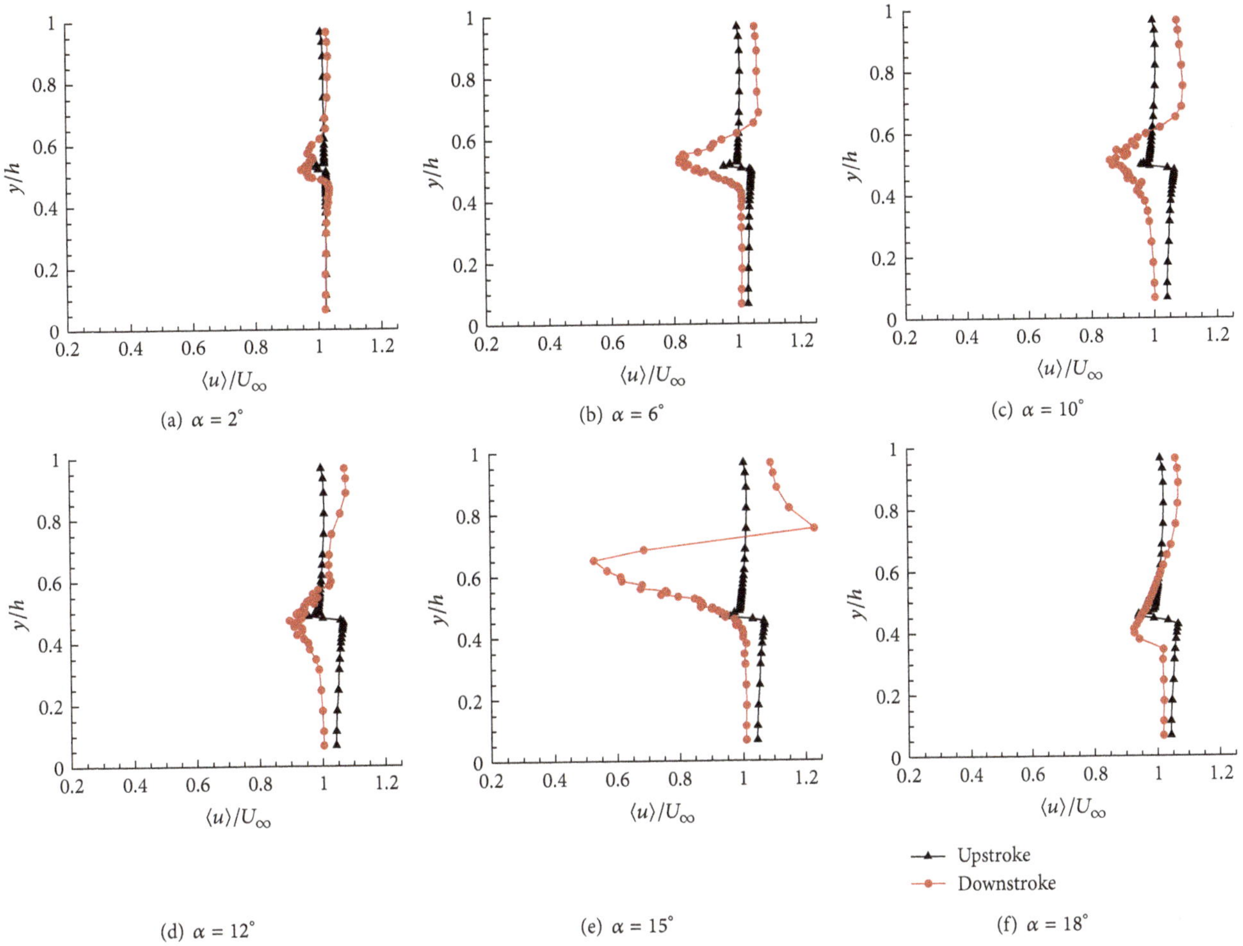

FIGURE 5: Streamwise velocity profiles for $\alpha = 10^\circ + 10^\circ \sin(\omega t)$ and $k = 0.1$ ($Re = 6 \times 10^5$ and $Ma = 0.09$).

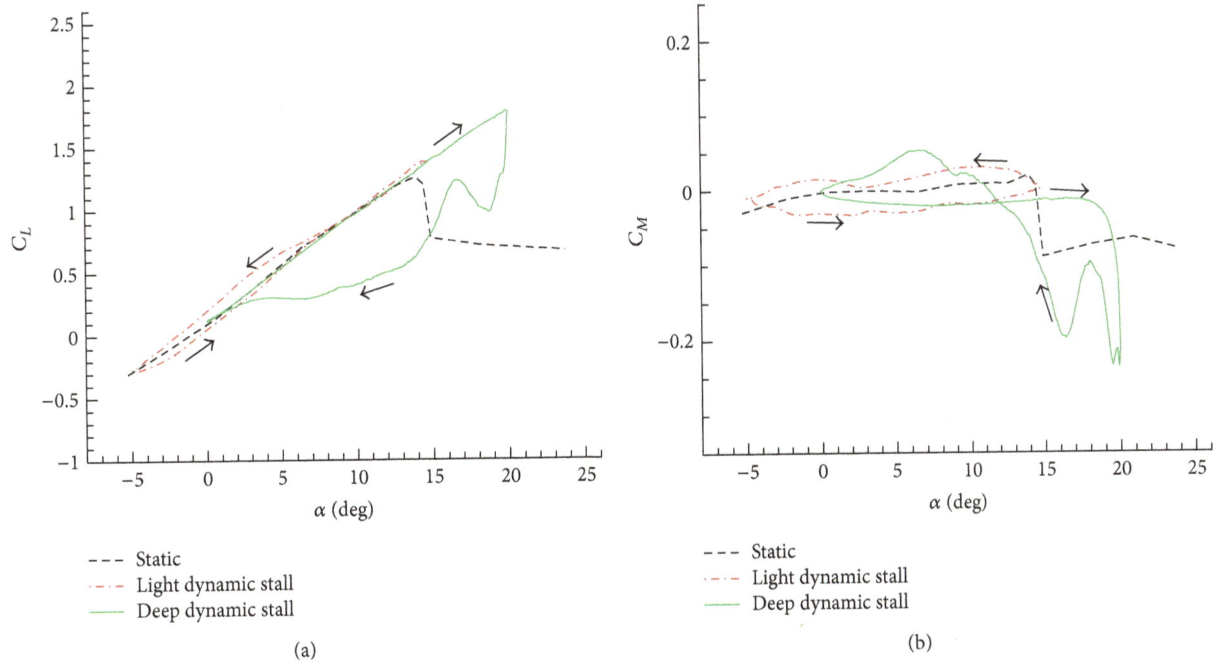

FIGURE 6: $C_L - \alpha$ and $C_M - \alpha$ curves for the NACA 23012 airfoil in static and dynamic stall conditions ($Re = 6 \times 10^5$ and $Ma = 0.09$).

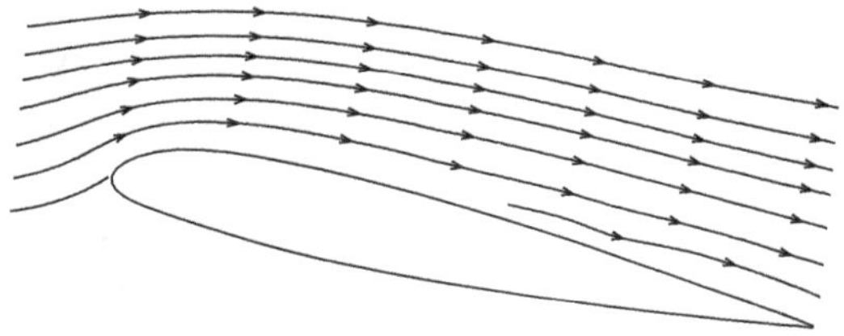

(a) $\alpha = 10^{\circ}$ upstroke

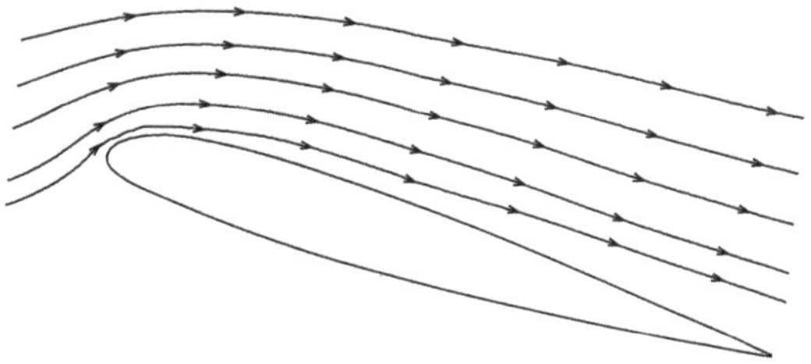

(b) $\alpha = 15^{\circ}$

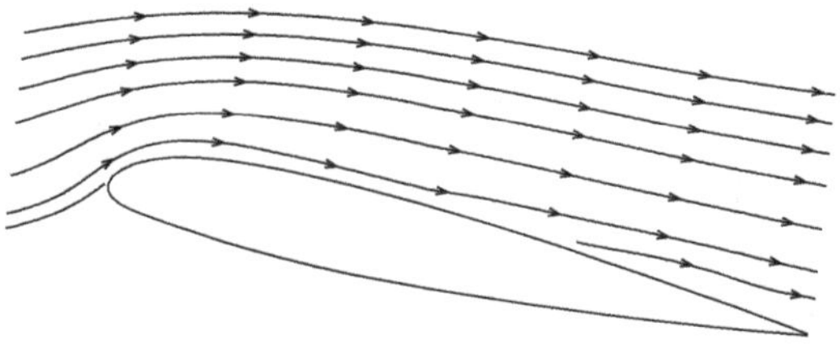

(c) $\alpha = 10^{\circ}$ downstroke

FIGURE 7: PIV results for $\alpha = 5^{\circ} + 10^{\circ}\sin(\omega t)$ and $k = 0.1$ (Re $= 6 \times 10^5$ and Ma $= 0.09$).

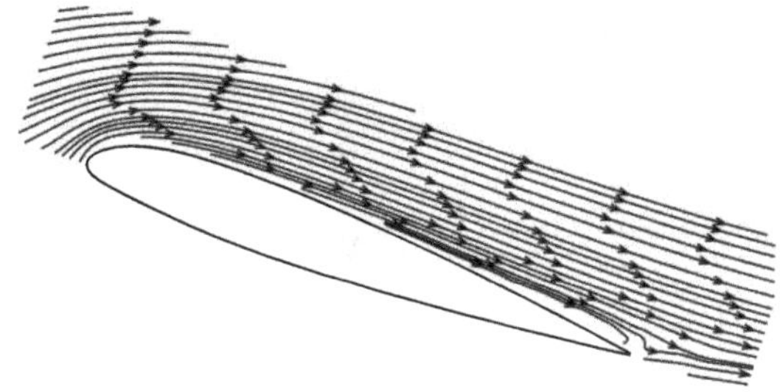

(a) $\alpha = 18^{\circ}$ upstroke

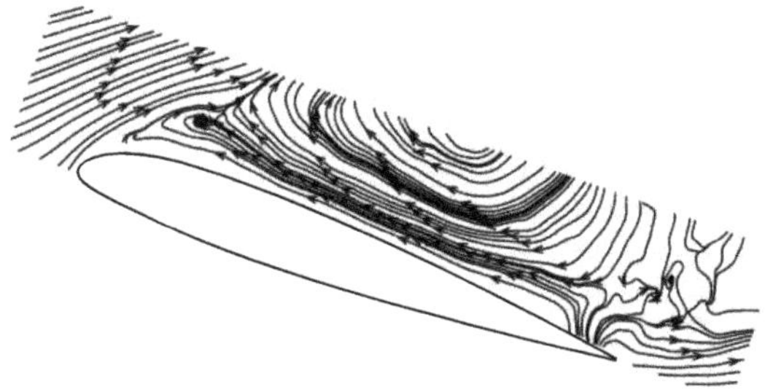

(b) $\alpha = 18^{\circ}$ downstroke

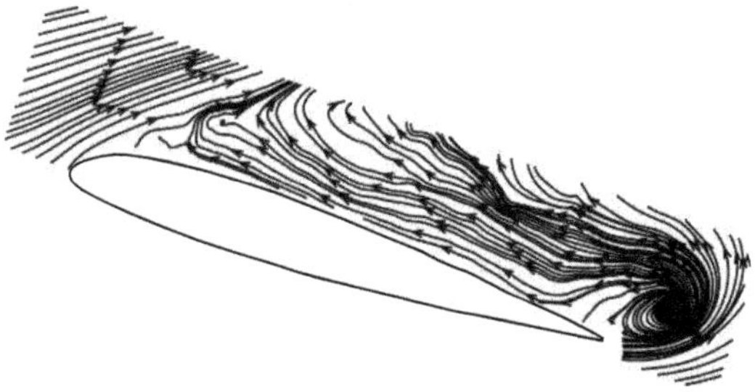

(c) $\alpha = 16^{\circ}$ downstroke

FIGURE 8: PIV results for $\alpha = 10^{\circ} + 10^{\circ}\sin(\omega t)$ and $k = 0.1$ (Re $= 6 \times 10^5$ and Ma $= 0.09$).

in Figure 5(e). The huge thickness of the wake evaluated at this angle of the attack is due to the passage of a strong vortical structure that starts on the airfoil upper surface at the beginning of the downstroke, as it can be observed from the PIV survey carried out at $\alpha = 18^{\circ}$ and $\alpha = 16^{\circ}$ in downstroke. At the lower angles of the attack in downstroke (see Figures 5(a) and 5(b)), the wake thickness tends to decrease according to the flow reattachment process start.

It is apparent from Figures 4 and 5 that the velocity profiles settle at two different values towards the wind tunnel floor and ceiling. A similar pattern is commonly observed in wind tunnel tests due to the flow circulation around the model interacting with the tunnel walls (see, e.g., [22]). In the present case, this up-down velocity difference is mainly due to the circulation in the wake [23]. In fact, for the case of a pitching airfoil with a high oscillation amplitude, a rapid circulation variation on the airfoil is associated with the lift variation, and consequently an opposite circulation is issued in the wake. Thus; in the upstroke phase, when the lift is increasing, a counterclockwise circulation is issued in the wake producing higher velocities on the floor with respect to the ceiling and vice versa in the downstroke phase.

This effect can also explain most of the difference between the ceiling and floor pressures highlighted by the pressure measurements results shown in Figure 9. In particular, Figure 9 presents the difference of the phase averaged pressures measured upstream and downstream the airfoil both on the ceiling and the floor of the test section for the different investigated dynamic stall conditions.

Further considerations about the pitching airfoil wakes in light and deep dynamic stall conditions can be deduced from the analysis of Figures 10 and 11, showing, respectively, the evolution of the adimensional freestream velocity defect $\widehat{U}_{\text{def}}$ and of the adimensional turbulent kinetic energy $\widehat{u}'$ during the pitching cycle, where

$$U_{\text{def}} = 1 - \frac{\langle U \rangle}{U_{\infty}}, \qquad e_k = \frac{1}{2}\frac{\langle u'u' \rangle \langle v'v' \rangle}{U_{\infty}^2}. \tag{5}$$

For the light dynamic stall condition, the defect velocity region moves along the vertical direction of the test section showing the wake oscillations. The more extended region of the velocity defect observed during the downstroke demonstrates a small thickening of the wake in this phase of the motion (see Figure 10(a)). In this test case, the level of the turbulent kinetic energy is quite low for the whole pitching cycle (see Figure 11(a)).

For the deep dynamic stall condition, the velocity defect evolution presents a similar behavior with respect to the light dynamic stall case. On the other hand in downstroke, the velocity defect behavior shows the conspicuous thickening of

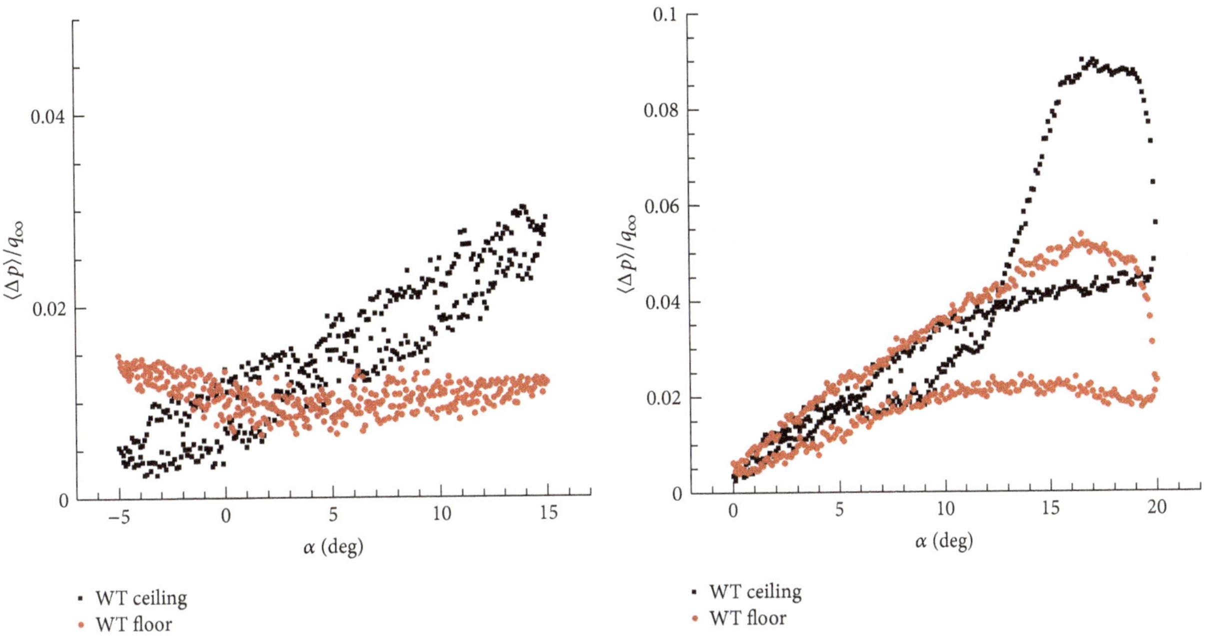

(a) Light dynamic stall

(b) Deep dynamic stall

FIGURE 9: Pressure differences measured upstream and downstream the airfoil; pressure taps located, respectively, 3 chords upstream the airfoil leading edge and 2 chords downstream the airfoil trailing edge.

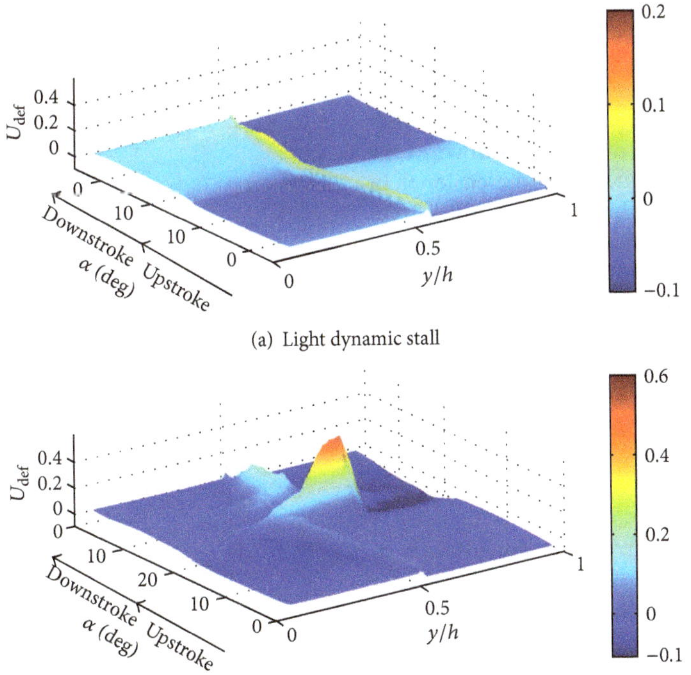

(a) Light dynamic stall

(b) Deep dynamic stall

FIGURE 10: Freestream adimensional velocity defect for the light dynamic stall (a) and the deep dynamic stall (b) conditions.

the wake due to the large vortical structures peculiar of this flow regime that are detached from the airfoil trailing edge (see Figure 10(b)). According to the occurrence of these flow phenomena, the turbulent kinetic energy in this phase of the motion is quite higher with respect to the light dynamic stall case, and in particular, its peak is about ten times greater (see Figure 11(b)).

In this phase of the pitching cycle, strong three-dimensional secondary flows occur as it can be deduced from Figure 12 that shows the behavior of the adimensional bulk

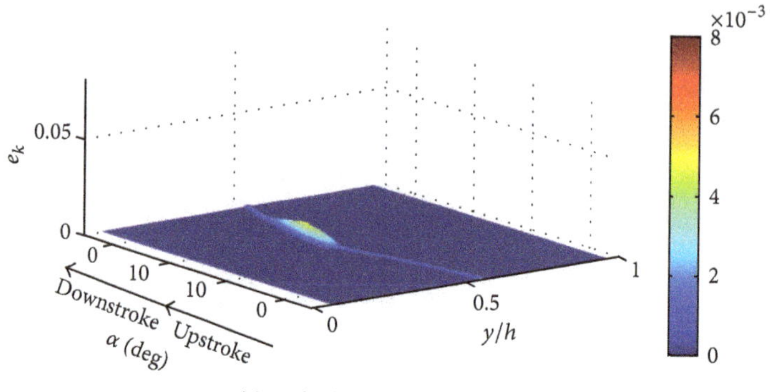

(a) Light dynamic stall

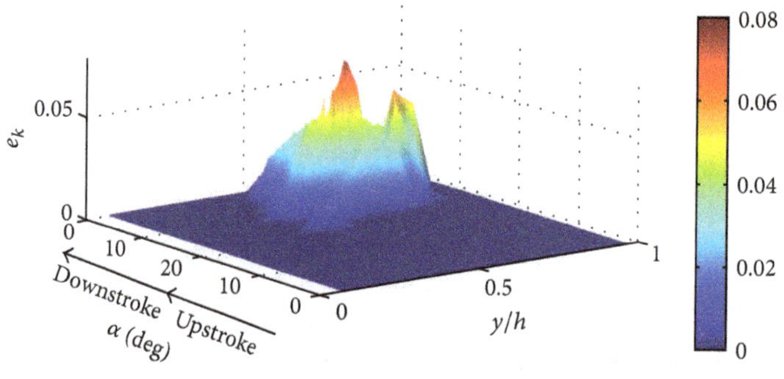

(b) Deep dynamic stall

FIGURE 11: Turbulent kinetic energy for the light dynamic stall (a) and the deep dynamic stall (b) conditions.

velocity for the light and the deep dynamic stall conditions. In particular, the bulk velocity defect observed in the deep dynamic stall curve from $12^\circ < \alpha < 18^\circ$ in downstroke demonstrates that the flow is quite three dimensional in this range of angles of attack. The results of the three-dimensional numerical simulations reported in Zanotti et al. [24, 25] carried out for the tested deep dynamic stall condition support this consideration. On the other hand, the quite flat behavior of the bulk velocity observed for the tested light dynamic stall condition suggests that in this regime three-dimensional flow effects are negligible during the whole pitching cycle.

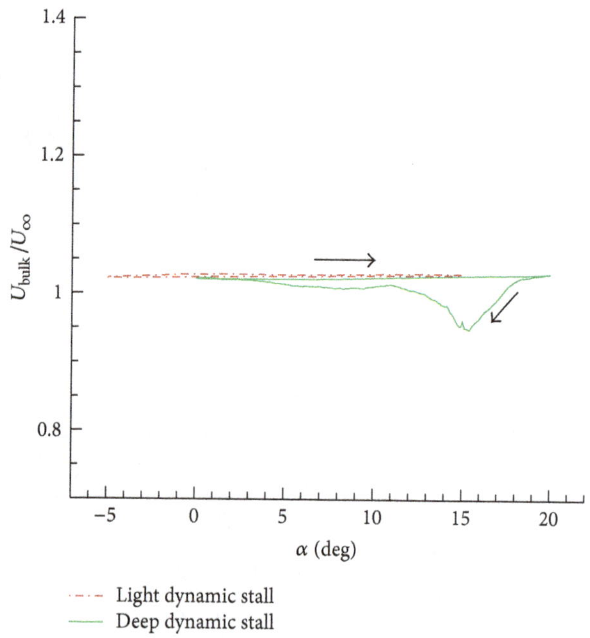

FIGURE 12: Wake bulk velocity for the light and deep dynamic stall conditions.

5. Conclusions

An extensive wake survey was carried out downstream of a NACA 23012 pitching airfoil in light and deep dynamic stall conditions. The analysis of the wake velocity profiles highlights the predominant aerodynamic features involved in the different regimes of the dynamic stall and gives interesting information about the turbulent kinetic energy in the wake of a pitching airfoil. The wake survey analysis was supported by PIV measurements carried out on the airfoil upper surface, by the measurement of the lift and pitching moments and by the pressure measurements carried out on the wind tunnel floor and ceiling. Therefore, the experimental investigation gives a complete overview about the dynamic stall process over an oscillating airfoil.

The comprehensive experimental database can be considered a reference for the development and validation of numerical tools. In particular, an interesting application would be the comparison of the wake velocity profiles with numerical simulations to evaluate the phase averaged total drag. Indeed, also on the side of the numerical simulation, the drag evaluation for a pitching airfoil in dynamic stall conditions is affected by a large uncertainty due to the severe unsteadiness that characterises this phenomenon. Thus, this problem is still quite open, and it represents a quite challenging activity.

Nomenclature

Symbols

α: Angle of attack
α_m: Mean angle of attack
α_a: Pitching oscillation amplitude
Δp: Pressure difference measured upstream and downstream the airfoil
ω: Circular frequency, $\omega = 2\pi f$
b: Blade section model span
c: Blade section model chord
C_D: Drag coefficient
C_L: Lift coefficient
C_M: Pitching moment coefficient about the airfoil quarter chord
e_k: Adimensional turbulent kinetic energy
f: Oscillation frequency
h: Test section height
k: Reduced frequency, $k = \pi f c / U_\infty$
q_∞: Freestream dynamic pressure
$\langle s \rangle$: Phase-averaged signal
U_∞: Freestream velocity
U_{bulk}: Wake bulk velocity
U_{def}: Adimensional freestream velocity defect
u: Streamwise velocity component
v: Vertical velocity component
u': Streamwise velocity component fluctuation
v': Vertical velocity component fluctuation
$\langle u \rangle$: Phase-averaged streamwise velocity component
x: Streamwise coordinate
y: Vertical coordinate
z: Spanwise coordinate.

Abbreviations

DAER: Dipartimento di scienze e tecnologie aerospaziali
HW: Hot wire
PIV: Particle image velocimetry
Ma: Mach number
Re: Reynolds number.

References

[1] W. J. McCroskey, "The phenomenon of dynamic stall," NASA TM 81264, 1981.

[2] J. G. Leishman, *Principles of Helicopter Aerodynamics*, Cambridge University Press, 2006.

[3] F. O. Carta, "An analysis of the stall flutter instability of helicopter rotor blades," *Journal of American Helicopter Society*, vol. 12, pp. 1–8, 1967.

[4] A. D. Gardner, K. Richter, H. Mai, and D. Neuhaus, "Experimental control of compressible OA209 dynamic stall by air jets," in *Proceedings of the 38th European Rotorcraft Forum*, Amsterdam, The Netherlands, September 2012.

[5] M. S. Chandrasekhara, P. B. Martin, and C. Tung, "Compressible dynamic stall control using a variable droop leading edge airfoil," *Journal of Aircraft*, vol. 41, no. 4, pp. 862–869, 2004.

[6] M. L. Post and T. C. Corke, "Separation control using plasma actuators: dynamic stall vortex control on oscillating airfoil," *AIAA Journal*, vol. 44, no. 12, pp. 3125–3135, 2006.

[7] K. Mulleners and M. Raffel, "The onset of dynamic stall revisited," *Experiments in Fluids*, vol. 52, pp. 779–793, 2012.

[8] M. Raffel, J. Kompenhans, B. Stasicki, B. Bretthauer, and G. E. A. Meier, "Velocity measurement of compressible air flows utilizing a high-speed video camera," *Experiments in Fluids*, vol. 18, no. 3, pp. 204–206, 1995.

[9] A. Zanotti and G. Gibertini, "Experimental investigation of the dynamic stall phenomenon on a NACA 23012 oscillating aerofoil," *Proceedings of the Institution of Mechanical Engineers, Part G: Journal of Aerospace Engineering*, 2012.

[10] A. D. Gardner, K. Richter, H. Mai, A. R. M. Altmikus, A. Klein, and C. H. Rohardt, "Experimental Inves- tigation of dynamic stall performance for the EDI-M109 and EDI-M112 airfoils," in *Proceedings of the 37th European Rotorcraft Forum*, Gallarate, VA, Italy, September 2011.

[11] J. G. Leishman, "Dynamic stall experiments on the NACA 23012 aerofoil," *Experiments in Fluids*, vol. 9, no. 1-2, pp. 49–58, 1990.

[12] L. W. Carr, K. W. McAlister, and W. J. McCroskey, "Dynamic stall experiments on the NACA 0012 airfoil," NASA TP-1100, 1978.

[13] Y. W. Jung and S. O. Park, "Vortex-shedding characteristics in the wake of an oscillating airfoil at low Reynolds number," *Journal of Fluids and Structures*, vol. 20, no. 3, pp. 451–464, 2005.

[14] J. W. Chang, "Near-wake characteristics of an oscillating NACA 4412 airfoil," *Journal of Aircraft*, vol. 41, pp. 1240–1243, 2004.

[15] H. Sadeghi, M. Mani, and M. A. Ardakani, "Effect of amplitude and mean angle of attack on wake of an oscillating airfoil," *Proceedings of World Academy of Science, Engineering and Technology*, vol. 33, pp. 125–129, 2008.

[16] H. Sadeghi, M. Mani, and M. A. Ardakani, "Unsteady wake measurements behind an airfoil and prediction of dynamic stall from the wake," *Aicraft Engineering and Aerospace Technology*, vol. 82, pp. 225–236, 2010.

[17] T. Lee and P. Gerontakos, "Investigation of flow over an oscillating airfoil," *Journal of Fluid Mechanics*, vol. 512, pp. 313–341, 2004.

[18] A. Zanotti, *Retreating blade dynamic stall [Ph.D. thesis]*, Politecnico di Milano, 2012.

[19] A. Zanotti, F. Auteri, G. Campanardi, and G. Gibertini, "An experimental set up for the study of the retreating blade dynamic stall," in *Proceedings of the 37th European Rotorcraft Forum*, Gallarate, VA, Italy, September 2011.

[20] F. Durst, S. Noppenberger, M. Still, and H. Venzke, "Influence of humidity on hot-wire measurements," *Measurement Science and Technology*, vol. 7, no. 10, pp. 1517–1528, 1996.

[21] P. Wernert and D. Favier, "Considerations about the phase averaging method with application to ELDV and PIV measurements over pitching airfoils," *Experiments in Fluids*, vol. 27, no. 6, pp. 473–483, 1999.

[22] T. Nishino and K. Shariff, "Numerical study of wind-tunnel sidewall effects on circulation control airfoil flows," *AIAA Journal*, vol. 48, no. 9, pp. 2123–2132, 2010.

[23] G. K. Batchelor, *An Introduction to Fluid Dynamics*, Cambridge University Press, 2000.

[24] A. Zanotti, S. Melone, R. Nilifard, and A. D'Andrea, "Experimental-numerical investigation of a pitching airfoil in deep dynamic stall," *Proceedings of the Institution of Mechanical Engineers, Part G: Journal of Aerospace Engineering*, 2013.

[25] A. Zanotti, R. Nilifard, G. Gibertini, A. Guardone, and G. Quaranta, "Experimental-numerical investigation of the dynamic stall phenomenon over the NACA 23012 airfoil," in *Proceedings of the 38th European Rotorcraft Forum*, Amsterdam, The Netherlands, September 2012.

9

A Numerical and Experimental Study of the Aerodynamics and Stability of a Horizontal Parachute

Mazyar Dawoodian,[1] Abdolrahman Dadvand,[2] and Amir Hassanzadeh[3]

[1] *Department of Mechanical Engineering, I. A. University of Takestan, Takestan, Iran*
[2] *Department of Mechanical Engineering, Urmia University of Technology, Urmia, Iran*
[3] *Department of Mechanical Engineering, Urmia University, Urmia, Iran*

Correspondence should be addressed to Mazyar Dawoodian; mazyar_dawoodian@yahoo.com

Academic Editors: J. López-Puente and R. K. Sharma

The flow past a parachute with and without a vent hole at the top is studied both experimentally and numerically. The effects of Reynolds number and vent ratio on the flow behaviour as well as on the drag coefficient are examined. The experiments were carried out under free-flow conditions. In the numerical simulations, the flow was considered as unsteady and turbulent and was modelled using the standard k-ε turbulence model. The experimental results reveal good agreement with the numerical ones. In both the experiments and numerical simulations, the Reynolds number was varied from 85539 to 357250 and the vent ratio was increased from zero to 20%. The results show that the drag coefficient decreases by increasing the Reynolds number for all the cases tested. In addition, it was found that at low and high Reynolds numbers, the parachutes, respectively, with 4% vent ratio and without vent are deemed more efficient. One important result of the present work is related to the effect of vent ratio on the stability of the parachute.

1. Introduction

The study of bluff bodies involves the consideration of complex aerodynamic phenomena such as semiwake and vortex shedding. The analysis of parachute dynamics is one of the most interesting problems, which would lie within this context. One important feature that may distinguish the computational fluid dynamics (CFD) modelling from the experimental analysis of the parachute behaviour is the geometry flexibility in the latter allowing large variations in the experiments [1]. A parachute is generally equipped with a venthole at the apex to allow some air to flow through the open canopy. The canopy can be represented by a hemispherical shell, which has been shown schematically in Figure 1. In the numerical simulations carried out in the present work, the canopy is considered solid and its direction of motion is from left to right (i.e., the flow direction is from right to left in the positive x direction). The flow structure near the wake of canopy is responsible for the aerodynamic forces and moments it experiences. The relationship between the Reynolds number and flow field structure around spherical bodies in incompressible flow regime has been studied extensively. The numerical study carried out by Natarajan and Acrivos [2] indicated that the wake of a sphere became unstable at a Reynolds number of 105. The experimental results of Sakamoto and Haniu [3] showed that the hairpin-shaped vortices begin to be periodically shading when the Reynolds number reaches about 350. Their research indicated that when the Reynolds number exceeds 6000, the vortex sheet separating from the surface of the sphere becomes completely turbulent.

According to the experimental work of Bakic and Peric [4] in wind tunnel and water channel, for the Reynolds numbers between 22000 and 50000, the near-wake recirculation region was observed to be large and the wake would perform a progressive wave motion. They also found that as the Reynolds number was increased from 22000 to 350000, the boundary layer separation point would move from about 82° to 132°.

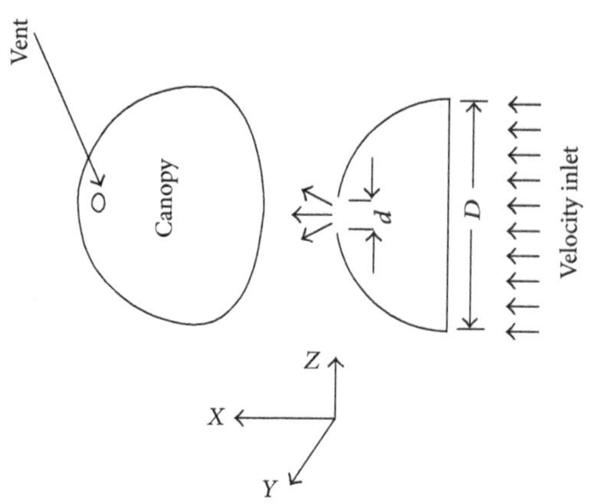

FIGURE 1: Hemispherical shell with a vent in the Cartesian coordinates system.

A circular disk can be regarded as a good representative of round parachutes since the separation line is fixed at the disk edge. Fuchs et al. [5] and Berger et al. [6] have studied the flow structure around the disk and identified three instability modes: an axisymmetric oscillation of the recirculation bubble at a low Strouhal number (≈0.05), a low-frequency mode at a Strouhal number of about 0.14, and a high-frequency mode at a Strouhal number equals to 1.6. The main recirculation bubble extends to about 2.5 diameters behind the disk. Flow visualization of Higuchi [7] confirmed the primary helical mode at a Reynolds number of 104 in the disk wake. However, close-up observations in a water tunnel revealed that the tilting of vortex rings results in asymmetric structures after the first diameter in the disk wake. Peterson et al. [8] have studied the characteristic features of parachute canopies during various stages from breathing to final developing.

Experimental investigation of the drag forces on flexible rectangular canopies was conducted by Filippone [9]. He showed that the drag coefficient decreased weakly by increasing the Reynolds number to 2000000 and decreased intensively by increasing the ratio of perforated tape area to plan form area. Izadi and Mohammadizadeh [10] and Izadi and Dawoodian [11] have numerically studied the Relationship between the drag coefficient and the vent diameter. Takizawa et al. [12] performed a comprehensive numerical calculation on the fluid-structure interaction (FSI) modelling of clusters of ringsail parachutes. They also presented a brief stability and accuracy analysis for the deforming-spatial-domain/stabilized space-time (DSD/SST) formulation, which is the core numerical technology of the SSTFSI technique. Cao and Jiang [13] calculated the drag coefficient for a parachute using a finite volume method and Spalart-Allmaras turbulence model. The separation flow, which truthfully reflects the characteristics during the terminal descent, accords with the fluid dynamics rules in their study. Cao and Xu [14] analyzed canopy shape and the parameters of parachute inflation by means of establishing parachute's flying physical model through four methods (i.e., tiny segment analysis method, inflating distance method, moment method and simulating canopy shape method). They concluded that moment method, and simulating canopy shape method are only fitted to the main phase of inflating in theory and inflating distance method can be applied to evaluate the dynamic load of parachute inflation.

A complete understanding of the flow physics around a parachute is still one of the most scientific challenging issues from both the theoretical and experimental viewpoints. Recent efforts that have successfully benefited from using CFD simulations could explain the basic principles behind this phenomenon. Thus far, a comprehensive research has not been performed to study the variations of drag coefficient of the flexible canopies with the Reynolds number and vent ratios. In this paper, the experimental tests and numerical simulations are conducted to obtain the drag coefficient on a hemispherical shell for different Reynolds numbers and vent ratios. The venthole is located at the shell apex. The main purpose is to find the effects of the Reynolds number and vent diameter on the drag coefficient as well as on the flow field. These findings may be useful in parachute designing and its applications.

2. Experimental Setup

The experiments were conducted under the free-flow conditions (i.e., at the atmospheric pressure and temperature). The test rig was mounted on a vehicle, which moved with the speed of 8.3 to 33.3 m/s. The inflated canopy setup mounted on the vehicle along with the supporting hardware is shown in Figure 2 schematically. The parachute assembly was positioned in a horizontal orientation. The canopy was attached to a stationary support by six threads. The forebody had a diameter of 1.4 cm and a length of 2 m. The cross-sectional area of the support was less than 2% of the projected area of the canopy.

By streamlining the support and reducing its diameter, the wakes around the support were kept to a minimum. Flow visualization confirmed that the wake of support rods had a negligible effect on the dynamics of the parachute canopy. The vent ratio of the parachute is varied from 0% to 20% (i.e., $d/D = 0$–0.2). The Reynolds number for a parachute canopy is commonly based on the diameter D. In this research, the Reynolds number ranges from 85539 to 357250. A dynamometer was used to record the force at every stage of the experiment. Figure 3 displays two typical parachutes with a venthole used in the experiments.

3. Numerical Simulation

3.1. Governing Equations. For the external flows around spherical obstacles, the critical Reynolds number is about 800, beyond which the wake flow becomes turbulent [15]. Since all the Reynolds numbers considered in the present work are much greater than this critical Reynolds, the flow around the parachute is assumed to be turbulent. Therefore, the three-dimensional unsteady Reynolds-averaged Navier-Stokes (RANS) equations combined with the standard k-ε turbulence model are employed to simulate the flow around

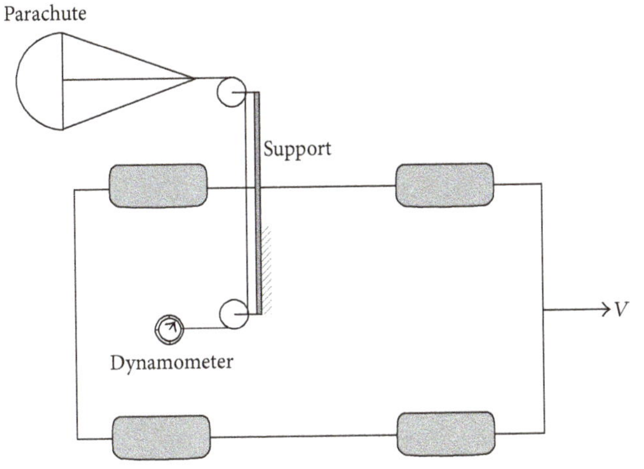

FIGURE 2: Top view of the experimental setup (not to scale).

FIGURE 3: Two typical parachutes utilized in the present experiments.

the parachute. It may be noted that the RNG k-ε turbulence model and more advanced turbulence models such as the Reynolds stress equations were also investigated, but these models could not give good numerical results. Since the experiments were performed under the free-flow conditions, the parachute would not have a significant effect on the temperature of air; so, in the simulations the density of air is assumed constant. In addition, since the minimum and maximum Mach numbers are equal, respectively, to 0.033 and 0.137, the flow was considered incompressible. The governing RANS equations (i.e., the continuity and momentum equations) are given as

$$\frac{\partial \bar{u}_i}{\partial x_i} = 0,$$

$$\rho\left(\frac{\partial \bar{u}_i}{\partial t} + \bar{u}_j \frac{\partial \bar{u}_i}{\partial x_j}\right) = -\frac{\partial \bar{p}}{\partial x_i} + \bar{B}_i + \frac{\partial}{\partial x_j}\left(\mu \frac{\partial \bar{u}_i}{\partial x_j} - \rho \overline{u_i' u_j'}\right). \quad (1)$$

$\bar{B}_i$ is the body force(s) in the ith direction. The normal velocity on all surfaces of the body is considered zero. The turbulence kinetic energy k and its rate of dissipation ε are obtained from the following transport equations:

$$\rho\frac{\partial k}{\partial t} + \rho\left(u_j \frac{\partial k}{\partial x_j}\right) = \frac{\partial}{\partial x_j}\left[\left(\mu + \frac{\mu_t}{\sigma_k}\right)\frac{\partial k}{\partial x_j}\right] + G_k + G_b - \rho\varepsilon - Y_M,$$

$$\rho\frac{\partial \varepsilon}{\partial t} + \rho\left(u_j \frac{\partial \varepsilon}{\partial x_j}\right) = \frac{\partial}{\partial x_j}\left[\left(\mu + \frac{\mu_t}{\sigma_\varepsilon}\right)\frac{\partial \varepsilon}{\partial x_j}\right] + C_{1\varepsilon}\frac{\varepsilon}{k}\left(G_k + C_{3\varepsilon}G_b\right) - C_{2\varepsilon}\rho\frac{\varepsilon^2}{k}. \quad (2)$$

In these equations, G_k, G_b, and Y_M represent the generation of turbulence kinetic energy due to the mean velocity gradients, the generation of turbulence kinetic energy due to buoyancy, and the contribution of the fluctuating dilatation in compressible turbulence to the overall dissipation rate, respectively. $C_{1\varepsilon}$ and $C_{2\varepsilon}$ are constants. σ_k and σ_ε are the turbulent Prandtl numbers for k and ε, respectively. The turbulent (or eddy) viscosity μ_t is computed as follows:

$$\mu_t = \rho C_\mu \frac{k^2}{\varepsilon}, \quad (3)$$

where C_μ is a constant. The constants $C_{1\varepsilon}$, $C_{2\varepsilon}$, C_μ, σ_k, and σ_ε take the following values: $C_{1\varepsilon} = 1.44$, $C_{2\varepsilon} = 1.92$, $C_\mu = 0.09$, $\sigma_k = 1.0$, and $\sigma_\varepsilon = 1.3$.

The flow equations, described above, are solved by FLUENT's pressure-based segregated solver. It is worth mentioning that when Re > 10^3, the inertial forces dominate and the drag force is approximately proportional to the square of the velocity; that is, $F_d = 0.5\rho C_d A U_\infty^2$.

Here C_d is the drag coefficient defined as

$$C_d = \frac{F_d}{0.5\rho A U_\infty^2}, \quad (4)$$

where $A = (\pi D^2/4)$ is the projected area of the hemispherical cup and U_∞ is the free stream velocity. The fluid properties are considered at standard condition. Therefore the air density ρ and viscosity μ are taken as 1.25 kg·m^{-3} and 1.78×10^{-5} kg·m^{-1}·s^{-1}, respectively. These values are kept constant during the simulation process.

3.2. Numerical Model Description and Boundary Conditions. Figure 4 depicts the 3D-CFD model of the parachute along with the boundary conditions. Note that, in all of the simulations carried out in the present work, the geometrical dimensions of the parachute are kept constant. However, the vent diameter and the Reynolds number are varied in order to study their effect on the drag coefficient as well as on the flow structure. In addition, to link/match similar conditions between the simulations and experiments, the numerical Reynolds numbers are chosen the same as those of the experiments. An unstructured (or irregular) T-Grid mesh with finer mesh points near the parachute body is used. As shown in Figure 5, the computational domain assumes

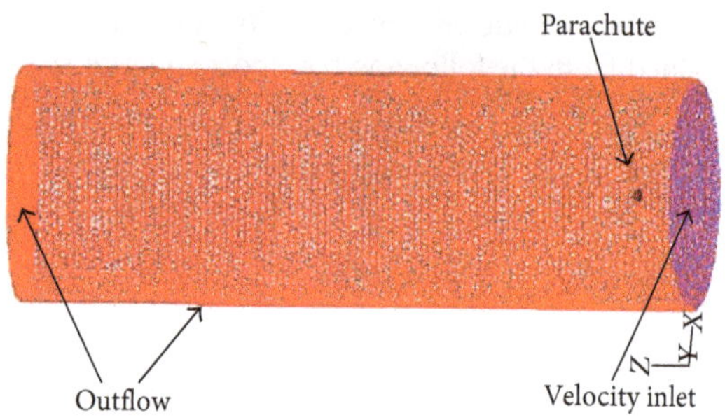

FIGURE 4: CFD mesh points and boundary conditions.

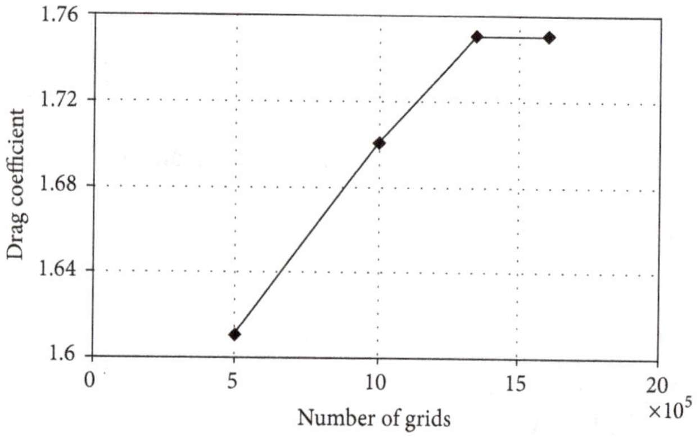

FIGURE 5: Grid size-dependence study of drag coefficient at different mesh densities.

a cylindrical shape, where A, B, and C distances are taken as 46, 10, and 10 times the parachute diameter *D*, respectively.

The first order upwind scheme is utilized to discretize the convective terms, and the SIMPLE algorithm is used for solving the governing equations. It must be noted that the calculations are allowed to continue for 8 seconds until the parameters like C_d remain constant; that is, the results are associated with the steady state condition. The outflow boundary condition has been applied at both the lateral and outlet boundaries, and the velocity inlet has been considered as the inlet boundary condition. The no-slip boundary condition is applied on the body of the hemispherical cup. The main boundaries are shown in Figure 4.

3.3. Grid Dependence Study. The numerical simulations have been carried out for different grid densities in order to find the grid independent solutions. For this purpose, first the grid independence study has been accomplished for Re = 85539 and $d/D = 0$. The variations of the drag coefficient as the key parameter against the number of grid cells have been depicted in Figure 5. According to this figure, a grid system of 1344000 tetrahedral cells was found to be sufficient to resolve the details of the flow field. Therefore, nominally a grid system of 1344000 tetrahedral cells was used in all the simulations carried out in the present work.

4. Results and Discussions

4.1. Drag Coefficient. In order to investigate the parachute performance under different flow conditions, six parachutes with various vent diameter ratios d/D of 0, 0.04, 0.08, 0.12, 0.16, and 0.20 each at different Reynolds numbers of 85539, 114000, 146330, 171100, 228070, 292660, and 357250 are studied in both the experimental and numerical cases. Figure 6 shows the numerical and experimental drag coefficient C_d variations as a function of the Reynolds number for different vent ratios of 0–20%.

It is clear that for all the vent ratios tested, C_d decreases by increasing the Reynolds number. In addition, the drag coefficient seems to experience a (relatively) sharp decrease at lower Reynolds numbers, but it takes a smooth change as the Reynolds number increases. One can also observe that the value of C_d reaches a constant value and the diagram finds an asymptote. Although the trend of all the diagrams in Figure 6 is roughly the same, this tendency is more evident for d/D = 0, 0.12, and 0.2. There is a good agreement between the experimental and numerical results for all the cases studied. The difference between the maximum and minimum values of C_d is about 0.2, which corresponds to the case with d/D = 0.20.

Figure 7 illustrates the experimental C_d changes versus vent ratio d/D for different Reynolds numbers. The results show that C_d decreases by increasing the Reynolds number. This may be attributed to the fact that the formation of the wakes behind the parachute causes the drag force to decrease. In all the Reynolds numbers tested except in the last two ones (i.e., Re = 292660 and 357250), the parachute with d/D ratio of 0.04 possesses the highest value of the drag coefficient as compared to the other vent ratios. According to this figure, the maximum amount of C_d achieved during the experiments is 1.77, which pertains to Re = 85539 and d/D = 0.04. For the d/D values greater than 0.04, C_d will be below this value for all the Reynolds numbers tested. Therefore, at high Reynolds numbers it is suggested to use parachutes without vent since such parachutes experience higher C_d in comparison with the other cases.

4.2. The Flow Field around the Parachute. Figure 8 shows the velocity vectors related to the unsteady turbulent flow around the parachute obtained from the numerical simulations.

According to this figure, it is evident that for all the vent ratios and the Reynolds numbers tested, two wakes both inside and outside the parachute are observed. In addition, the outer wakes are responsible for increasing the drag force.

The size of these wakes seems to increase moderately as the Reynolds number is increased. In general, the vent ratio is the main cause of the pressure difference decrease across the wall of the parachutes (i.e., between the outer and inner regions of it). In fact, by increasing the vent diameter the outflow flux increases, hence, decreasing the pressure difference between the inner and outer regions of the parachute.

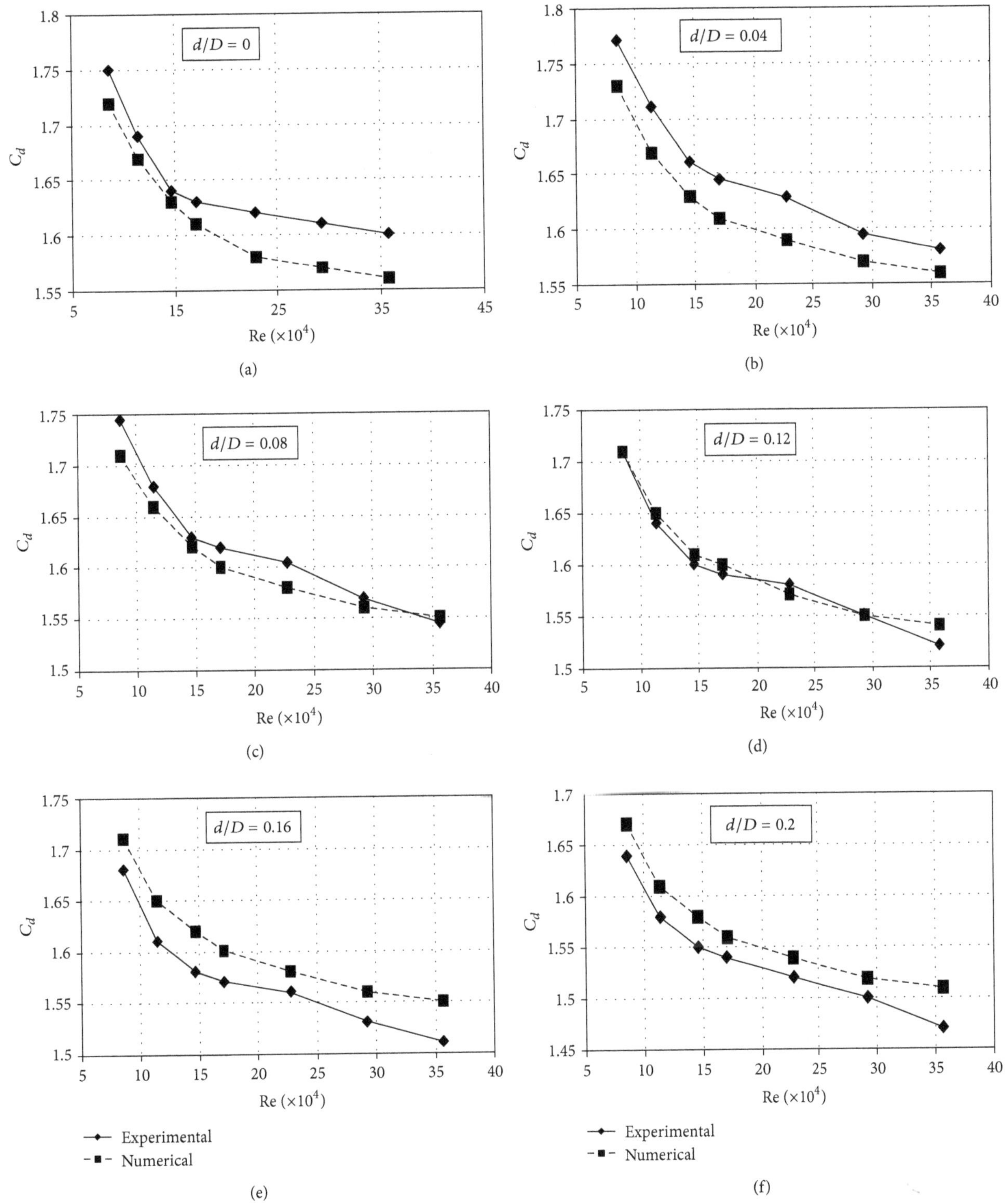

FIGURE 6: Drag coefficient variations as a function of the Reynolds number for different vent ratios.

4.3. Stability Analysis. During the experimental tests, the parachute was observed to be oscillating when the Reynolds number exceeded some certain values.

At low Reynolds numbers, the parachute was found to be stable without any oscillations. However, as the Reynolds number was increased beyond a certain value, the parachute became unstable. So transitions from stable to unstable condition were observed for all vent ratios. Figure 9 shows the stable and unstable regions observed in the experiments.

It is clear that as the vent diameter is increased, the transition from stable to unstable conditions is delayed. For the parachute without venthole, at Re = 85539 no oscillation

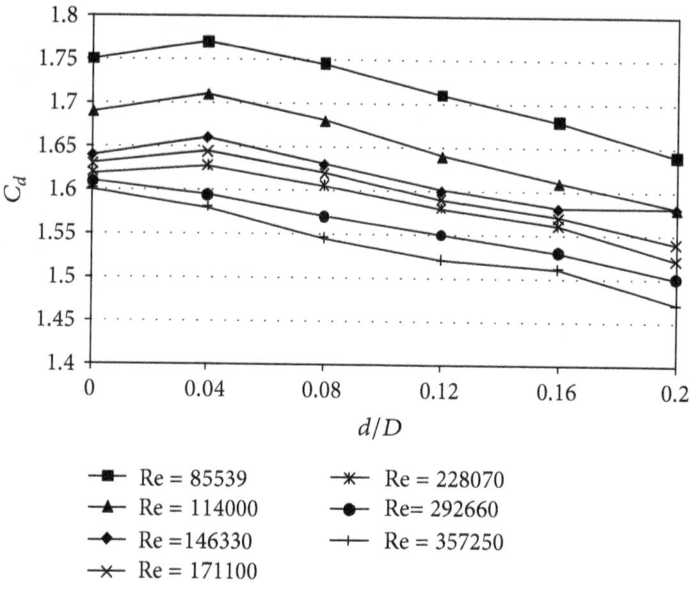

FIGURE 7: Experimental drag coefficient variations against vent ratios for different Reynolds numbers.

is observed. However, for Re = 114000 we observed very intensive oscillations. For the parachute with vent of 4%, the oscillations were commenced at Re = 114000.

It may be noted that by increasing the vent ratio, the critical Reynolds number at which the parachute started to oscillate is increased. For parachutes with vent ratios of 8% and 12%, these critical Reynolds numbers are 171100 and 292600, respectively.

5. Conclusion

In this paper, the dynamic behaviour of a parachute is studied both experimentally and numerically. Different cases with various vent ratios are tested at different Reynolds numbers. In the numerical simulations, the flow is considered turbulent and the standard k-ε turbulence model is used to model the flow field. In addition, an unsteady state condition is considered for the numerical models, but the calculations are continued until the results remain constant. The CFD simulations have been done by using FLUENT software. The effects of the vent ratio and the Reynolds number on the drag coefficient are examined. The results show that for all cases as the Reynolds number is increased the drag coefficient is decreased. In addition, in all the simulations carried out in the present work, two wakes are observed behind the parachute. The maximum value of the drag coefficient measured during the experiments is about 1.77, which is associated with Re = 85539 and d/D = 0.04. In addition, at high Reynolds numbers the parachute without vent experiences a higher drag force as compared with the other cases. Comparison between the simulations and experiments evidences good agreements. Finally, the experiments show that the parachute begins oscillating when the Reynolds numbers exceed a certain critical value, which is different for different vent ratios. This would imply that every case has a stable and unstable region. As the vent ratio increases, the oscillations would occur at higher Reynolds numbers.

Abbreviations

Nomenclature

A: Area (m^2)
C_d: Drag coefficient
d: Vent diameter (m)
D: Canopy diameter (m)
F_d: Drag force (N)
k: Turbulence kinetic energy ($m^2 \cdot s^{-2}$)
M: Mach number
p: Pressure ($N \cdot m^{-2}$)
Re: Reynolds number
U_∞: Free stream velocity ($m \cdot s^{-1}$).

Greek Symbols

ε: Turbulence dissipation rate ($m^2\, s^{-3}$)
ρ: Density ($kg\, m^{-3}$)
μ: Dynamic viscosity ($kg\, m^{-1}\, s^{-1}$)
μ_t: Turbulent viscosity ($kg\, m^{-1}\, s^{-1}$).

Subscripts

i: Direction
max: Maximum
min: Minimum.

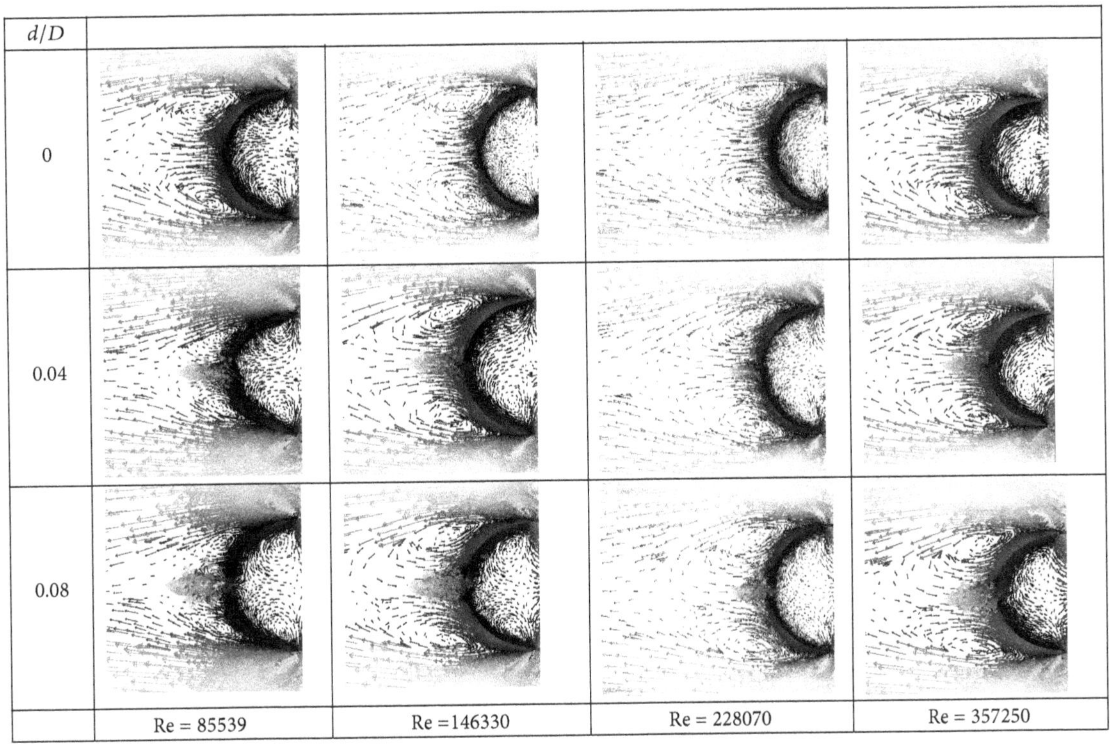

(a)

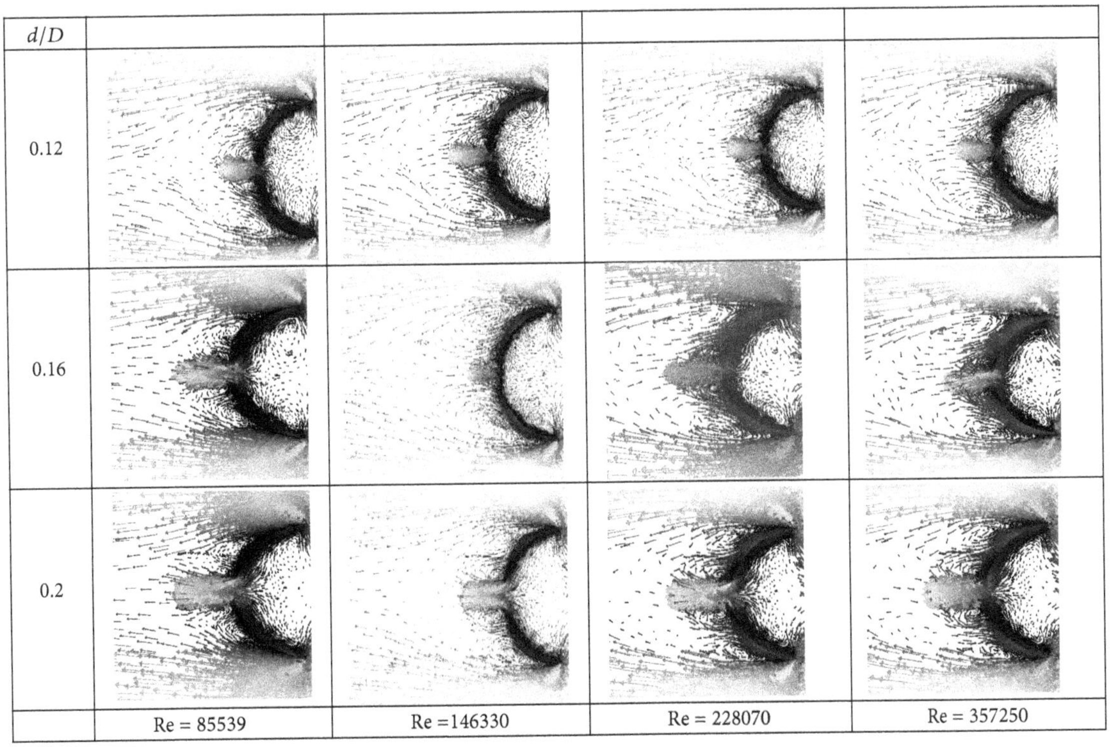

(b)

FIGURE 8: Velocity vectors around a hemispherical shell parachute under different conditions.

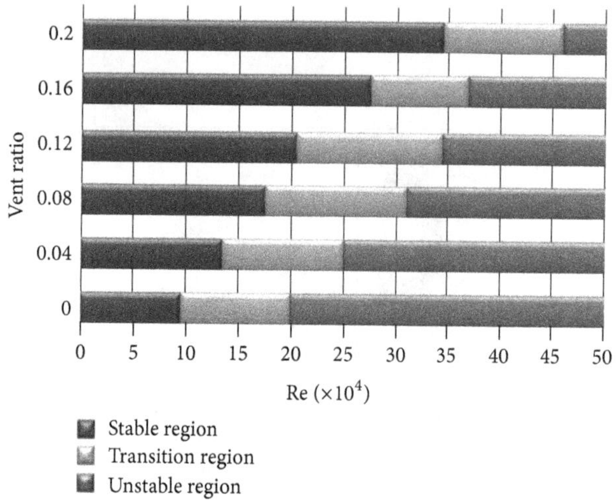

FIGURE 9: Stable and unstable zones obtained from the experimental analysis.

References

[1] A. Al Musleh and A. Frendi, "On the effects of a flexible structure on boundary layer stability and transition," *Journal of Fluids Engineering*, vol. 133, no. 7, Article ID 071103, 6 pages, 2011.

[2] R. Natarajan and A. Acrivos, "The instability of the steady flow past spheres and disks," *Journal of Fluid Mechanics*, vol. 254, pp. 323–344, 1993.

[3] H. Sakamoto and H. Haniu, "A study on vortex shedding from spheres in a uniform flow," *Journal of Fluids Engineering*, vol. 112, no. 4, pp. 386–392, 1990.

[4] V. Bakic and M. Peric, "Visualization of flow around sphere for Reynolds numbers between 22,000 and 400,000," *Thermophysics and Aeromechanics*, vol. 12, no. 3, pp. 307–315, 2005.

[5] H. V. Fuchs, E. Mercker, and U. Michel, "Large-scale coherent structures in the wake of axisymmetric bodies," *Journal of Fluid Mechanics*, vol. 93, no. 1, pp. 185–207, 1979.

[6] E. Berger, D. Scholz, and M. Schumm, "Coherent vortex structures in thewake of a sphere and a circular disk at rest and under forced vibrations," *Journal of Fluids and Structures*, vol. 4, no. 3, pp. 231–257, 1990.

[7] H. Higuchi, "Visual study on wakes behind solid and slotted axisymmetric bluff bodies," *Journal of Aircraft*, vol. 28, no. 7, pp. 427–430, 1991.

[8] C. W. Peterson, J. H. Strickland, and H. Higuchi, "The fluid dynamics of parachute inflation," *Annual Review of Fluid Mechanics*, vol. 28, pp. 361–387, 1996.

[9] A. Filippone, "On the flutter and drag forces on flexible rectangular canopies in normal flow," *Journal of Fluids Engineering*, vol. 130, no. 6, Article ID 061203, 8 pages, 2008.

[10] M. J. Izadi and M. Mohammadizadeh, "Numerical study of a hemi-spherical cup in steady, unsteady, laminar and turbulent flow conditions with a vent of air at the top," *Fluids Annual Review of Fluid Mechanics*, vol. 28, pp. 361–387, 2008.

[11] M. J. Izadi and M. Dawoodian, "CFD analysis of drag coefficient of a parachute in steady and turbulent conditions in various Reynolds numbers," in *Proceedings of the Fluids Engineering Division Summer Conference (FEDSM '09)*, pp. 2285–2293, ASME, Colorado, Colo, USA, August 2009.

[12] K. Takizawa, S. Wright, C. Moorman, and T. E. Tezduyar, "Fluid-structure interaction modeling of parachute clusters," *International Journal for Numerical Methods in Fluids*, vol. 65, no. 1–3, pp. 286–307, 2011.

[13] Y. Cao and C. Jiang, "Numerical simulation of the flow field around parachute during terminal descent," *Aircraft Engineering and Aerospace Technology*, vol. 79, no. 3, pp. 268–272, 2007.

[14] Y. Cao and H. Xu, "Parachute flying physical model and inflation simulation analysis," *Aircraft Engineering and Aerospace Technology*, vol. 76, no. 2, pp. 215–220, 2004.

[15] D. Ormières and M. Provansal, "Transition to turbulence in the wake of a sphere," *Physical Review Letters*, vol. 83, no. 1, pp. 80–83, 1999.

10

Experimental Evaluation of the Density Ratio Effects on the Cooling Performance of a Combined Slot/Effusion Combustor Cooling System

Antonio Andreini, Gianluca Caciolli, Bruno Facchini, and Lorenzo Tarchi

DIEF, Department of Industrial Engineering Florence, University of Florence, Via S. Marta 3, 50139 Florence, Italy

Correspondence should be addressed to Bruno Facchini; bruno.facchini@htc.de.unifi.it

Academic Editors: K. A. Sallam and H. Xiao

The purpose of this study is to investigate the effects of coolant-to-mainstream density ratio on a real engine cooling scheme of a combustor liner composed of a slot injection and an effusion array with a central dilution hole. Measurements of heat transfer coefficient and adiabatic effectiveness were performed by means of steady-state thermochromic liquid crystals technique; experimental results were used to estimate, through a 1D thermal procedure, the Net Heat Flux Reduction and the overall effectiveness in realistic engine working conditions. To reproduce a representative value of combustor coolant-to-mainstream density ratio, tests were carried out feeding the cooling system with carbon dioxide, while air was used in the main channel; to highlight the effects of density ratio, tests were replicated using air both as coolant and as mainstream and results were compared. Experiments were carried out imposing values of effusion blowing and velocity ratios within a range of typical modern engine working conditions. Results point out the influence of density ratio on film cooling performance, suggesting that velocity ratio is the driving parameter for the heat transfer phenomena; on the other hand, the adiabatic effectiveness is less sensitive to the cooling flow parameters, especially at the higher blowing/velocity ratios.

1. Introduction

In the course of last years, the increase of performances for the gas turbine for aeronautics has been achieved by increasing the pressure ratio and the maximum cycle temperature. These working conditions are not bearable with the materials employed in the components exposed to high thermal loads; hence, the development of effective cooling schemes is fundamental to match the increasing trend of gas turbine operating temperature. On the other hand, the development of aeroengine combustor is driven also by the effort to reduce NO_x emissions, in order to meet stricter legislation requirements.

To satisfy future ICAO standards concerning NO_x emissions, main engine manufacturers have been updating the design concept of combustors. Future aeroengines combustion devices will operate with very lean mixtures in the primary combustion zone, switching as much as possible to premixed flames. Whatever detailed design will be selected, the amount of air in the primary zone will grow significantly at the expense of liner cooling air, which thus will be reduced. Consequently, important attention must be paid to the appropriate design of the liner cooling system in order to optimize coolant consumption and guarantee an effective liner protection. In addition, further goals need to be taken into account: reaction quenching due to a sudden mixing with cooling air should be accurately avoided, whilst temperature distribution has to reach the desired levels in terms of OTDF.

In recent years, the improvement of drilling capability has allowed making a large quantity of extremely small cylindrical holes, whose application is commonly referred to as effusion cooling. Alternative solutions to the typical 2D-slot combustor cooling systems, like the full coverage film cooling or multihole film cooling, still relies on the generation of a high effective layer of film cooling; on the contrary, an effusion cooling system permits to lower the wall temperature mainly through the so-called "heat sink effect," which is the wall cooling due to the heat removed by the passage of the coolant through the holes [1, 2]. A high number of small tilted holes homogeneously distributed over the whole surface of the liner allows, if accurately designed, a significant improvement in lowering wall temperature, despite a slight reduction of the wall protection at least in the first part of the liner. Even if early effusion cooling schemes were developed to be an approximation of transpiration cooling, the design of effusion cooling in current combustor is usually based on very shallow injection holes ($\alpha \leq 30^{\circ}$) with high coolant jet momentum; this solution allows to greatly increase the heat sink effect (higher holes Reynolds number and higher exchange areas) without excessive detriment to film effectiveness. With this design approach, the analysis and the characterization of the heat transfer and the wall protection due to the injection of coolant become a fundamental issue in order to estimate the entire cooling system performance.

Many studies of full coverage film cooling have been focused on measuring or estimating the film effectiveness generated by coolant jets and the heat transfer of effusion cooling. Scrittore et al. [3] studied the effects of dilution hole injection on effusion behaviour; they found relevant turbulence levels downstream dilution holes, thus leading to an increased spreading of coolant jets. Scrittore et al. [4] measured velocity profiles and adiabatic effectiveness of a full coverage scheme with blowing ratios from 3.2 to 5.0, finding the attainment of a fully developed effectiveness region at the 15th row and a very low effect of blowing ratio on cooling performance. Metzger et al. [5] studied the variation of heat transfer coefficient for full-coverage film cooling scheme with normal holes, founding an augmentation of 20–25% in the local heat transfer with blowing ratio 0.1 and 0.2. Crawford et al. [6] experimentally determined Stanton number for an effusion cooling geometry. Martinez-Botas and Yuen [7] measured heat transfer coefficient and adiabatic effectiveness of a variety of geometries in a flat plate to test the influence of the injection angle by varying blowing ratio from 0.33 to 2.0. They measured the variation of the heat transfer coefficient h with respect to a reference case h_0; results show that there is a maximum of h/h_0 close to the hole and further downstream with highest heat transfer augmentation for 30° injection angle. Kelly and Bogard [8] investigated an array of 90 normal holes and found that the largest values for h/h_0 occur immediately downstream of the film cooling holes and the levels of h/h_0 are similar for the first 9 rows. They explained that this could be due to an increase in the local turbulence levels immediately downstream of the holes, created by the interaction between the cooling jet and the mainstream flow. Another reason could be the creation of a new thermal boundary layer immediately downstream of the cooling jets. In the open literature, none of the previous studies investigates the effect that a high blowing ratio has on adiabatic effectiveness, heat transfer coefficient, and Net Heat Flux Reduction. As reported by Kelly and Bogard [8], increases in heat transfer coefficient due to high blowing ratios could potentially be replaced by an increase in heat transfer coefficient due to high mainstream turbulence.

More recently Facchini et al. [9, 10] measured the overall effectiveness and the heat transfer coefficient at variable blowing ratios on a real engine cooling scheme to evaluate the combined effects of slot, effusion, and a large dilution hole; they found that an increase in BR leads to lower values of effectiveness. On the other hand, they found that high BR values enhance the heat transfer phenomena. Facchini et al. [11] investigated also the influence of a recirculating area in the mainstream on the same geometry; they highlight that the presence of the recirculation leads to a general reduction of effectiveness, while it does not have significant effects on the heat transfer coefficient.

Despite many studies deal with the investigation of the effusion cooling performance, most of them were conducted by using air as coolant and mainflow, precluding the possibility to point out the effects of density ratio between the two flows. Density ratio is, however, a key parameter for the design of a liner cooling system, mainly because of the actual large temperature difference between coolant and burned gases inside the core of the combustor. Ekkad et al. [12, 13] measured effectiveness and heat transfer coefficient distribution over a flat surface with one row of injection holes inclined streamwise at 35° for several blowing ratios and compound angles; tests were carried out by using air and carbon dioxide as coolant, finding that both heat transfer and effectiveness increase with blowing ratio. They also pointed out the effects of density ratio, showing how these effects are more evident with increasing the compound angle and the momentum flux ratio. This experimental survey was, however, oriented for turbine blade applications rather than combustors. More recently, Lin et al. [14, 15] investigated both experimentally and numerically adiabatic film cooling effectiveness of four different 30° inclined multihole film cooling configurations; the survey, which was specific for combustor liner applications, was performed by using a mixture of air and CO_2 as coolant, but it was mainly focused on studying the influence of hole geometrical parameters and blowing ratio on film cooling rather than on the effects of density ratio. Andreini et al. [16] performed a CFD analysis on the a test article which replicated a slot injection and an effusion array; they simulated the behaviour of the cooling system both with air and CO_2. Numerical results show that the entity of local heat transfer enhancement in the proximity of effusion holes exit is due to gas-jets interaction and that it mainly depends on effusion velocity ratio; furthermore, a comparison between results obtained with air and with CO_2 as coolant pointed out the effects of density ratio, showing the opportunity to scale the increase in heat transfer coefficient with effusion jets velocity ratio.

In the present study, the effects of density ratio on heat transfer coefficient, adiabatic effectiveness, NHFR, and overall effectiveness are investigated on a test rig which replicates a real cooling system for a combustor liner application, made up of a slot, an effusion array, and a dilution hole. In order to reproduce a representative value of combustor coolant-to-mainstream density ratio, tests were carried out by feeding the cooling system with carbon dioxide (CO_2), while air was used in the main channel; the test plate was tested imposing several values of blowing and velocity ratios within the range of typical modern engine working conditions. To highlight the effects of density ratio and, as a consequence, to distinguish between the influence of blowing ratio and velocity ratio, tests were replicated by using air both as coolant and mainstream and results were compared. Results point out the influence of DR on the performance of the cooling scheme; moreover, they give useful indications on how to take into account the density ratio effects without using a foreign gas in a low temperature lab-scaled facility.

2. Experimental Facility

This investigation was aimed at pointing out the dependence of film cooling performance on coolant-to-mainstream density ratio. In order to achieve this goal, measurements on a test rig which represents a specific cooled combustor liner were carried out by using air and carbon dioxide (CO_2) as cooling flows and results were compared in terms of heat transfer coefficient (HTC), adiabatic effectiveness (η_{aw}), Neat Heat Flux Reduction (NHFR), and overall effectiveness (η_{ov}).

The test rig, depicted in Figure 1, consists of an open-loop suction type wind tunnel which allows the complete control of three separate flows: the hot mainstream, the slot cooling, and the effusion cooling flows. The vacuum system is made up of two rotary vane vacuum pumps with a capacity of 900 m^3/h each dedicated to the extraction of the mainstream mass flow.

The mainstream flow rate is set up by guiding the speed of the pumps and using a calibrated orifice located at the beginning of the wind tunnel (throttle). The mainstream temperature is set up by using a 24.0 kW electronically controlled electric heater, placed at the inlet of the rig.

Slot and effusion coolant flows reach the test rig crossing two different lines that connect the wind tunnel with a pressure tank which stores the cooling fluid up to a maximum pressure of 1 MPa. Flow rates are set up by throttling two separated valves. Heaters for a total power of 1.5 kW are placed along the lines which connect the tank to the rig in order to set the desired inlet coolant temperature.

The mass flow rate is measured at three different locations of the rig: according to the standard EN ISO 5167-1 one orifice measures the flow rate blown by the pumps, while two orifices measure the slot and the effusion mass flow rates.

Two pressure scanners Scanivalve DSA 3217 with temperature compensated piezoresistive relative pressure sensors measure the static pressure in 32 different locations with a maximum accuracy of 6.9 Pa. Several T type thermocouples connected to a data acquisition/switch unit (HP/Agilent 34970A) measure the mainstream and the coolant static temperatures.

The main channel has a constant cross-section of 100 × 150 mm and is 1000 mm long. In the first part of the channel the mainstream flow crosses a honeycomb and three screens which allow to set an uniform velocity profile. A 6.0 mm square hole grid (hole pitch 7.6 mm, plate thickness 0.7 mm) is placed 125 mm upstream the slot coolant injection, so as to set turbulence level at $x/S_x = 0$ around 5%, with a macroscopic length scale of 2.8 mm, according to correlations proposed by Roach [17].

Heat transfer coefficient and effectiveness are determined by a steady-state technique, measuring wall temperatures from a heated surface by means of TLC paint. Wide band TLC $30C20W$ supplied by Hallcrest and active from ~30°C to 50°C are used. Crystals are thinned with water and sprayed with an airbrush on the test surface after the application of a black background paint. TLC were previously calibrated following the steady-state gradient method [18]. The calibration setup is made by a 4.5 mm thick aluminium rectangular plate, which houses seven thermocouples and is sprayed with black background paint and then TLC. One of its edges is heated by an electric heater, while the other is cooled by air. The whole apparatus is housed into an insulating basis. Camcorder and illuminating system are placed at the same distance and inclination of the real test, so as to replicate the exactly alike optic conditions. A linear temperature gradient will appear on TLC surface: once steady conditions are reached, a single picture is sufficient for a precise measurement of color-temperature response, with the latter parameter measured through thermocouples. Several tests have been carried out, so as to increase global precision; moreover, the calibration has been checked directly on the test article before each experiment.

A digital camera (Sony XCD-SX90CR) records color bitmap images (1280 × 960 pixel) from the TLC painted surface on a PC. The illuminating system (Shott-Fostec KL1500 LCD) uses an optical fiber goose-neck to ensure a uniform illumination on the test surface and it allows to keep both color temperature and light power constant. The test article is completely made of transparent PMMA to allow the required optical access for TLC measurements; the effusion plate only was made of PVC.

2.1. Geometry of the Test Sample. Figure 2 reports a sketch of the test article, which represents the cooling system of the combustor prototype developed within the European Integrated Project NEWAC. A picture of the prototype is shown in Figure 3. The slot coolant representing the starter film cooling is injected in the mainstream from a 6.0 mm high channel, with a lip thickness of 3.0 mm. The effusion array and the dilution hole are fed by an annulus with a rectangular 30.0 mm high and 120.0 mm wide cross-section.

The effusion geometry consists of a staggered array of 272 circular holes ($d = 1.65$ mm), with an inclination angle of $\alpha = 30°$, drilled in a 4.5 mm thick PVC plate and with a length to diameter ratio of $L/D = 5.5$. The spanwise and the streamwise pitches are, respectively, $S_y = 9.9$ mm and $S_x = 12.6$ mm. The first row is located 22.25 mm ($1.77S_x$) after the slot injection, while the last row 375 mm downstream. The origin of the coordinate system ($x = 0$) was set in order to have $x/S_x = 1$ at the first row and $x/S_x = 29$ at the last row, while the slot injection is located at $x/S_x = -0.77$. The dilution hole ($D = 18.75$ mm) is located immediately after the 14th row, at $x/S_x = 14.16$.

3. Measurement and Test Conditions

The experimental survey was formed by two main campaigns: the first campaign was aimed at measuring the heat transfer

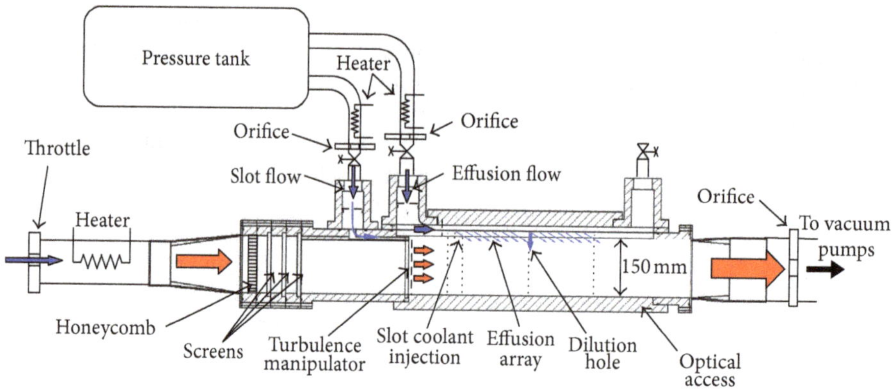

FIGURE 1: Sketch of the test rig.

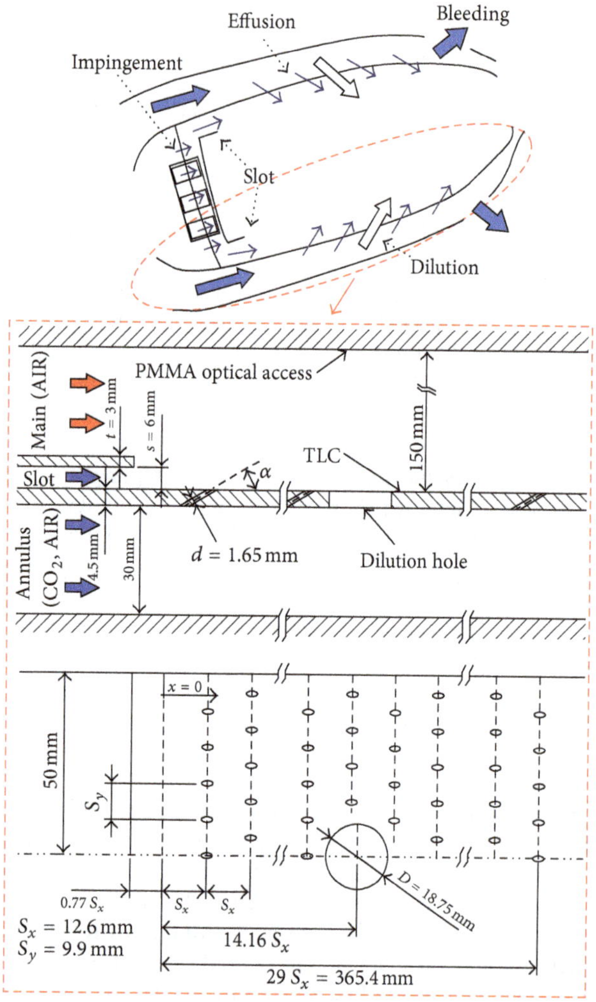

FIGURE 2: Liner geometry.

FIGURE 3: Combustor prototype of European Project NEWAC.

coefficient over the effusion plate, which was tested by imposing values of blowing ratio and velocity ratio within a typical range of an aeroengine combustor. Afterwards the same fluid conditions were replicated in the second campaign in order to estimate the adiabatic effectiveness of the film cooling generated by the system. Results of the two campaigns were finally combined to calculate the Net Heat Flux Reduction and the overall effectiveness.

In modern combustor, the temperature differences between the cooling air and the hot gases lead to a coolant-to-mainstream density ratio which usually falls within the range 1.5–3.0. To reproduce the effects of DR in a low temperature lab-scaled facility, two main approaches are adopted in the literature: cool down the cooling flow or use a foreign gas with a molecular weight greater than the air one. In this work, measurements were carried out by feeding the cooling system with carbon dioxide (CO_2): including the typical temperature differences required to perform experiments with TLC paints, the use of this foreign gas leads to a DR $\approx$ 1.7.

To highlight the effects of density ratio, the test matrix was duplicated and each fluid dynamic condition of the campaign was tested twice: the cooling system was first fed with air and then with CO_2, while air was used for the mainflow. Main investigation parameters are defined as follows:

$$\mathrm{BR}_{\mathrm{eff}} = \frac{1}{N_{\mathrm{row}}} \cdot \sum_{k=1}^{N_{\mathrm{row}}} \mathrm{BR}_k, \qquad (1)$$

$$\mathrm{BR}_k = \frac{\mathrm{Cd}_k \cdot \left(4 \cdot m_{\mathrm{is},k} / \left(\pi d^2\right)\right)}{m_{\mathrm{main},k} / A_{\mathrm{main}}}, \quad (2)$$

$$\mathrm{VR}_{\mathrm{eff}} = \mathrm{BR}_{\mathrm{eff}} \cdot \frac{\rho_{\mathrm{main}}}{\rho_{\mathrm{cool}}} = \frac{\mathrm{BR}_{\mathrm{eff}}}{\mathrm{DR}}. \quad (3)$$

$\mathrm{BR}_{\mathrm{eff}}$ is the averaged blowing ratio of the effusion rows (the dilution hole was excluded). BR of the kth row was evaluated by using the actual mass flow rate through the holes and the correspondent mainstream mass flow (inlet mainstream mass flow and coolant mass flow injected by the previous $(k-1)$th rows); the amount of coolant crossing each effusion row was calculated by using hole discharge coefficient, which is

$$\mathrm{Cd} = \frac{m_{\mathrm{real}}}{m_{\mathrm{is}}} = \left(m_{\mathrm{real}}\right) \times \left(p_{\mathrm{Tc}} \left(\frac{p_{\mathrm{main}}}{p_{\mathrm{Tc}}}\right)^{(\gamma+1)/2\gamma} \times \sqrt{\frac{2\gamma}{(\gamma-1)\,\mathrm{RT}_{\mathrm{Tc}}}\left(\left(\frac{p_{\mathrm{Tc}}}{p_{\mathrm{main}}}\right)^{(\gamma-1)/\gamma} - 1\right)} \frac{\pi}{4} d^2 \right)^{-1}. \quad (4)$$

Starting from the cooling and main mass flow and the several pressure values measured along the annulus and in main channel, the procedure uses the previous equations to evaluate the average effusion Blowing Ratio ($\mathrm{BR}_{\mathrm{eff}}$). The isentropic mass flow rate of coolant, calculated row by row, and the measured mass flow were employed to estimate a mean value of the effusion discharge coefficient, which is then used to calculate the actual mass flow through each row and consequently the blowing ratio (2). It was estimated that Cd $\approx$ 0.73, almost constant for all the tested conditions. Other parameters are

$$\mathrm{BR}_{\mathrm{sl}} = \frac{m_{\mathrm{sl}} / A_{\mathrm{sl}}}{m_{\mathrm{main}} / A_{\mathrm{main}}},$$
$$\mathrm{BR}_{\mathrm{dil}} = \frac{\mathrm{Cd}_{\mathrm{dil}} \cdot \left(m_{\mathrm{is,dil}} / \pi D^2/4\right)}{m_{\mathrm{main},14} / A_{\mathrm{main}}}. \quad (5)$$

A_{main} is the mainstream channel cross-section ($150 \times 100\,\mathrm{mm}^2$), A_{sl} is the slot cross-section ($6 \times 100\,\mathrm{mm}^2$), and $\mathrm{Cd}_{\mathrm{dil}}$ was imposed equal to 0.6 [19]. When air is used both as cooling and mainstream flows, the temperature differences of the experiments cannot raise the density ratio over DR $\approx$ 1.1 and, as a consequence, tests carried out by imposing the desired values of VR coincide with tests with the correspondent values of BR imposed.

For a better comprehension of slot and effusion influence on the cooling performance, some tests were performed activating only the effusion cooling flow; when the two cooling systems were tested together, the slot flow was set in order to keep a constant value of $\mathrm{BR}_{\mathrm{sl}} \approx 1.5$. Mainstream absolute pressure was kept constant at about $p_{\mathrm{main}} = 50000$ Pa ($\mathrm{Re}_{\mathrm{main}} \approx 75000$, $\mathrm{Ma}_{\mathrm{main}} \approx 0.04$–$0.05$), while coolant pressure was varied to ensure the desired values of coolant velocity inside the holes.

Table 1: Test matrix.

Flow type (coolant/mainstream)	$\mathrm{BR}_{\mathrm{eff}}$ ($\mathrm{VR}_{\mathrm{eff}}$)	$\mathrm{VR}_{\mathrm{eff}}$ ($\mathrm{BR}_{\mathrm{eff}}$)
AIR/AIR	1.5 (1.5)	—
	3.0 (3.0)	—
	5.0 (5.0)	—
	7.0 (7.0)	—
CO_2/AIR	1.5 (0.9)	1.5 (2.6)
	3.0 (1.8)	3.0 (5.1)
	5.0 (2.9)	5.0 (8.5)
	7.0 (4.1)	7 (11.9)

Table 1 sums up the test matrix of the campaign. The effusion plate was first tested using air as coolant, and then using CO_2; 4 different $\mathrm{BR}_{\mathrm{eff}}$ and 4 different $\mathrm{VR}_{\mathrm{eff}}$ were investigated. The full test matrix was made up of 48 experiments: each point of Table 1 was tested twice, feeding or not feeding the slot cooling system (8 AIR/AIR and 16 AIR/CO_2 experiments). The resulting 24 experiments matrix was finally performed twice in order to measure HTC and adiabatic effectiveness. It is important to underline that, in reference to the classification introduced by L'Ecuyer and Soechting [20], the effusion jets work within the penetration regime ($\mathrm{VR}_{\mathrm{eff}} > 0.8$) in all testing conditions.

All the tests were run after steady conditions were reached by all the measured quantities: flow rates, pressures, and temperatures. The uncertainty analysis was performed following the standard ANSI/ASME PTC 19.1 [21] based on the Kline and McClintock method [22]. Temperature accuracy is ±0.5 K, differential pressure ±6.9 Pa, and mass flow rate ±3–5%; the estimated error for the heat transfer coefficient calculation is ±10%, while it is ±0.05 for the adiabatic effectiveness.

4. Data Post Process

4.1. Heat Transfer Measurements. Heat transfer coefficients were determined by a steady-state technique, using TLC paint to measure the wall temperature from a heated surface. The heating element was a 25.4 μm thick Inconel Alloy 600 foil; it was laser drilled with the same array pitches of the PVC plate, and then applied on the test plate with a double sided tape. Surface heat flux was generated by Joule effect, fed by a DC power supply (Agilent N5763A) which is connected to the Inconel sheet through two copper bus bars fixed on lateral extremities of the test plate.

The mainstream heat transfer coefficient is defined as

$$\mathrm{HTC}_{\mathrm{main}} = \frac{q_{\mathrm{conv}}}{T_w - T_{\mathrm{main}}}, \quad (6)$$

where T_{main} is the mainstream static temperature, measured by means of three thermocouple located one pitch upstream the slot injection. T_w is the wall temperature measured by means of TLC while q_{conv} represents the heat rate exchanged by convection between the effusion plate and the mainstream flow. Due to the presence of the effusion and dilution holes, heat generated by the Inconel foil is not uniform on the surface of the plate; in addiction, despite the low thermal conductivity of the PVC, test sample is not ideally adiabatic and heat losses due to the conduction through the plate and the convective heat removed by coolant both in the annulus and inside the holes have to be taken into account. As a consequence, in order to have an accurate evaluation of the net heat flux transferred from the surface to the mainstream, q_{conv} was estimated implementing an iterative procedure based on a complete 3D thermal-electric FEM simulation. The procedure evaluates the nonuniform heat locally generated on the surface, allowing to obtain an accurate estimate of q_{conv}. Moreover, heat losses are taken into account too: depending on the fluid dynamics conditions of the tests, they are approximately 2%–5%. A detailed description of the iterative procedure can be found in Facchini et al. [10].

Heat transfer experiments were carried out with coolant and mainstream at room temperature. Likewise effectiveness measurements, the mainstream absolute pressure was kept constant at about $p_{\text{main}} = 50000$ Pa, while coolant pressure was varied in order to ensure the desired values of BR_{eff} and VR_{eff}.

According to Jones [23], the use of a foreign gas requires a special correction during the postprocess of the experimental data (both for heat transfer and effectiveness measurements); this correction allows to take into account the difference in specific heat and thermal conductivity between the foreign gas (CO_2 in this campaign) and the actual cooling flow of a real application (air). Jones assesses that a little correction is necessary in the case of CO_2 injection; in particular, the correction becomes smaller with increasing the coolant velocity (high BR-VR), due to the fact that the transport of species, momentum, and enthalpy becomes mainly dependent on the turbulent flow field rather than on the concentration gradients and the viscosity. Concerning the results of this campaign, whose tests were performed at high BR-VR, it was estimated that the entity of this correction is almost negligible and falls within the error due to the experimental uncertainties.

4.2. Adiabatic Effectiveness Measurements. Effectiveness measurements were carried out by heating both the coolant and the mainflow, in order to obtain temperature of about 300 K and 350 K, respectively. Likewise HTC measurements, the mainstream absolute pressure was kept constant at about $p_{\text{main}} = 50000$ Pa, while coolant pressure was varied in order to ensure the desired values of coolant velocity.

Adiabatic effectiveness is defined as

$$\eta_{\text{aw}} = \frac{T_{\text{main}} - T_{\text{aw}}}{T_{\text{main}} - T_{\text{cool}}}. \tag{7}$$

Three thermocouples located one pitch upstream the slot injection acquired mainstream static temperature T_{main}. Three additional probes were dedicated to measure coolant flow static temperature and were inserted into the annulus, at $x/S_x = 0; 14; 29$; one further probe was located inside the slot channel at $x/S_x = -1$. T_{aw} was evaluated through a post-processing procedure which takes into account the thermal fluxes across the plate due to conduction and due to the coolant inside the annulus and the holes. This procedure is based on an 1D approach and considers the following equation:

$$T_{\text{aw}} = T_w - \frac{q}{\text{HTC}_{\text{main}}}, \tag{8}$$

where T_w is the wall temperature measured with TLC. Heat flux across the plate (q) is evaluated through TLC wall temperature and coolant temperature and using the Colburn correlation $\text{Nu} = 0.023\text{Re}^{0.8}\text{Pr}^{1/3}$ to estimate heat transfer coefficients inside the holes and on the annulus side of the plate; Reynolds and Nusselt numbers were evaluated with the hole diameter and with the annulus cross-section hydraulic diameter, respectively. Values of HTC_{main} were directly taken from results of the dedicated experimental campaign. Conduction through the PVC was taken into account too.

5. Results

5.1. Heat Transfer Coefficient. Figure 4 shows heat transfer coefficient maps for the experiments carried out by imposing $\text{BR}_{\text{eff}} - \text{VR}_{\text{eff}} = 3$ (due to the small coolant to mainstream density ratio, $\text{VR}_{\text{eff}} \approx \text{BR}_{\text{eff}}$ in AIR tests); results are displayed dividing the local HTC_{main} by a constant reference value (HTC_{ref}). In white areas close to the dilution hole HTC was not measured because the local low/high surface heat generation did not allow TLC paints working properly within their activation range. Maps displays the overall trend of HTC, showing that it increases up to the 14th row and then remains nearly constant. Difference between tests with and without slot is restricted to the first 2-3 rows, where coolant coming out from the slot mitigates the heat transfer; after the 5th row, the presence of the slot flow does not alter significantly the behaviour of the effusion cooling.

Imposing the same mainstream conditions of all the other experiments, a reference test was carried out in order to evaluate the heat transfer coefficient without film cooling (HTC_0); map of HTC_0 is displayed in Figure 4. Values of HTC_0 were spanwise averaged and the resulting trend along the centerline was used as the reference.

Figures 5(a) and 5(b) show trends of spanwise averaged heat transfer coefficient along the plate with effusion coolant only and with both slot and effusion flows for $\text{BR}_{\text{eff}} - \text{VR}_{\text{eff}} = 3$; data are plotted in terms of ($\text{HTC}_{\text{main}}/\text{HTC}_0$) in order to highlight the increase of heat transfer due to coolant injections. Figure 5(a) shows that HTC remains constant in the first five rows, even if it is enhanced compared to the reference case ($\text{HTC}_{\text{main}}/\text{HTC}_0 > 1$); after the 5th row, it increases up to the dilution hole, where it reaches an asymptotic value. The beginning of the rising trend of HTC is brought forward to the 2-3th row in presence of the slot cooling flow (Figure 5(b)); however, as it was already shown in the maps, after the 5th row, the slot flow has only a slight influence on the heat transfer. Results show

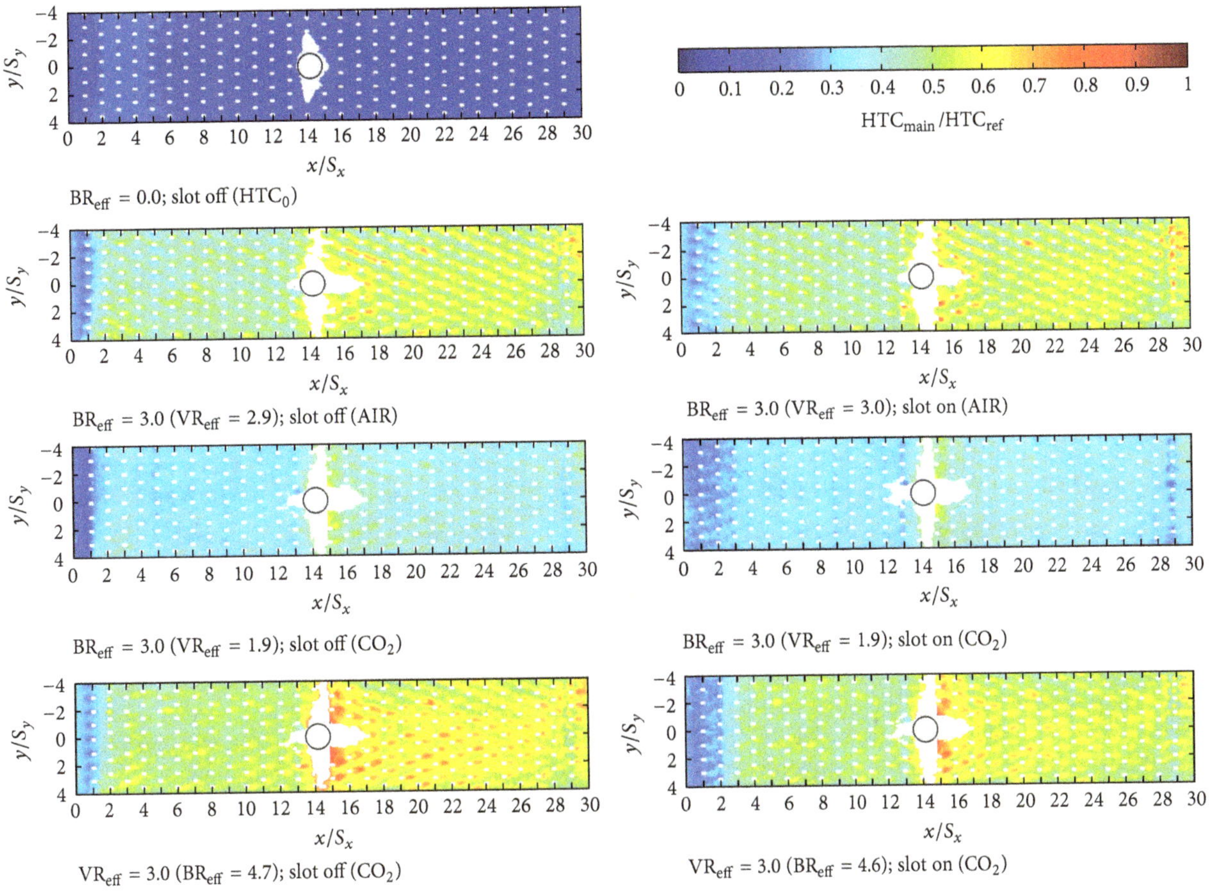

FIGURE 4: Heat transfer coefficients maps (BR$_{eff}$ − VR$_{eff}$ = 3).

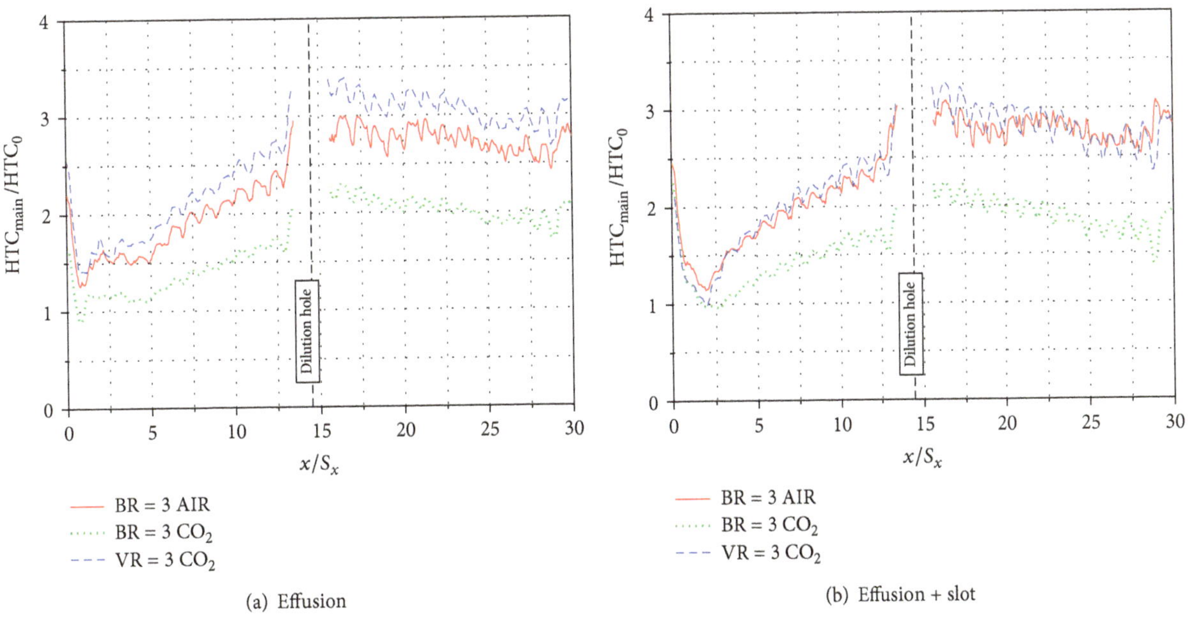

FIGURE 5: Spanwise averaged HTC (BR$_{eff}$ − VR$_{eff}$ = 3).

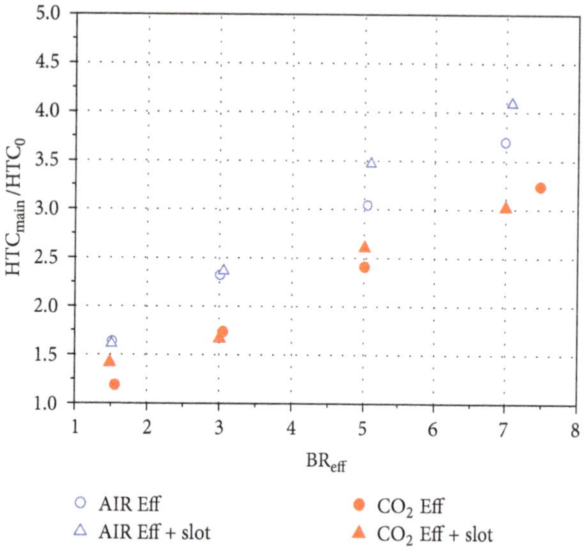

(a) Averaged value of HTC/HTC$_0$ versus BR$_{eff}$

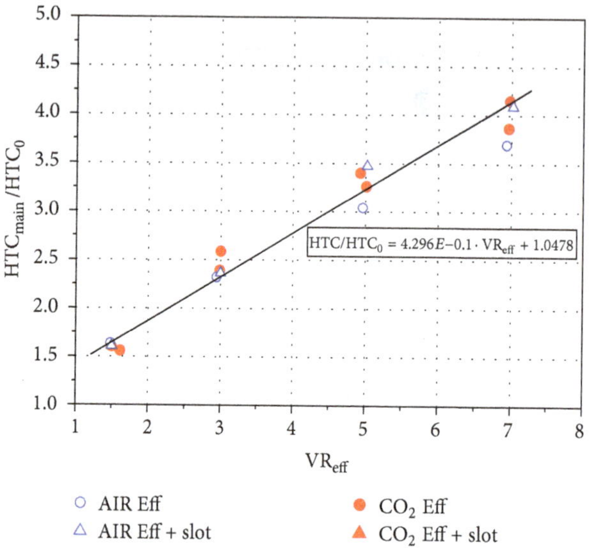

(b) Averaged value of HTC/HTC$_0$ versus VR$_{eff}$

FIGURE 6: Heat transfer coefficient results.

that for a constant blowing ratio, heat transfer decreases with increasing density ratio; on the other hand, results of tests carried out by imposing the same value of velocity ratio are almost coincident. Maps and trends are here shown only for one point of the test matrix, which significantly represents the typical behaviour of the system in each testing condition. A more detailed description of the behaviour of heat transfer coefficient over this effusion plate can be found in Facchini et al. [10]. Results of the full test matrix are summarized in Figure 6: it shows the average value of HTC_{main}/HTC_0 of the whole plate with and without the slot flow, plotted versus the actual BR_{eff} Figure 6(a) and VR_{eff} Figure 6(b). Figures clearly display how the HTC linearly increases with increasing BR-VR; furthermore, it is possible to highlight that air tests are in good agreement with CO_2 tests with the same velocity ratio (Figure 6(b)). This means that, within the effusion jets penetration regime, VR acts as the driving parameter of the phenomena instead of BR; this confirms the results numerically found by Andreini et al. [16]. Figure 6(b) includes also the equation of the linear fitting of the experimental data: the maximum estimated relative error was around 10%, while the averaged value was around 4.5%. Even if the outcomes of the campaign are affected by a low turbulence level with respect to an actual combustor, results give useful information about the behaviour of the effusion system and, moreover, show that the use of velocity ratio in a low temperature facility allows reproduction of the effects of DR without employing a foreign gas.

5.2. Adiabatic Effectiveness. Figure 7 shows the adiabatic effectiveness maps for $BR_{eff} - VR_{eff} = 3$ test points. Maps display the effects of effusion and slot coolant injections and highlight both how the wake generated by the dilution hole and the presence of the slot flow influence the film cooling distribution over the surface. Results point out that, without the slot flow, the effusion system does not guarantee a sufficient protection of the first part of the liner by itself. On the other hand, maps with both effusion and slot coolant show that a very efficient protection of the liner can be obtained combining the two cooling systems.

Figure 8 shows trends of spanwise averaged adiabatic effectiveness along the plate: it is possible to observe that, when only effusion is activated, the film cooling superposition increases η_{aw} quite linearly until the 15th row, where the dilution hole is located. In the following rows, the wake generated by the dilution hole deeply affects the film distribution: a slight decrease of effectiveness can be observed immediately downstream the hole, after which the dilution jet draws the coolant from the effusion holes towards the center of the test palate and generates a high effectiveness area. The effective level in this area is almost asymptotic until the end of the plate.

Focusing on the effects of DR on test, it is possible to observe that an increase of VR_{eff} leads to a slight increase in η_{aw}; moreover, even if the differences among the curves are quite small, BR_{eff} seems to be the parameter which allows to take into account the effects of density ratio even in air-to-air tests.

The presence of the slot coolant strongly enhances the adiabatic effectiveness; after the first three rows where the η_{aw} remains nearly constant, there is a lower effectiveness area due to the detrimental interaction between the two cooling flows: here the highly penetrating effusion jets partially destroy the high effectiveness film cooling layer generated by the slot [9]. This behaviour can be directly related to the velocity ratio of the jets, which can be used to scale the effects of DR; after the dilution hole, η_{aw} reaches an asymptotic value which can still be related to the velocity ratio.

Finally Figure 9 shows the adiabatic effectiveness results for the whole test matrix, plotted versus BR_{eff} and VR_{eff}; each point represents the averaged value of the entire test sample.

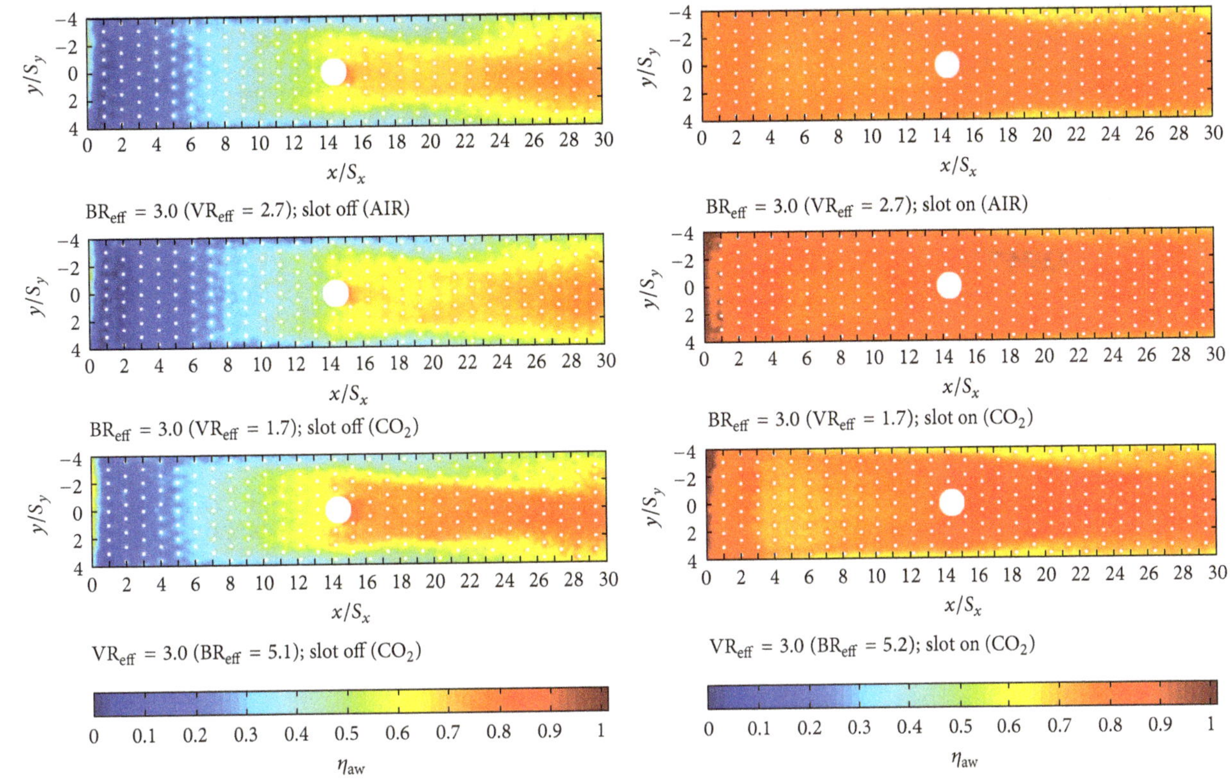

FIGURE 7: Adiabatic effectiveness maps ($BR_{eff} - VR_{eff} = 3$).

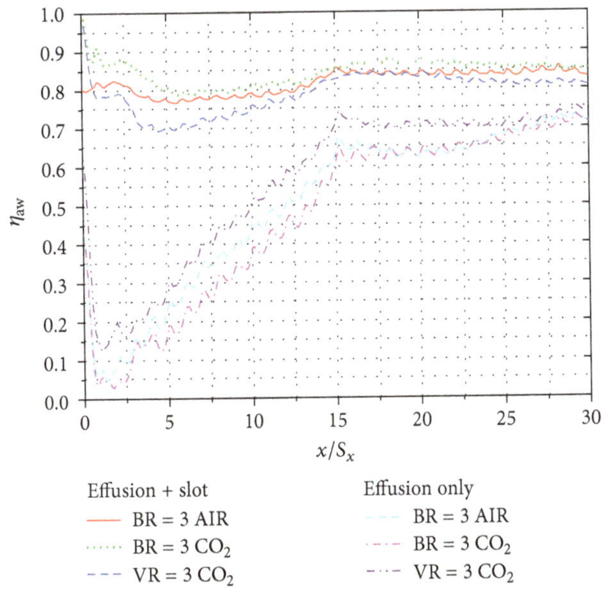

FIGURE 8: Spanwise averaged η_{aw} ($BR_{eff} - VR_{eff} = 3$).

Tests without slot flow (circles) confirm the outcomes from the analysis of Figure 8: experimental results show that within the penetration regime, the effects of coolant to mainstream density ratio are small thus, as commonly found in the literature, BR have to be used to scale the effects of DR. In addition, results allow the highlight of the behaviour of the system at high values of BR-VR: here η_{aw} is only weakly affected by those parameters since the high effectiveness is mainly due to the coolant mass addiction. In fact, even if the penetration of effusion jets increases, η_{aw} does not fall because the large amount of coolant mass flow injected in the mainstream grows row by row and guarantees the good protection of the liner.

Concerning tests with both cooling systems (triangles), results indicate that η_{aw} decreases with increasing BR-VR (due to the increasing penetration of effusion jets), but only slight differences were found changing the coolant to mainstream density ratio. Focusing on tests with the same BR_{eff} Figure 9(a), it is possible to note that an increase in DR causes a small enhancement in η_{aw}; on the other hand, results indicate that when both the slot flow and the effusion are active and the jets work within the penetration regime, VR_{eff} has to be used to take into account the effects of density ratio.

5.3. Net Heat Flux Reduction and Overall Effectiveness. Net Heat Flux Reduction (NHFR) is a commonly used parameter to evaluate the reduction of heat flux across a cooled surface. This parameter was defined by Sen et al. [24] as

$$\mathrm{NHFR} = 1 - \frac{q}{q_0} = 1 - \frac{\mathrm{HTC_{main}}}{\mathrm{HTC_0}}(1 - \eta\theta), \quad (9)$$

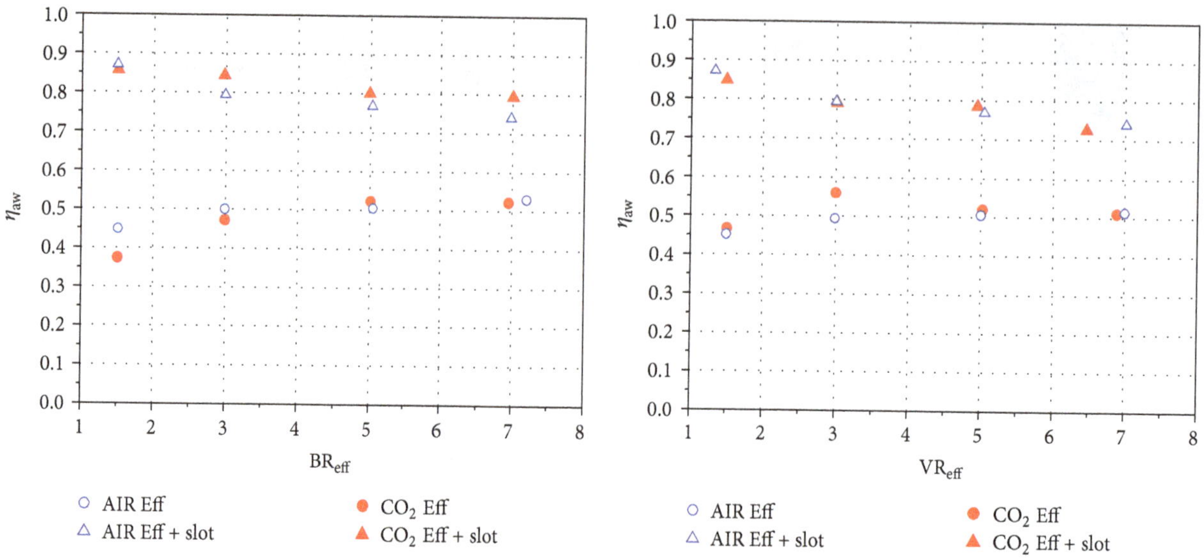

(a) Averaged value of η_{aw} versus BR_{eff}

(b) Averaged value of η_{aw} versus VR_{eff}

FIGURE 9: Adiabatic effectiveness results.

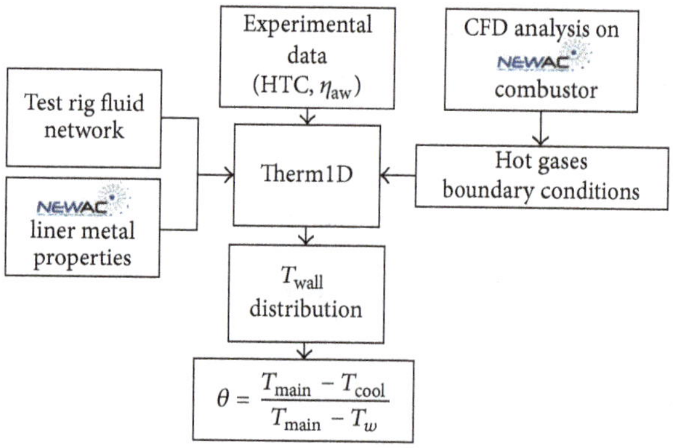

FIGURE 10: Procedure to evaluate θ.

where θ represent the dimensionless temperature:

$$\theta = \frac{T_{main} - T_{cool}}{T_{main} - T_w}. \quad (10)$$

In the open literature [24–27], NHFR was mainly used to evaluate turbine endwall and blades cooling systems and the dimensionless temperature was set within the range of $\theta = 1.5$–1.6. Since the overall effectiveness of a combustor cooling system is generally higher than the one of a turbine airfoil, more recently Facchini et al. [10, 11] updated this parameter to evaluate the NHFR of a cooled liner, imposing a value of $\theta = 1.2$. In the present study, NHFR was evaluated by using the experimental results within a one-dimensional thermal procedure (*Therm1d*) in order to estimate an engine representative distribution of θ. *Therm1d* is an in-house procedure which solves heat conduction inside a combustor liner and provides its temperature distribution by using an 1D Finite Difference Model. On the coolant side, the procedure solves the coolant fluid network of the system, taking into account the different cooling techniques of the specific combustor architecture and the heat exchange with metal surfaces; on the hot gas side, it estimates the convective heat load and the luminous and the nonluminous radiation through a correlative approach, mainly following the one-dimensional approach suggested by Lefebvre [19]. The final temperature distribution is obtained considering also the film cooling, usually estimated through correlations, and the heat sink effect due to the presence of cooling holes. Further details on the procedure can be found in Andreini et al. [28, 29].

Figure 10 illustrates the procedure followed to estimate the NHFR of the cooling scheme in engine representative working conditions. *Therm1d* was used to set up the cooling fluid network of the test rig, including both the slot and the effusion system; test plate material was specified to be the real steel of NEWAC combustor prototype instead of PVC as in the experiments to more realistically model the conduction through the plate itself. Moreover, hot gas boundary conditions (i.e., gas temperature, pressure, etc.), taken from previous CFD analysis on the NEWAC combustor, were imposed in order to simulate the behaviour of a realistic combustor diffusion flame and, consequently, to estimate realistic the heat loads [28]. As an example, Figure 11(a) shows the distribution of mainstream gas temperature inside the core of the combustor (temperatures are adimensionalized with a reference value): it is possible to identify the position of the flame front, which causes the temperature rise downstream the second row. Coolant side inlet pressure was varied in order to set the desired averaged BR-VR through the effusion holes and reproduce the experimental test matrix, while inlet coolant temperature and outlet pressure were kept

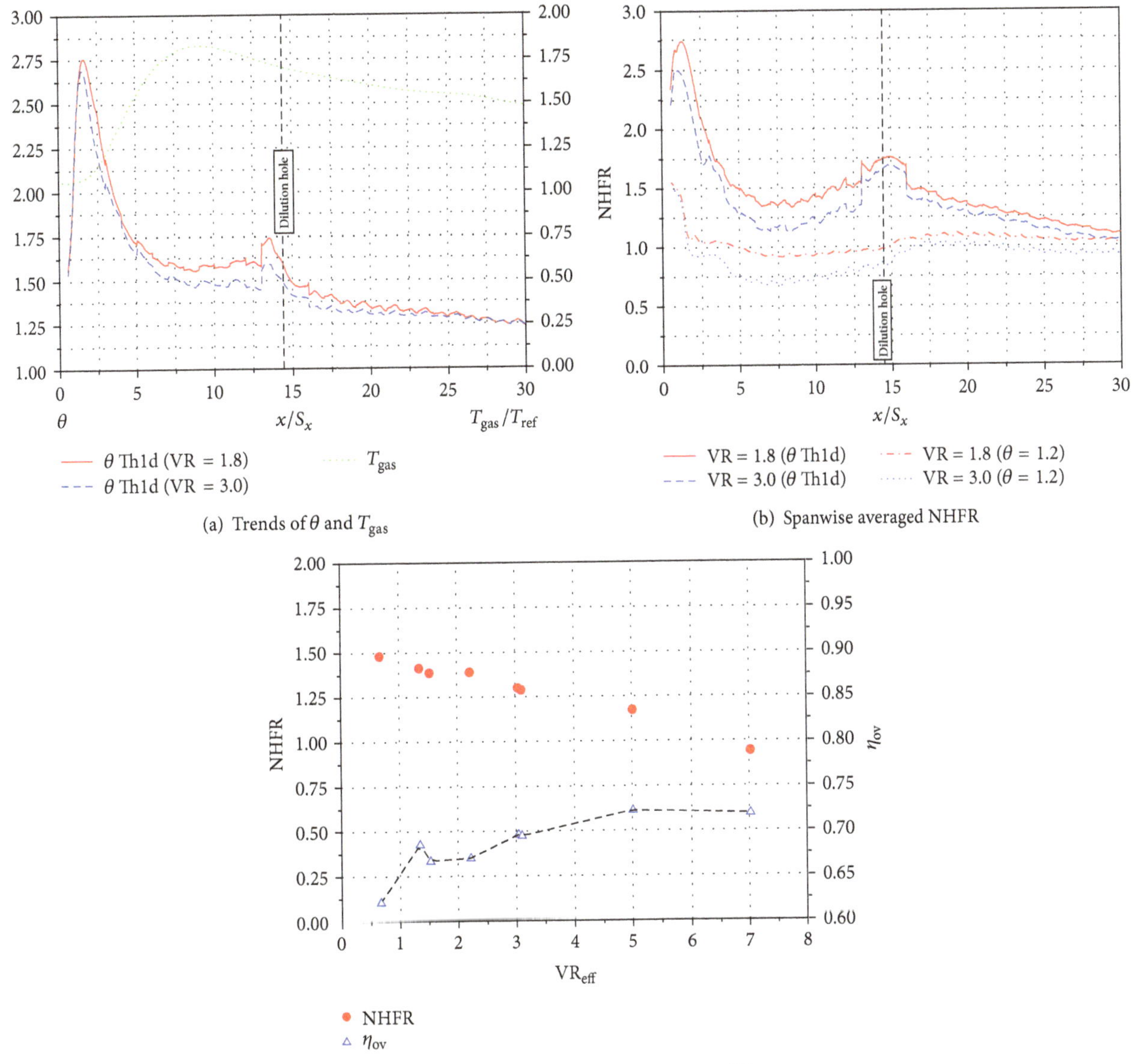

(a) Trends of θ and T_{gas}

(b) Spanwise averaged NHFR

(c) Averaged value of NHFR and η_{ov} versus VR_{eff}

FIGURE 11: Net Heat Flux Reduction and η_{ov} results.

constant. Experimental film cooling and convective gas side HTC distributions obtained in this work were imposed as boundary conditions too: in order to take into account the effects of density ratio due to the temperature difference between coolant and mainstream, data were taken only from CO_2 tests (only tests with both effusion and slot cooling system). This procedure allowed simulation of the behaviour of the cooling system under realistic operating condition and to finally estimate the distribution of NHFR along the liner: in fact, each run provided the liner surface temperature distributions (T_w) and the local temperature of the coolant coming out from each effusion row (T_{cool}), including its heating through the hole due to the heat removal by heat sink effect.

Figure 11(a) shows trends of the θ parameter for cases $BR_{eff} = 3$ ($VR_{eff} \simeq 1.8$) and $VR_{eff} = 3$. It is possible to observe how θ varies along the liner as a consequence of a nonuniform mainstream temperature distribution and of the resulting T_{cool} and T_w; it can be noticed that an increase in VR_{eff} leads to a slight decrease of θ.

Figure 11(b) displays trends of NHFR for the two previous test points: results evaluated through *Therm1d* are compared with those estimated by imposing a constant value of $\theta = 1.2$. As a consequence of the θ behaviour, NHFR calculated considering the heat sink effect is much higher than the simple θ-imposed case (the averaged value is almost 30% higher). Obviously trends evaluated with *Therm1d* are highly affected by the hot gases imposed boundary conditions (e.g., the drop after the 2th row is related to the gas temperature rise which represents the flame front), but what is important to highlight is the influence of effusion velocity ratio on NHFR. Even if NHFR linearly decreases with VR_{eff}, Figure 11(c) points out that the cooling system always brings to a reduction of the heating flux towards the liner (NHFR > 0).

Despite previous considerations, NHFR is not properly representative for an effusion cooling system since its definition (9) does not explicitly take into account the heat sink effect, which instead plays a major role in this type of cooling technique. To overcome this aspect and give a complete description of the cooling performance of the system, wall temperatures estimated through *Therm1d* were finally employed to calculate also the overall effectiveness of the test plate in real engine conditions. This further parameter indicates the overall cooling capability of a cooling system and is defined as:

$$\eta_{ov} = \frac{T_{main} - T_w}{T_{main} - T_{cool}}. \quad (11)$$

Results depicted in Figure 11(c) show that, except for very low values of velocity ratio, η_{ov} remains high and almost constant for a wide range of operative conditions; even if results do not take into account the effects of turbulence, whose level is rather lower than a real combustor, they give useful indications for the designer: in fact, this study points out that the combined effusion and slot system is very effective and robust from a cooling point of view for a wide range of operative conditions. As a consequence, this flexibility of the system allows, during the design phase, to focus its optimization taking into account other requirements, like combustion issues, aeroacoustic or simply off-design working conditions.

6. Conclusions

The aim of the present study is the investigation of the effects of coolant-to-mainstream density ratio on the cooling performance of a real engine cooling scheme of a combustor liner. The cooling scheme consists of a slot injection, followed by a flat plate with 29 effusion rows and a single large dilution hole. Values of effusion blowing ratio and velocity ratio typical of modern engine working conditions were imposed in order to measure the heat transfer coefficient and the adiabatic effectiveness; tests were carried out by using a steady-state technique with wide band thermochromic liquid crystals. To obtain a value of density ratio which is representative for a combustor, tests were carried out by feeding the cooling system with a foreign gas (CO_2).

HTC results show that, for a constant blowing ratio, heat transfer is reduced with increasing the density ratio; on the other hand, within the effusion jets penetration regime, velocity ratio is the driving parameter of the phenomena in order to scale the effects of DR. Concerning the adiabatic effectiveness, experiments show that after $VR_{eff} = 3$, η_{aw} generated by the effusion jets is weakly affected by BR-VR; furthermore, effects of density ratio can be neglected within the penetration regime. When both slot and effusion system are activated, results point out that, for a constant velocity ratio, effectiveness increases with increasing density ratio and that VR_{eff} can be used to take into account the effects of DR.

Finally, NHFR and the overall effectiveness were estimated combining heat transfer and effectiveness results: real engine working conditions were simulated by using an in-house 1D thermal procedure. Results point out a linear decrease of NHFR with VR_{eff}, even if the system always brings a reduction of the heating flux towards the liner; results in terms of η_{ov} indicate that the combined effusion and slot cooling system has a very effective and robust behaviour over a wide range of operative conditions.

Nomenclature

A: Reference area (m^2)
BR: Blowing ratio (–)
Cd: Hole discharge coefficient (–)
d: Effusion hole diameter (m)
D: Dilution hole diameter (m)
DR: Density ratio (–)
HTC: Heat transfer coefficient (W/m^2K)
L: Hole length (m)
Ma: Mach number (–)
m: Mass flow rate (kg/s)
p: Pressure (Pa)
q: Heat flux (W/m^2)
Re: Reynolds number (–)
s: Slot lip thickness (m)
S_x: Streamwise pitch (m)
S_y: Spanwise pitch (m)
T: Temperature (K)
VR: Velocity ratio (–)
x: Abscissa along the plate (m)
y: Spanwise location (m).

Greeks

α: Effusion hole injection angle (deg)
η: Effectiveness (–)
γ: Ratio of specific heat (–)
θ: Dimensionless temperature (–)
ρ: Density (kg/m^3).

Subscript

aw: Adiabatic wall
c: Coolant
conv: Convection
eff: Effusion
is: Isentropic
main: Mainstream
ov: Overall
ref, 0: Reference
sl: Slot
T: Total
w: Wall.

Acronyms

CO_2:	Carbon dioxide
FEM:	Finite Element Method
ICAO:	International Civil Aviation Organization
NEWAC:	NEW Aeroengine Core concept
NHFR:	Net Heat Flux Reduction
OTDF:	Outlet Temperature Distribution Factor
PMMA:	Poly(methyl methacrylate)
PVC:	Polyvinyl chloride
TLC:	Thermochromic Liquid Crystal.

Acknowledgments

The authors wish to express their gratitude to F. Simonetti and A. Picchi for their useful suggestions and support. The present work was supported by the European Commission as part of FP6 IP NEWAC (NEW Aero engine Core concepts) research program (FP6-030876), which is gratefully acknowledged together with consortium partners.

References

[1] G. E. Andrews, F. Bazdidi-Tehrani, C. I. Hussain, and J. P. Pearson, "Small diameter film cooling hole heat transfer: the influence of hole length," ASME Paper 91-GT-344, 1991.

[2] G. E. Andrews, I. M. Khalifa, A. A. Asere, and F. Bazdidi-Tehrani, "Full coverage effusion film cooling with inclined holes," ASME Paper 95-GT-274, 1995.

[3] J. J. Scrittore, K. A. Thole, and S. W. Burd, "Experimental characterization of film-cooling effectiveness near combustor dilution holes," ASME Turbo Expo GT2005-68704, 2005.

[4] J. J. Scrittore, K. A. Thole, and S. W. Burd, "Investigation of velocity profiles for effusion cooling of a combustor liner," ASME Turbo Expo GT2006-90532, 2006.

[5] D. E. Metzger, D. I. Takeuchi, and P. A. Kuenstler, "Effectiveness and heat transfer with full-coverage film cooling," *ASME Journal of Engineering For Power*, vol. 95, no. 3, pp. 180–184, 1973.

[6] M. E. Crawford, W. M. Kays, and R. J. Moffat, "Full-coverage film cooling-2. Heat transfer data and numerical simulation," *Journal of engineering for power*, vol. 102, no. 4, pp. 1006–1012, 1980.

[7] R. F. Martinez-Botas and C. H. N. Yuen, "Measurement of local heat transfer coefficient and film cooling effectiveness through discrete holes," ASME Turbo Expo 2000-GT-243, 2000.

[8] G. B. Kelly and D. G. Bogard, "An investigation of the heat transfer for full coverage film cooling," ASME Turbo Expo GT2003-38716, 2003.

[9] A. Ceccherini, B. Facchini, L. Tarchi, and L. Toni, "Combined effect of slot injection, effusion array and dilution hole on the cooling performance of a real combustor liner," ASME Turbo Expo GT2009-60047, 2009.

[10] B. Facchini, F. Maiuolo, L. Tarchi, and D. Coutadin, "Combined effect of slot injection, effusion array and dilution hole on the heat transfer coefficient of a real combustor liner-part 1 experimental analysis," ASME Turbo Expo GT2010-22936, 2010.

[11] B. Facchini, F. Maiuolo, L. Tarchi, and D. Coutadin, "Experimental investigation on the effects of a large recirculating area on the performance of an effusion cooled combustor liner," *Journal of Engineering For Gas Turbines and Power*, vol. 134, no. 4, Article ID 041505, 2012.

[12] S. V. Ekkad, D. Zapata, and J. C. Han, "Film effectiveness over a flat surface with air and CO2 injection through compound angle holes using a transient liquid crystal image method," *Journal of Turbomachinery*, vol. 119, no. 3, pp. 587–593, 1997.

[13] S. V. Ekkad, D. Zapata, and J. C. Han, "Heat transfer coefficients over a flat surface with air and CO2 injection through compound angle holes using a transient liquid crystal image method," *Journal of Turbomachinery*, vol. 119, no. 3, pp. 580–586, 1997.

[14] Y. Lin, B. Song, B. Li, G. Liu, and Z. Wu, "Investigation of film cooling effectiveness of full-coverage inclined multihole walls with different hole arrangements," ASME Turbo Expo GT2003-38881, 2003.

[15] Y. Lin, B. Song, B. Li, and G. Liu, "Investigation of film cooling effectiveness of full-coverage inclined multihole walls with different hole arrangements," *ASME Journal of Heat Transfer*, vol. 128, no. 6, pp. 580–585, 2006.

[16] A. Andreini, A. Ceccherini, B. Facchini, and D. Coutadin, "Combined effect of slot injection, effusion array and dilution hole on the heat transfer coefficient of a real combustor liner-part 2 numerical analysis," ASME Turbo Expo GT2010-22937, 2010.

[17] P. E. Roach, "The generation of nearly isotropic turbulence by means of grids," *International Journal of Heat and Fluid Flow*, vol. 8, no. 2, pp. 82–92, 1987.

[18] T. L. Chan, S. Ashforth-Frost, and K. Jambunathan, "Calibrating for viewing angle effect during heat transfer measurements on a curved surface," *International Journal of Heat and Mass Transfer*, vol. 44, no. 12, pp. 2209–2223, 2001.

[19] A. H. Lefebvre, *Gas Turbine Combustion*, Taylor & Francis, London, UK, 1998.

[20] M. R. L'Ecuyer and F. O. Soechting, "A model for correlating flat plate film cooling effectiveness for rows of round holes," in *AGARD Heat Transfer and Cooling in Gas Turbines 12p (SEE N86-29823 21-07)*, 1985.

[21] The American Society of Mechanical Engineers, "Measurement uncertainty," in *Instrument and Apparatus, Volume ANSI/ASME PTC 19.1-1985 of Performance Test Code*, The American Society of Mechanical Engineers, New York, NY, USA, 1985.

[22] S. J. Kline and F. A. McClintock, "Describing uncertainties in single sample experiments," *Mechanical Engineering*, vol. 75, pp. 3–8, 1953.

[23] T. V. Jones, "Theory for the use of foreign gas in simulating film cooling," *International Journal of Heat and Fluid Flow*, vol. 20, no. 3, pp. 349–354, 1999.

[24] B. Sen, D. L. Schmidt, and D. G. Bogard, "Film cooling with compound angle holes: heat transfer," *Journal of Turbomachinery*, vol. 118, no. 4, pp. 800–806, 1996.

[25] M. Gritsch, A. Schulz, and S. Wittig, "Film-cooling holes with expanded exits: near-hole heat transfer coefficients," *International Journal of Heat and Fluid Flow*, vol. 21, no. 2, pp. 146–155, 2000.

[26] J. R. Christophel, K. A. Thole, and F. J. Cunha, "Cooling the tip of a turbine blade using pressure side holes-part ii: heat transfer measurements," *ASME Turbo Expo*, no. 2004-GT-53254, 2005.

[27] J. D. Piggush and T. W. Simon, "Measurements of net change in heat flux as a result of leakage and steps on the contoured endwall of a gas turbine first stage nozzle," *Applied Thermal Engineering*, vol. 27, no. 4, pp. 722–730, 2007.

[28] A. Andreini, C. Carcasci, A. Ceccherini et al., "Combustor liner temperature prediction: a preliminary tool development and its application on effusion cooling systems," in *Proceedings of the 1st CEAS European Air and Space Conference Century Perspectives*, Paper n.026, 2007.

[29] A. Andreini, A. Ceccherini, B. Facchini, F. Turrini, and I. Vitale, "Assesment of a set of numerical tools for the design of aeroengines combustors: study of a tubular test rig," ASME Turbo Expo GT2009-59539, 2009.

11

Novel Multipoint Relays Scheme Based on Hybrid Cost Function

Ali Ouacha,[1] Bachir Bouamoud,[2] Ahmed Habbani,[1,2] and Jamal El Abbadi[1]

[1] *Laboratoire d'Electronique et de Communications (LEC), Ecole Mohammadia d'Ingénieurs (EMI), Université Mohammed V-Agdal (UM5A), BP 765, Avenue Ibn Sina, Agdal, 10000 Rabat, Morocco*
[2] *Laboratoire SIME, Ecole Nationale Supérieure d'Informatique et d'Analyse des Systèmes (ENSIAS), Université Mohammed V-SOUISSI (UM5S), BP 713, Madinat Al Irfane, Avenue Mohammed ben Abdallah Regragui, Agdal, 10000 Rabat, Morocco*

Correspondence should be addressed to Bachir Bouamoud; bouamoud.bachir@gmail.com

Academic Editors: V. G. M. Annamdas, F. Lu, and Z. Mazur

When evaluating the performance of QoS protocols, a number of factors have a major impact on the results. Notably, QoS is emphasized when mobile ad hoc networks (MANETs) are employed into aerospace fields. Some of these parameters are a particular manifestation of characteristics of the MANET environment, such as mobility. Indeed, our proposal is a novel multipoint relays scheme based on hybrid cost function taking into account QoS criteria and avoiding mobility effect of nodes, especially those selected as MPRs. A comprehensive simulation study was conducted to evaluate the performance of the proposed scheme. Performance results show that RQMPR outperforms existing MPR heuristic adopted in the ad hoc routing protocols OLSR and QOLSR, in terms of packet delivery and average end-to-end delay.

1. Introduction

Ad hoc network is wireless network composed of autonomous individual nodes. It is easy to install and deploy and provides point to point communications between nodes without any infrastructure network. In an ad hoc network, since there is no central coordinator, for example, an access point (AP) or base station (BS), all nodes are supposed to work as terminals and routers at the same time. Thus, a routing protocol will play a major role in an ad hoc network to connect nodes that cannot communicate with each other directly and does not stop to be a subject of research work to improve the performance of wireless networking solutions.

Due to the dynamic changes of the factors that affect the performance in a mobile ad hoc network, it would be convenient that any proposed optimizations should consider the dynamics that act on nodes and links which interconnect them. In this context, knowledge of network must also have the same character in terms of taking into account factors such as available bandwidth, delays, and the lifetime of nodes in the process of selection of multi point relays.

In proactive routing mechanism, the use of relays aims to reduce the broadcast messages senders and then the number of flooded messages; here we highlight the importance of relays in the OLSR protocol case since they become the only responsible for broadcasting Topology control messages. While this approach is pleasing to the eye, beside some control functions that are necessary to prevent an eternal duplication of broadcast messages, it is required to selected relays in a reliable manner; indeed, defects in the reception of broadcast packets from MPR nodes can greatly affect the rate of delivery of packets across the network.

QoS routing relies on the state of parameters specifying resource availability at network nodes or links and uses them to find paths with enough free resources, and this concept can be applied also on the selection of relays, by choosing among the candidates nodes those maximizing a composite constraint based on the path characteristics to reach a 2-hop farther nodes.

In Figure 1, we present a case that highlights the process of MPR selection in node V_1, where V_2 is the only intermediate node to reach V_5. Besides V_6 and V_7 are reachable through both nodes V_3 and V_4; thus, V_1 must make a decision to select which one of them to be in its MPR set. Although in the standard OLSR (RFC 3626) the selection is made randomly

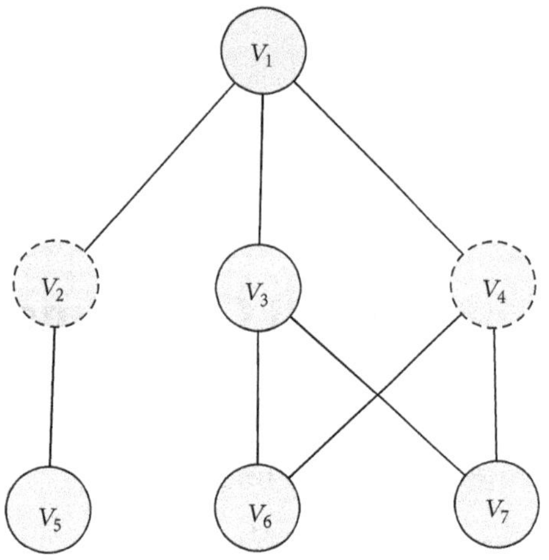

FIGURE 1: MPR selection case study.

among all the candidates, node is to be selected as a relay. Related to this situation, the question is how can we avoid a random choice and select the most convenient node between V_3 and V_4?

In the rest of the paper, we refer to our proposal as RQMPR, and it is organized as follows. In Section 2, we present some previous works whose aim is to improve routing reliability. In Section 3, we introduce a new composite metric. In Section 4, we present a new MPR selection heuristic. Section 5, contains a set of simulations and results which concern the evaluation of our approach. In Section 6, we evaluate the performance of the proposed scheme. Lastly in Section 7, we conclude and discuss future works.

2. Related Work

Multipoint relay nodes publish inside the TC messages the links which could constitute the paths from source to destination; then the set of MPRs form a kind of backbone, in the mobile ad hoc network. Thus, one promising issue of the routes selection optimization is to carefully select MPR that meets a given requirement to improve the targeted network performance. Indeed, an analysis of MPR selection in the OLSR Protocol [1] concluded that routes performance can be increased by adopting some supplementary criteria on MPR mechanism. In other words, the routing metrics can be also taken into consideration to choose the nodes relays, to face up to the nonadvertisement of potential link [2]. Most of the literature of routing optimization in OLSR aims to find other efficient metrics rather than the default one defined in the RFC3626, where the path quality is measured by the number of hops.

Some proposed routing metrics are based on Mac layer information such as the queue length, bandwidth, or Bit Error Rate (BER) [3]. QOLSR [4] is one of the major QoS extension in the proactive routing category. It enforces restrictions on bandwidth availability and/or delay by evaluating these metrics using a heuristic method that provides an option of best effort routing by searching for a so-called shortest-widest path. However, a mean drawback of this approach is that the number of generated MPRs is increased since the criteria of available bandwidth are a prior condition to reachability for choosing a node as a relay indeed, and this situation causes control overhead due to excess of topology control messages, particularly in a very mobile context where links are broken frequently, so it is not efficient to select MPR whose links that provide a high QoS, but it is lost in few coming moments.

MPRs utilization decreases the number of retransmissions to disseminate a broadcast packet, and a relay selection method has been proposed in [5] to further reduce retransmissions by aggregating broadcast packets at MPR even if reducing the number of retransmissions may not be necessarily advantageous for reliable packet delivery due to instability of radio support. A reverse approach has been proposed in [6] based on choosing a redundant MPR coverage for relays in 2-hop neighbourhood of a given node in this manner; routing overhead is generated as consequence.

In cross-layer measurement, an early approach called signal stability routing in [7] consists of prioritizing the paths whose links have the strongest signal. In a similar way, authors present preemptive routing techniques in [8] to calculate approximately the time of link breakage so that failure can be announced in advance. Moreover, ETX (expected transmission count) metric [9] has been proposed as an MANET internet draft, and it is in the way to become a standard. Thereby in [10], a comparison work has been done among OLSR and RFC with default hop count metric and OLSR-ETX metric in a mesh network. Their results reveal the ETX metric to be fundamentally imperfect when assessing optimal routes in large dense mesh network scenario and worse than the OLSR-RFC standard. Despite of that in the various literature [11–13], a different approach is adopted based on seeking the well-suited parameter configurations of existing mobile ad hoc network protocols in order to get better performances. In the same way, others [14] propose a cross-layer design that jointly considers routing and topology control and take mobility and interference into account.

Nevertheless, many works discuss routing optimization based on online nodes measurements in order to categorize paths which are preferentially used for routing. But these works have a common weakness, where they cannot prevent possible change in links status occurring in the future. A link qualified as reliable based on past or current measurements may become unreliable with time because of dynamic nature of mobile environments.

In some works, authors draw attention to the stability of routes; indeed, in [15], the objective is to seek for stable paths between source and destination that also have lower hop count using the "predicted link expiration time (LET)" concept [16] employed for the Flow-oriented routing protocol (FORP) [17]. The FORP has been observed to look for stable routes that have even twice the highest lifetime and the minimum-hop routes discovered using the well-known dynamic source routing (DSR) protocol [18]. However, the trade off is that the hop count of FORP routes is significantly larger (twice or more) than the minimum hop count [19].

So far, many foregoing studies focus on statistical examination of link availability. The study made in [20] showed

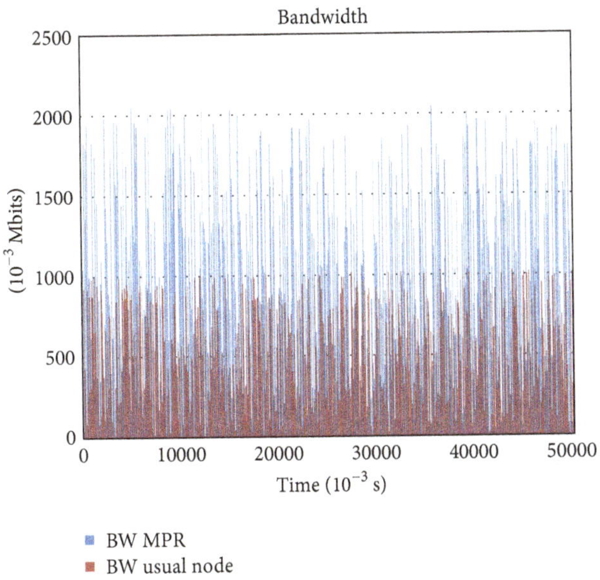

FIGURE 2: Bandwidth consumption.

that the link lifetime has a multimodel distribution when the node's speed is slow, and the path availability duration can be approximated by an exponential distribution at moderate and high velocities. On the other hand, the solutions provided by these studies are valid only for some specific situations; they could not be completely extended to universal ad hoc networks and practical MANETs applications. In [21, 22], a prediction method has been proposed and investigated with random walk mobility model based on link availability estimation. The algorithm aims to predict the probability of a link available with a continuously manner for a certain period, which is obtained based on the current node's movement. However, this algorithm is inaccurate when calculating the link availability; it makes the tendency of a given link availability known. Additionally, another approach [23] tries to predict the link availability during a time span based on a rough estimation of the distance between two nodes.

Routes rediscovering causes significant data loss, communication overheads especially with reactive routing behaviour, and jitter, so, routing protocol for mobile ad hoc networks must match the mobility character. In this way, our reflexion aims to adopt a composite metric taking into account the online measurement such as bandwidth/delay and links lifetime in the selection of relays.

Different mobility models can be used to evaluate MANET routing protocols performance. They can be classified into two categories: entity and group mobility models. Detailed reviews of these models can be found in [24–26]. In this paper, we will use random waypoint (RWP) [27].

3. Background

Metrics must accurately capture the triggered link characteristic, based on direct measurement or even on prediction computing. An obvious approach is to express the resource availability at a given node by a unique measure (total cost) and then use it as a single metric. For example, we can express the metrics of the bandwidth and the delay of a path by a function proportional to the bandwidth and inversely proportional the delay. However, this approach does not take into account the possible change in future time.

3.1. Cross-Layer Measurement

3.1.1. Bandwidth. Bandwidth metrics are popular, especially for QoS applications. Indeed, it indicates the capacity of data which can be sent through a link within a time spanner. From the perspective of a node, this is equal to the transfer rate of a link. Many factors other than theoretical physical bandwidth have a significant effect on this metric such as packet loss ratio. It Even refers to the measurement of bandwidth that is reduced to the determination of the number of free communication slots.

Firstly, we highlight the fact that nodes selected as MPRs introduce more traffic (context of OLSR). Indeed they generate TC messages on which links with the MPR selector set are published by relay. Beside that, Figure 2 presents bandwidth consumption by MPR, within simulation environment according to Table 2.

Nonetheless, bandwidth graph reveal; the debit consumed afterwards simulation but to make decision the mechanism of the latter must instantaneously give the available bandwidth.

Definition 1 (capacity of link). For a pair of nodes within transmission range of each other, we define the capacity C of the link between them as the physical transmission bit rate of the source node.

Definition 2 (bandwidth of link). In absence of competing stations, the time to get and release the medium in a one-hop transmission is a random variable T. The time required to transmit an L-bit long packet at a link transmission rate of C bps will be $T + L/C$, which means that if the link is completely available as follows the link bandwidth is

$$\mathrm{BW} = \frac{L}{T + L/C}. \tag{1}$$

In fact, L/C quantity represents the physical transmission time and T represents, fading, internal, external noise, and contention; let us aggregate the formula (1) and express it in 802.11 way, considering the packet size and the time spanner between sending and acknowledgement,

$$\mathrm{BW} = \frac{L}{t_q + (t_S + t_{\mathrm{CA}} + t_{\mathrm{Over}}) \times R + \sum_{r=1}^{R} B_{\mathrm{off}}^{r}}, \tag{2}$$

where t_q is the Mac layer queuing time, t_S is the transmission time of the L bits, t_{CA} is the collision avoidance phase time, t_{Over} is the control overhead time such as ACK and RTS/CTS, R is the necessary retransmissions, and B_{off}^{r} is the back-off time for a retransmission r.

This formula reveals some undesirable characteristics that are however common to measurements such as packet size dependence. Thereby, in our calculation, the BW is averaged

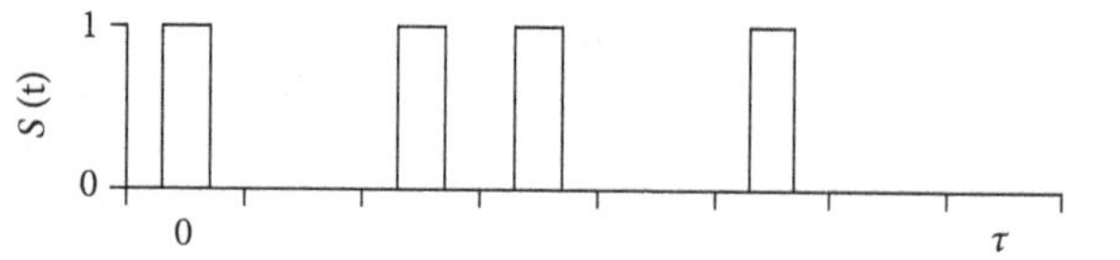

FIGURE 3: Link utilisation during a time span estimation.

TABLE 1: Cost comparison of QoS-based MPR schemes.

Schemes	I.R	S.D	T.C	M.C
RQMPR	2 hops	Yes	$O(2\|N_1\|\|N_2\|)$	$O(\|N_1\|+\|N_2\|)$
QMPR-1	2 hops	Yes	$O(3\|N_1\|\sigma+\|N_2\|)$	$O(\|N_1\|+\|N_2\|)$
QMPR-2	2 hops	Yes	$O(3\|N_1\|\sigma+\|N_2\|)$	$O(\|N_1\|\sigma+\|N_2\|)$
QMPR-3	2 hops	Yes	$O(2\|N_1\|\|N_2\|)$	$O(\|N_1\|\sigma+\|N_2\|)$

$|N_1|$: the maximum number of 1-hop neighbours of a node.
$|N_2|$: the maximum number of 2-hop neighbours of a node.
σ: the maximum MPR number of a node.

TABLE 2: Simulation parameters.

Simulation environment	Option and parameter
Flat size	1000 m × 1000 m
Max number of nodes	70 nodes
Radio scoop	100 m
MAC layer	IEEE.802.11.peer to peer mode
Transport layer	User datagram protocol (UDP)
Traffic model used	CBR
Package size	512 bytes
Rate	0.4
The number of connections	1/5 of the number of nodes
Mobility model	Random waypoint (RWP)
Pause time	0 second
Maximum speed of nodes	5, 10, 15, 20, and 30 m/s
Simulation time	200 sec

according to past values, which mitigates the foregoing dependences.

Definition 3 (available bandwidth (ABW)). It is defined as the unused bandwidth over the time interval τ. Here τ is referred to as the estimation period (namely, the time needed for estimating ABW once). It is not a constant value and can be changed in different estimation tools, or even in a tool according to the network scenario.

The state of a given link i is at time t is as follows:

$$S_i(t) = \begin{cases} 0, & \text{link is idle,} \\ 1, & \text{link is busy.} \end{cases} \tag{3}$$

As it is shown in Figure 3, then the average utilisation of a link is as follows:

$$\overline{u}_i = \frac{1}{\tau}\int_{t-\tau}^{t} S_i(t)\,dt. \tag{4}$$

The available bandwidth in a time interval of $[t-\tau, t]$ is as follows:

$$\text{ABW}_i = (1-\overline{u}_i)\,\text{BW}_i. \tag{5}$$

In the rest of paper, the available bandwidth is referred to as bandwidth.

3.1.2. Delay. The average end-to-end delay is the average time taken between packet sending and successful message receiving from the source to the destination. Similar to the QOLSR [4], each node includes in the *Hello* message the creation time of this message. When a neighbour node receives this message, it calculates the difference between such time and the current time; this is done in a synchronized network. Due to the characteristics of sparse ad hoc networks, classical clock synchronization algorithms are not applicable. For the moment, the aim of the current work is not to answer synchronization issues. Time synchronization in ad hoc networks is a wide subject of research, such as the work presented in [28]. This metric is important in delay sensitive applications such as video and voice transmission:

$$\text{Average End to End Delay} = \frac{\sum \text{received\_Time} - \text{sent\_Time}}{\text{Total\_Data\_packets\_received}}. \tag{6}$$

3.2. Remaining Time to Quite (RTTQ). In a previous work, [29] we have introduced a new metric called RTTQ remaining time to quite which estimates the remaining time of each node to be unreachable by the node executing the MPR selection process, based on the distance and the radio scoop for a given speed. We assume that between each pair of nodes, there is only one link, and for a given node, the remaining lifetime of the peer node is equivalent to the link lifetime between them. Here we talk about prediction because of the prior idea that we get, by knowing when the link status is susceptible to being invalid.

The Remaining time to quite (RTTQ) (Figure 4) of each neighbour node referred to as V_{peer} (peer node) is susceptible to leaving the neighbourhood of a reference node V_{ref}; running MPRs selection procedure is estimated based on travelled distances at two consecutive messages receptions (positions P_t and $P_{t+\tau}$). These positions are computed based on nodes abscises and axis coordinates and the elapsed time during this travel $[\![t, t+dt]\!]$ and radio scoop (RANGE), where the instants t and $t+dt$ are the time duration to reach the positions P_t and $P_{t+\tau}$, respectively. Indeed, the sign of $\Delta_{t,t+\tau}$ gives an idea about the direction of each neighbouring node relatively. So the positive value ($\Delta_{t,t+\tau} \geq 0$) indicates that distance between the pair of nodes is getting larger and a negative value ($\Delta_{t,t+\tau} < 0$) indicates they are closer. Indeed, in this case, link failure is unlikely; thus, RTTQ is set to its maximum value. However, in the opposite case, we predict the moment of the connection loss. Then, in time interval $[\![t, t+dt]\!]$, we calculate the travelled distance relative speed ϑ (7) the node V_{peer} (assuming the V_{reef} node as a reference) to reach the radio scoop edge. The estimated RTTQ is given by (9) as follows:

$$\Delta_{t,t+\tau} = D_{t+\tau} - D_t, \quad (7)$$

$$\vartheta = \frac{\Delta_{t,t+\tau}}{\tau}, \quad (8)$$

$$\text{RTTQ}(t + dt) = \frac{\text{RANGE} - D_{t+\tau}}{\vartheta}. \quad (9)$$

3.3. Network Model. We assume that a network has been a direct graph $G(V, E)$, where V is the set nodes and E the set of links $l = (x, y)$ where the node y is within the transmission range of x. In general cases, routing metrics have an attributed values which we call weighs associated to each edge. Consider

dist(x, y): number of hops between node x and node y

$N_1(x) = \{y \mid \text{dist}(x, y) = 1\}$, set of 1-hop adjacent nodes

$N_2(x) = \{y \mid \text{dist}(x, y) = 2\}$, set of 2-hop adjacent nodes.

Definition. Consider the following.

$B_2(x)$ is bandwidth of 2-hop farther path; it is a concave metric equal to the minimum of link bandwidth composing the path to a given node in $N_2(x)$.

$\text{Del}_2(x)$ is delay of 2-hop farther path; it is an additive metric equal to the delay sum of intermediate links.

$\text{RTTQ}_c(x)$ is the remaining time to quite for a node in $N_1(x)$, by which node x reaches a given node in $N_2(x)$.

$F_i(x)$ is the function cost of each available i path between source x and y node in $N_2(x)$, in which we aggregate three objectives by maximizing bandwidth and RTTQ and minimizing the delay, where

$$F_i(x) = B_2(x) + \text{RTTQ}_c(x) - \text{Del}_2(x). \quad (10)$$

For accuracy care, the cost function is achieved by a normalized weighted additive utility function. Indeed, each metric is as follows:

$$B_2(x) = \frac{b_2(x)}{\max_{x\in V} b_2(x)}, \quad (11)$$

$$\text{del}_2(x) = \frac{\text{del}_2(x)}{\max_{x\in V} \text{del}_2(x)}, \quad (12)$$

$$\text{RTTQ}_c(x) = \frac{\text{RTTQ}_c(x)}{\max_{x\in V} \text{RTTQ}_c(x)}, \quad (13)$$

where $b_2(x)$ is measured bandwidth, $\text{del}_2(x)$ is measured delay, and $\text{RTTQ}_c(x)$ is measured RTTQ.

4. Proposal Heuristic

In general, all the MPR broadcast schemes based on cross-layer information aim to revise the original MPR selection heuristic to achieve QoS awareness. However, it also has more relays with other MPR schemes, therefore growing retransmissions in the networks.

In [30] two schemes based on QoS measurement are proposed. First heuristic referred to as QMPR-1 still has the same initialization steps as the original MPR heuristic, except it changes procedure in order to provide QoS priorities. Instead of higher degree, a node with maximum bandwidth is chosen in case of multiple choices. If equal solution still exists, a node with minimum delay is selected. Then, likely MPRs with large bandwidth are selected, but the improvement is insignificant. Second heuristic, referred to as QMPR-2, is similar to the first one but selects nodes with higher bandwidth as MPRs, and the delay is used when there is a tie. And in case of multiple node with maximum delay reachability is choises. This heuristic highlights QoS criteria in the MPR selection; thus, MPRs are chosen based on QoS conditions, so the optimal links are published between a given pair of source and destination. Another heuristic referred to as QMPR-3 has been proposed in [31] and surveyed by [32], based on the idea that lets all 2-hop neighbours have an optimal bandwidth path through MPRs to the source node. For each 2-hop neighbours y, *a* node x chooses from its 1-hop neighbour node as the MPR if it covers y, and the path is the largest from y to x. Each 2-hop node has to go through this calculation until it finds an optimal path to the source node.

A survey [32] has established a comparative QoS-based MPR heuristics such as QMPR-1, QMPR-2, and QMPR-3 based on the following.

Time Complexity (TC). The maximum number of steps required in the worst case of a heuristic.

Message Complexity (MC). The maximum number of messages used in the worst case for a heuristic to obtain necessary information.

Information Range (IR). The number of hops of neighbor's information is required.

Source Dependent (SD). Before forwarding, node needs to know whether or not messages it received are from its MPR selectors.

Beside that in Table 1, we extendc this comparison work by including our proposal MPR schemes referred to as RQMPR based on a cost function. It still follows the same steps as the original MPR heuristic but modifies priorities action done in step (4) (Algorithm 1) where, for each 2-hop node, the cost function $F_i(x)$ of all of its available paths to the source node and reachability of candidate relay are calculated. This step takes $O(|N_1|)$ time to be complete in the worst case when a 2-hop node is reachable by all one-hop nodes. Then, for each two-hop node, add to the MPR set a node that can provide the maximum cost function. This step takes $O(|N_1|)$ time to run. Since these two steps have to be operated for all two-hop neighbours, the total time complexity of the heuristic can be $O(2|N_2||N_1|)$. Our proposal is source dependent because it need information to be included in broadcast messages in order to decide whether or not a node will relay the message. Furthermore, it requires the information range of 2 hops; thereby, the message complexity is $O(|N_2| + |N_1|)$.

Initial: A node k, $N_1(k)$, $N_2(k)$.
Return: MPR_k; MPR set of k
begin
(1) Add to MPR_k the node in $N_1(k)$ which is the only one to reach a node in $N_2(k)$
(2) Remove the nodes from $N_2(k)$ which are covered by a node in MPR_k
(3) **While** ($N_2(k)$ not empty) do
  For each node in $N_1(k)$,
    (i) Calculate the number of nodes in $N_2(k)$ that it can reach, that is, reachability
    (ii) Calculate $F_i(k)$ to reach $N_2(k)$
(4) Add to MPR_k the nodes which has the highest $F_i(k)$, If multiple choices, add which provide the highest reachability. If multiple choices, select node with highest degree, that is, number of 1-hop neighbors.
(5) Remove the nodes from $N_2(k)$ which are covered by a node in MPR_k
end

Algorithm 1: RQMPR broadcast schema (RQMPR selection).

5. Simulation Environments

Simulations are done in NS2 [33] (network simulator) version 2.35 in which we have integrated a standard version of OLSR. (UM-OLSR-0.8.8 [34, 35]), which is developed by MANET Simulation and Implementation at the University of Murcia (MASIMUM).

Our simulation parameters are as follows: for all simulations, our network consists of a maximum numbers of mobile nodes (70) whose radio scoop is 100 m, moving in an area of $1000 \times 1000\,\text{m}^2$. Each node moves according to the random waypoint (RWP) mobility model [36] with pause time fixed to 0 second and maximum speed varying between 5 and 30 meter/second with step (5). The scenario that defines the nodes movement is regenerated at the beginning of each simulation. To generate traffic in the network, in each simulation, 1/5 of nodes are randomly selected to be a source of constant bit rate (CBR) traffic. And these selected nodes use user datagram protocol (UDP) connections to send Packets with 512 bytes of size in the order of one packet every 2.5 seconds. Table 2 summarizes all the parameters used during simulations.

Within the simulations set, we distinguish all the possible attributes of the network, such as node mobility and the node density according to the following.

Varying Number of Nodes. We vary the number of nodes in the network. Our objective is to investigate the impact of node density on the protocol's performance. We use the same simulation area as in our previous simulations and gradually increase the number of nodes in the network. A desirable property of a protocol is to have stable behavior regardless of the number of nodes in the network. However, due to wireless medium limitations, we do not place an inadequate number of nodes in the simulation area. A small number of nodes in a large simulation area will result in low connectivity due to the large distances between nodes. In contrast, a large number of nodes in a small simulation area will result in signal interference, as nodes are located very close to each other.

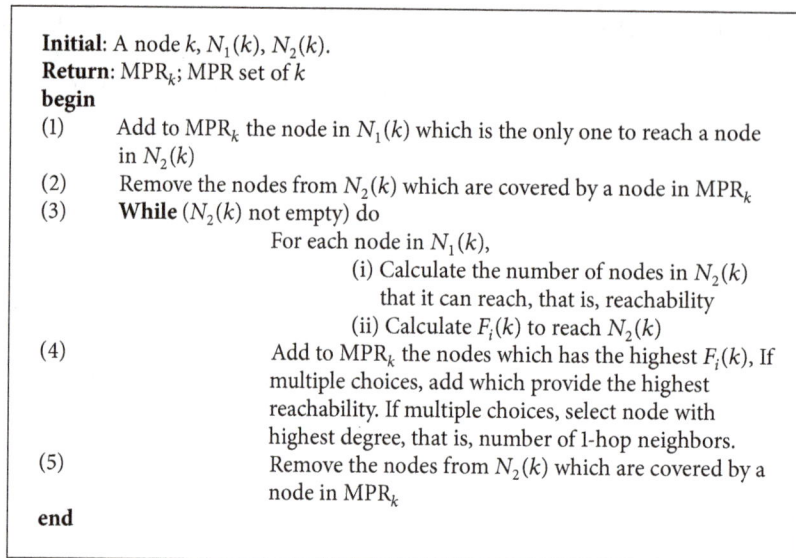

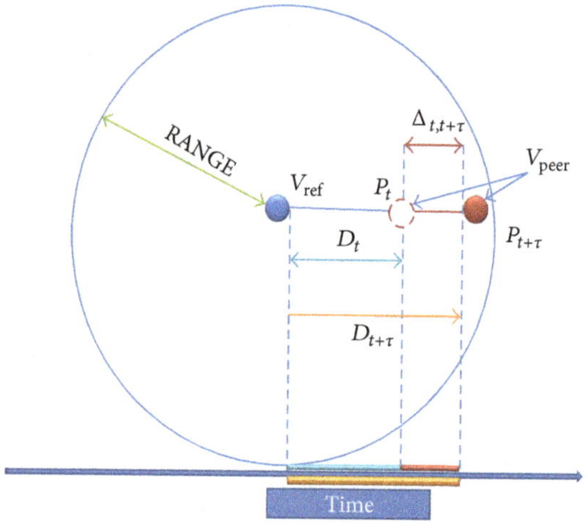

Figure 4: Comprehensive case of RTTQ computing.

Varying Network Mobility. We vary the mobility of nodes. We start with a mobility scenario in which the nodes have a low velocity of 5 m/s (18 km/h). We then increase the node velocity up to 30 m/s (108 km/h). Our intention is to investigate the behaviour of protocols in networks with varied mobility although the high mobility (108 km/h) cannot be easily found in military movements. We observed that, at higher data rates with increasing mobility, the performance of the protocols decreases. The decreased performance is large, mainly due to network congestion, in a way that makes any comparison meaningless.

All the metrics of performances should be based on same network attributes, such as mobility, network density, data density, bandwidth, energy resources, transmission and

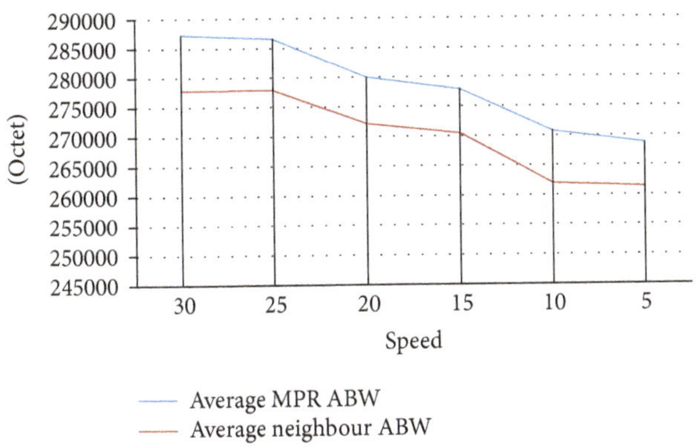

FIGURE 5: Average ABW for neighbours and MPRs.

receiving power, antenna types, and any other "component" that affects the performance of a routing protocol. In our performance evaluation, we follow the general ideas described in RFC 2501 [37] and use some quantitative metrics to evaluate our proposal RQMPR using as cost function $F_i(x)$, the standard one in OLSR (MPR-1), and QOLSR referred to as QMPR-2.

6. Evaluation of Performance

Previously, we have highlighted the need of bandwidth nodes selected as MPRs in OLSR where the link attributes are not considered in the selection of those relays. In this step, we check the effect of our proposal to deal with that problem in Figure 5.

Indeed, it reveals that the selected MPRs have more available bandwidths, and it scales well for different velocities.

6.1. Packet Delivery Ratio. The packet delivery ratio is defined as the fraction of all the received data packets at the destinations over the number of data packets sent by the sources (14). This is an important metric in networks. Indeed, packet loss at the intermediate nodes will result in retransmissions by the sources, which may result in network congestion,

$$\text{Packet Delivery Ratio} = \frac{\text{Total Data packets received}}{\text{Total Data packets sent}}. \tag{14}$$

6.2. Throughput. It indicates the capacity of data which can be sent over the network within a given time. From the perspective of a node, this is equal to the transfer rate of a link. Many factors other than theoretical physical bandwidth have a significant effect on this metric, for example, packet loss ratio.

Figure 6(a) presents the packet delivery ratio for both protocols MPR-1 and QMPR-2, and we see that it deteriorates while speed is getting higher. Despite that, QMPR-2 dominates MPR-1 by almost 15% of more successfully delivered packets.

Figure 6(b) presents the latter measure of performance for both protocols in different densities, where there is no meaningful difference until the number of nodes reaches 60 nodes, we see that QMPR-2 scales better than MPR-1.

So from both cases, we conclude that QMPR-2 offers higher success ratio of packets transmission, and it scales well for high velocities within dense networks. These results confirm our hypothesis and reveal the positive effect of our proposal such as selected MPR are those nodes which maximize the utility function. Where each node publishes inside the TC messages all links with its MPRs which could constitute the paths from source to destination, then the set of MPRs forms a kind of backbone, in the mobile ad hoc network.

In Figure 7 we see that the average throughput per nodes behaves in the same way as the ratio of delivered packet. This comes to confirm the opposite dominance of our proposal that is explained by its ability of finding robust paths between destinations and sources in context of high mobility and dense network.

Figure 8 presents the average delay measured in packets transmission where there is a clear dominance of MPR-1. From the side of RQMPR, this bad performance is due to the increased number of MPR because of the features of the method used in the selection of relays. But this observation is valid until the number of nodes reaches 45 node; indeed, starting from this value, the delay of RQMPR performs better than MPR-1.

7. Conclusion

Generally, in proactive broadcasting scheme, the use of relays aims to reduce the broadcast messages senders and then the number of flooded messages. MPRs have been the core of broadcasting scheme and it is required to selected relays in reliable manner; indeed, our proposal is based on a multicriteria selection method of the latter node. Indeed, it is in a form of aggregated multicriteria problem, taking into account bandwidth, delay, and RTTQ. The innovative idea of the current work is to be aware of a set of criteria applied not just to the 1-hop neighbourhood but also to the 2-hop neighbourhood of a given node.

So to each of the criteria aforementioned, we have associated a metric and a technique of measurement, aiming to establish tradeoffs between each objective of those criteria. By summarizing the outcomes of our proposal, we conclude that it performs well in high mobility environment and scales better in dense network. Beside the performance that our approach realizes, namely, for PDR and throughput, some improvements still have been possible by using statistical techniques that allows correction of the new recorded values based on previous ones. Also we plan to create more composite metric using several techniques of measurements that have a direct impact on network performance such as residual energy of nodes. Further optimization can be done by monitoring the current network characteristics, based on some metrics, mapping these metrics to related routing parameters, and adjusting parameters if necessary.

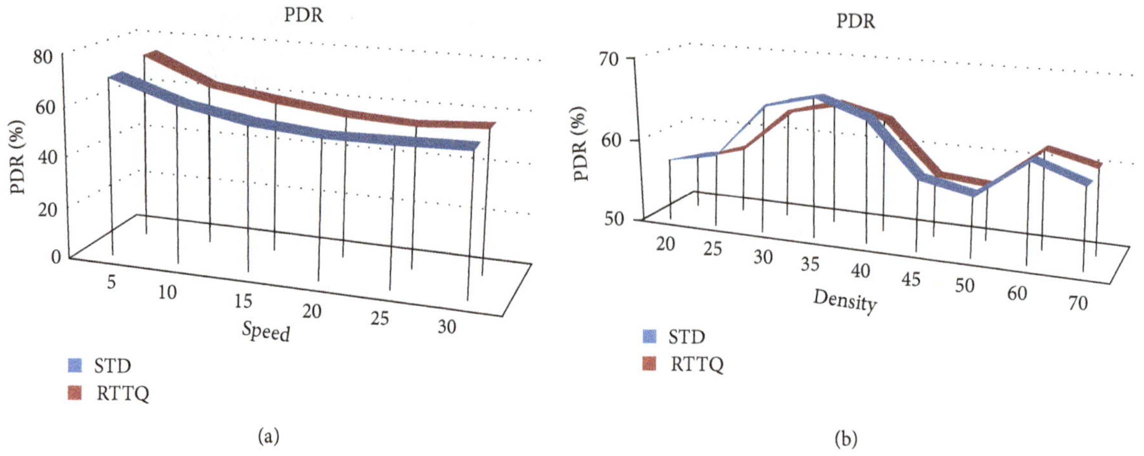

FIGURE 6: (a) PDR varying velocity. (b) PDR varying density.

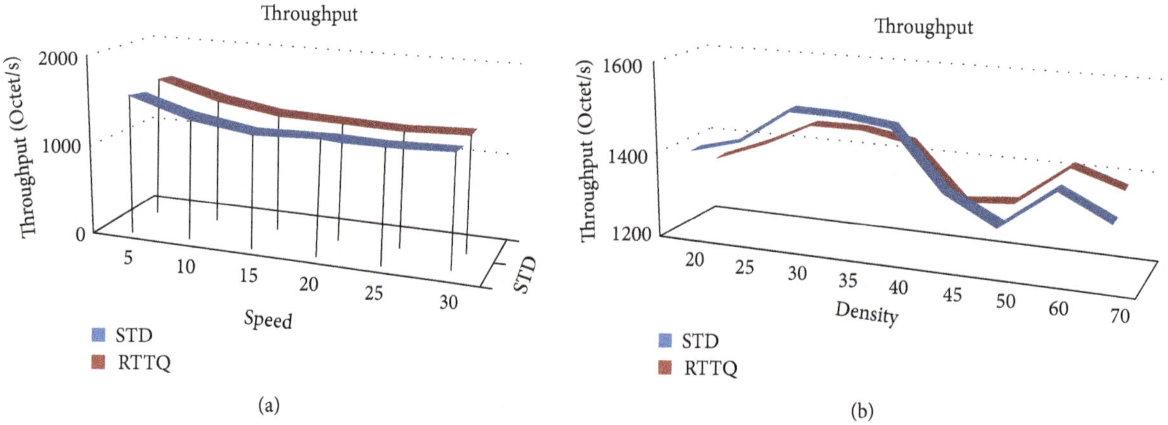

FIGURE 7: (a) Throughput varying velocities. (b) Throughput varying number of nodes.

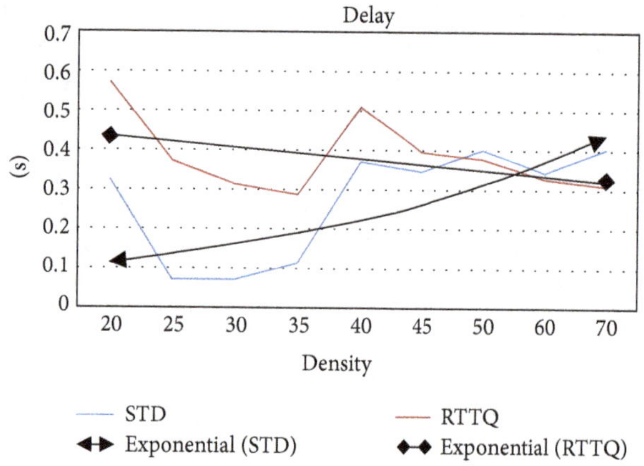

FIGURE 8: Delay measurement.

This adaptive process itself may also be dynamic, which means the node may receive feedback from the network and realise an interactive optimization.

References

[1] D. Nguyen and P. Minet, "Analysis of MPR selection in the OLSR protocol," in *Proceedings of the 21st International Conference on Advanced Information Networking and Applications Workshops/Symposia (AINAW '07)*, pp. 887–892, May 2007.

[2] J. Leguay, V. Conan, and T. Friedman, "QoS routing in OLSR with several classes of service," in *Proceedings of the 4th Annual IEEE International Conference on Pervasive Computing and Communications Workshops (PerCom Workshops '06)*, pp. 420–425, March 2006.

[3] A. M. Poussard, W. Hamidouche, R. Vauzelle, Y. Pousset, and B. Parrein, "Realistic SISO and MIMO physical layer implemented in two routing protocols for vehicular ad hoc network," in *Proceedings of the 9th International Conference on Intelligent Transport Systems Telecommunications (ITST '09)*, pp. 393–397, October 2009.

[4] H. Badis and K. A. Agha, "QOLSR, QoS routing for ad hoc wireless networks using OLSR," *European Transactions on Telecommunications*, vol. 16, no. 5, pp. 427–442, 2005.

[5] K. Yamada, T. Itokawa, T. Kitasuka, and M. Aritsugi, "Cooperative MPR selection to reduce topology control packets in OLSR," in *Proceedings of the IEEE Region 10 Conference, TENCON 2010*, pp. 293–298, November 2010.

[6] J. H. Ahn and T. J. Lee, "A multipoint relay selection method for reliable broadcast in ad hoc networks," in *Proceedings of the International Conference on ICT Convergence (ICTC '11)*, pp. 616–617, September 2011.

[7] R. Dube, C. D. Rais, K. Y. Wang, and S. K. Tripathi, "Signal stability-based adaptive routing (SSA) for ad hoc mobile networks," *IEEE Personal Communications*, vol. 4, no. 1, pp. 36–45, 1997.

[8] T. Goff, N. B. Abu-Ghazaleh, D. S. Phatak, and R. Kahvecioglu, "Preemptive routing in ad hoc networks," in *Proceedings of the 7th Annual International Conference on Mobile Computing and Networking*, pp. 43–52, Rome, Italy, July 2001.

[9] D. Djenouri and I. Balasingham, "LOCALMOR: LOCALized Multi-Objective Routing for wireless sensor networks," in *Proceedings of the IEEE 20th Personal, Indoor and Mobile Radio Communications Symposium (PIMRC '09)*, pp. 1188–1192, September 2009.

[10] D. Johnson and G. Hancke, "Comparison of two routing metrics in OLSR on a grid based mesh network," *Ad Hoc Networks*, vol. 7, no. 2, pp. 374–387, 2009.

[11] Y. Fu, X. Wang, and S. Li, "Performance comparison and analysis of routing strategies in mobile ad hoc network," in *Proceedings of the International Conference on Computer Science and Software Engineering (CSSE '08)*, pp. 505–510, December 2008.

[12] J. Toutouh, J. Garcia-Nieto, and E. Alba, "Intelligent OLSR routing protocol optimization for VANETs," *IEEE Transactions on Vehicular Technology*, vol. 61, pp. 1884–1894, 2012.

[13] C. Gomez, D. Garcia, and J. Paradells, "Improving performance of a real ad-hoc network by tuning OLSR parameters," in *Proceedings of the 10th IEEE Symposium on Computers and Communications (ISCC '05)*, pp. 16–21, June 2005.

[14] K. Ghada, J. Li, and Y. Ji, "Cross-layer design for topology control and routing in MANETs," *Wireless Communications and Mobile Computing*, vol. 12, no. 3, pp. 257–267, 2012.

[15] S. Marwaha, D. Srinivasan, C. K. Tham, and A. Vasilakos, "Evolutionary fuzzy multi-objective routing for wireless mobile ad hoc networks," in *Proceedings of the 2004 Congress on Evolutionary Computation (CEC '04)*, vol. 2, pp. 1964–1971, June 2004.

[16] W. Su, S. J. Lee, and M. Gerla, "Mobility prediction and routing in ad hoc wireless networks," *International Journal of Network Management*, vol. 11, pp. 3–30, 2001.

[17] W. Su and M. Gerla, "IPv6 flow handoff in ad hoc wireless networks using mobility prediction," in *Proceedings of the IEEE Global Telecommunication Conference (GLOBECOM '99)*, vol. 1, pp. 271–275, December 1999.

[18] D. B. Johnson, D. A. Maltz, and J. Broch, "DSR: the dynamic source routing protocol for multihop wireless ad hoc networks," in *Ad Hoc Networking*, pp. 139–172, Addison-Wesley Longman Publishing, 2001.

[19] N. Meghanathan, "Exploring the stability-energy consumption-delay-network lifetime tradeoff of mobile ad hoc network routing protocols," *Journal of Networks*, vol. 3, no. 2, pp. 17–28, 2008.

[20] X. M. Zhang, F. F. Zou, E. B. Wang, and D. K. Sung, "Exploring the dynamic nature of mobile nodes for predicting route lifetime in mobile Ad hoc networks," *IEEE Transactions on Vehicular Technology*, vol. 59, no. 3, pp. 1567–1572, 2010.

[21] S. Jiang, "An enhanced prediction-based link availability estimation for MANETs," *IEEE Transactions on Communications*, vol. 52, no. 2, pp. 183–186, 2004.

[22] S. Jiang, D. He, and J. Rao, "A prediction-based link availability estimation for routing metrics in MANETs," *IEEE/ACM Transactions on Networking*, vol. 13, no. 6, pp. 1302–1312, 2005.

[23] Q. Han, Y. Bai, L. Gong, and W. Wu, "Link availability prediction-based reliable routing for mobile ad hoc networks," *IET Communications*, vol. 5, no. 16, pp. 2291–2300, 2011.

[24] T. Camp, J. Boleng, and V. Davies, "A survey of mobility models for ad hoc network research," *Wireless Communications and Mobile Computing*, vol. 2, no. 5, pp. 483–502, 2002.

[25] F. Bai, N. Sadagopan, and A. Helmy, "The IMPORTANT framework for analyzing the impact of mobility on performance of RouTing protocols for Adhoc NeTworks," *Ad Hoc Networks*, vol. 1, no. 4, pp. 383–403, 2003.

[26] N. Sadagopan, F. Bai, B. Krishnamachari, and A. Helmy, "PATHS: analysis of PATH duration statistics and their impact on reactive MANET routing protocols," in *Proceedings of the 4h ACM International Symposium on Mobile Ad Hoc Networking and Computing*, pp. 245–256, Annapolis, Md, USA, June 2003.

[27] J. Broch, D. A. Maltz, D. B. Johnson, Y. C. Hu, and J. Jetcheva, "A performance comparison of multi-hop wireless ad hoc network routing protocols," in *Proceedings of the 4th Annual ACM/IEEE International Conference on Mobile Computing and Networking*, Dallas, Tex, USA, 1998.

[28] S. Mo, J. Hsu, J. Gu, M. Luo, and R. Ghanadan, "Network synchronization for distributed MANET," in *Proceedings of the IEEE Military Communications Conference (MILCOM '08)*, pp. 1–7, November 2008.

[29] A. Ouacha, N. Lakki, B. Bouamoud, A. Habbani, J. E. ABBADI, and M. Elkoutbi, "Reliable MPR selection based on link lifetime-prediction method," in *Proceedings of the 10th IEEE International Conference on Networking, Sensing and Control (ICNSC '13)*, pp. 11–16, Paris-Evry University, Evry, France, April 2013.

[30] H. Badis, A. Munaretto, K. Al Agha, and G. Pujolle, "Optimal path selection in a link state QoS routing protocol," in *Proceedings of the IEEE 59th Vehicular Technology Conference (VTC '04)*, vol. 5, pp. 2570–2574, May 2004.

[31] Y. Ge, T. Kunz, and L. Lamont, "Quality of service routing in Ad-Hoc networks using OLSR," in *Proceedings of the 36th Annual Hawaii International Conference on System Sciences (HICSS '03)*, vol. 9, 2003, Track 9.

[32] O. Liang, Y. A. Sekercioglu, and N. Mani, "A survey of multipoint relay based broadcast schemes in wireless ad hoc networks," *Communications Surveys & Tutorials*, vol. 8, pp. 30–46, 2006.

[33] H. Alwan and A. Agarwal, "Multi-objective reliable multipath routing for wireless sensor networks," in *Proceedings of the IEEE Globecom Workshops (GC '10)*, pp. 1227–1231, December 2010.

[34] R. L. Gomes, W. A. Moreira, J. J. H. Ferreira, and A. J. G. Abele, "LatinCon14—providing QoE and QoS in wireless mesh networks through dynamic choice of routing metrics," *IEEE Latin America Transactions*, vol. 8, no. 4, pp. 454–462, 2010.

[35] G. Koloniari and E. Pitoura, "A game-theoretic approach to the formation of clustered overlay networks," *IEEE Transactions on Parallel and Distributed Systems*, vol. 23, no. 4, pp. 589–597, 2012.

[36] D. Mahjoub and H. El-Rewini, "Adaptive constraint-based multi-objective routing for wireless sensor networks," in *Proceedings of the IEEE International Conference on Pervasive Services (ICPS '07)*, pp. 72–75, July 2007.

[37] S. Corson and J. Macker, *Routing Protocol Performance Issues and Evaluation Considerations, RFC 2501*, Mobile ad hoc Networking (MANET), 1999.

A Model Matching STR Controller for High Performance Aircraft

Adel A. Ghandakly[1] and Jason A. Reed[2]

[1] *Department of ECE, California State University, Chico, Chico, CA 95929-0888, USA*
[2] *Principal Engineer AREVA NP Inc., 1345 Ridgeland Pkwy, Alpharetta, GA 30004, USA*

Correspondence should be addressed to Adel A. Ghandakly; aghandakly@csuchico.edu

Academic Editors: M. V. Predoi and Z. Qin

This paper presents a development, as well as an investigation of a Model Matching Controller (MMC) design based on the Self-Tuning Regulator (STR) framework for high performance aircraft with direct application to an F-16 aircraft flight control system. In combination with the Recursive Least Squares (RLS) identification, the MMC is developed and investigated for effectiveness on a detailed model of the aircraft. The popular robust Quantitative Feedback Theory (QFT) controller is also outlined and used to represent a baseline controller, for performance comparison during four simulated test flight maneuvers. In each of the four maneuvers, the proposed MMC provided consistently stable and satisfactory performance, including the challenging pull-up and pushover maneuvers. The baseline stationary controller has been found to become unstable in two of the four maneuvers tested. It also performs satisfactorily-to-arguably poorly in the remaining two as compared to the MMC. Simulation results presented in this investigation support a clear argument that the proposed MMC provides superior performance in the realm of automatic flight control.

1. Introduction

Challenges in automatic flight control are predominant over those in many systems due to the uncertainties that are involved in the aircraft itself, as well as its surroundings [1–5]. Nonlinearities are found in the dynamics of the plane and the actuators that control it. In addition, atmospheric conditions can always be given credit to the uncertainties in flight control. An aircraft's velocity, altitude, and orientation are all factors that decide how the plane will perform. Differences in these factors along with varying atmospheric conditions throughout the flight envelope can result in a less than optimum, or even unstable system. For the purpose of stability and control, the ability to cope with these different conditions cannot be compromised.

Current methods of flight control include dynamic inversion, gain scheduling, and QFT, among others [6–12]. These are stationary controllers, in which they incorporate a design that does not adapt to the many changes that an aircraft can encounter. Beyond the design phase, their behavior is fixed. For these reasons, the focus of most designs is robustness. This can prove successful, but maneuverability of the aircraft is usually sacrificed to some extent. The flight envelope may even be bounded by the restrictions of the controller itself. Additionally, the aircraft and its surroundings are modeled only in the design phase. The drawback to a design whose primary focus is robustness is that a more desirable response possibly could have been achieved for many situations. The benefit is a wider range of conditions by which the system can be controlled successfully. However, in the case of a high performance aircraft, both optimum maneuverability and stability are critical in most situations. Many studies have made an attempt to implement a linear controller with robustness that is sufficient enough to stabilize a flight control system [7, 8].

Nonlinear dynamic inversion is a commonly used method of controller design in the application of flight control and is quite different from QFT [8]. The basis of nonlinear dynamic inversion is to use feedback to linearize the system being controlled and to provide a desired response.

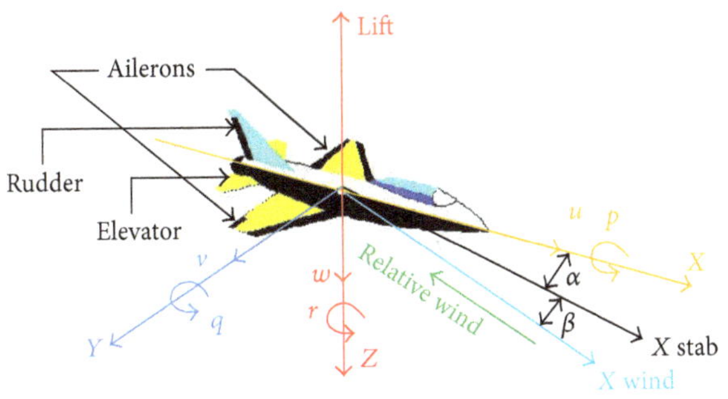

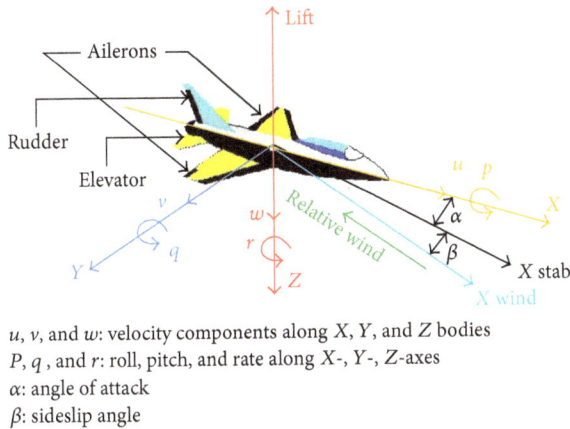

FIGURE 1: Aircraft three-dimensional representation with axes of control.

The method made several advancements, but its nature of linearizing the process meant that it was faced with the challenge of model uncertainties, which can prevent exact linearization. For this reason, dynamic inversion appears in many hybrid controllers that utilize some other types of control. These generally consist of Quantitative Feedback Theory, gain scheduling, and μ-synthesis.

Adaptive control methods, including Gain Scheduling, Self Tuning Regulators (STR), Model Reference Adaptive Controllers (MRAC) and Intelligent Adaptive Controllers have been recently proposed [1–3]. The complexity of the resulting algorithms and hence their control computational requirements have presented a limiting factor for their implementation. Because of the low computational requirements found in stationary designs, an increase in robustness has been their primary focus. Gain scheduling has also been implemented in hybrid designs [7]. This form of control has been a popular method and can be found in many finished designs in flight control today. While mildly related to adaptive methods, gain scheduling is another method that utilizes linear models in the design phase and is in many ways fixed. It does not directly face the nonlinearities that an aircraft has in its dynamics and encounters in its environment. Robustness is sometimes obtained by including high loop gains, which can become a problem in the presence of unmodeled dynamics, sensor noise, and unpredictable changes to the system or its environment. The use of an adaptive controller can be a significant improvement to automatic flight control. While preserving the robustness of other control methods, stability of the aircraft can be maintained while giving a faster and more accurate response by using a control scheme that features adaptability. Quantitative Feedback Theory has been proposed as a means of robust controller design. In addition to this, hybrid controllers have been discussed that include dynamic inversion, a nonlinear control method [6, 8]. QFT involves the selection of a feedback compensator and prefilter to ensure a response within acceptable limits. This is a stationary design whose structure inhibits the control from any adaptability once the design phase is complete. When applied to a nonlinear system, numerous models are formed to aid in the design process. These models are linear and time invariant in nature and do not fully represent the plant nonlinearities. For this reason, a response can only be predicted to the extent by which the designer tests the control and by the number of linear models that are used. QFT was originally devised for applications involving linear, time-invariant systems. However, efforts have been made to demonstrate the usefulness of this method of control for other systems that are nonlinear and time varying [6]. The objective of the controller design is to select two transfer functions, one to process the command and another to perform the closed loop performance requirements. The goal of both transfer functions is to change the elevator deflection angle such that the process output will track the desired angle of attack command. During the design, the plant is represented by a series of linear models. The linear models are used to determine the parameters for the two transfer functions. This is where a disadvantage of QFT control can be found for a flight control application. Just as in many other control methods, linearized models are used in the design phase, not in real time when the aircraft is in motion. In regards to an aircraft flight control system, the nonlinear parameters will vary throughout the flight envelope. It is conceivable that this system could be regarded as having an infinite number of parameter variations, thus creating many questions about the design of a controller of this type.

A real-time Model Matching Control (MMC) algorithm is proposed in this paper for implementation within the Self-Tuning Regulator (STR) structure of adaptive control framework. The proposed controller will be investigated on an F-16 aircraft simulated by a detailed dynamic model that has been developed and will be outlined in the paper. The popular QFT controller will be used as a baseline controller for the purpose of comparison and assessment of effectiveness of the proposed MMC.

2. Development of the Aircraft Model

A detailed model of the F-16 aircraft is given in the appendix. The aircraft model is based on the aircraft representation in Figure 1 and consists of a number of subsystems, each of which represent a significant part of the overall plant and its environment [1, 2]. First, the vehicle or aircraft itself has a manner in which it moves throughout the medium irrespective of any influential external action. These behaviors are defined by the flight dynamics. The flight dynamics have two components. They are governed by Newton's law, which identifies the translational dynamics experienced by the aircraft, and Euler's law, which relates the attitude dynamics or orientation of the aircraft.

There also exists a summation of forces that join with the equations of motion to deliver the resultant set of parameters that are meaningful to an analysis, such as speed, altitude, position, and orientation. All objects inherent to the model system other than the equations of motion can be linked to and have some relation to the forces. The primary components of these are propulsion, gravitation and aerodynamics.

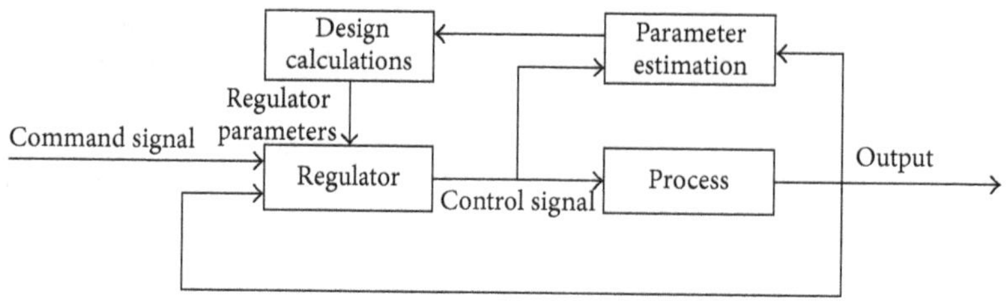

Figure 2: Self-Tuning Regulator framework.

The following nomenclature applies to the aircraft model for the states, variables, and parameters:

a_n = normal acceleration, gravity units
$\bar{c}$ = wing mean aerodynamic chord, m
b = wing span, m
C_{xt} = total aerodynamic x-axis force coefficient
ρ = air density, kg/m^3
C_{zt} = total aerodynamic z-axis force coefficient
C_{mt} = total aerodynamic pitching moment coefficient
F_X = aerodynamic force in x-direction, N
F_Z = aerodynamic force in z-direction, N
g = acceleration due to gravity, m/s^2
h = altitude, m
I_Y = moment of inertia about y-axis, kg·m^2
m = mass, kg
M = mach number
M_Y = aerodynamic pitching moment
q = pitch rate, rad/s
$\bar{q}$ = dynamic pressure, N/m^2
r = reference command, rad
S = wing area, m^2
T = thrust, N
u = velocity x-component, m/s
V = velocity magnitude, m/s
w = velocity z-component, m/s
x_{cg} = center of gravity location, fraction of $\bar{c}$
α = angle of attack, rad/s
δ_h = elevator deflection angle, rad
δ_{lef} = leading edge flap deflection angle, rad
η = elevator effectiveness factor
θ = pitch att angle, rad.

In this study, the angle of attack (AOA) of the aircraft will be controlled by manipulating the elevator deflection angle in the investigated flight maneuvers.

For that purpose, the following reduced state space model that is relevant to those specific dynamics of the longitudinal aircraft model is developed from the overall detailed model of the appendix:

$$\begin{aligned}
\dot{u} &= -qw - g_c \sin(\theta) + \frac{\bar{q}\,SC_{xt} + T}{m}, \\
\dot{w} &= qu + g_c \cos(\theta) + \frac{\bar{q}\,SC_{zt}}{m}, \\
\dot{q} &= \frac{\bar{q}\,S\bar{c}C_{mt}}{I_Y}, \\
\dot{\theta} &= q, \\
\dot{h} &= u \sin(\theta) - w \cos(\theta), \\
\dot{\alpha} &= q + \frac{mg \cos(\theta) + \bar{q}\,SC_{zt}}{mu}.
\end{aligned} \tag{1}$$

3. Development of the STR MMC

A real-time Model Matching Control algorithm is proposed in this paper for implementation within the Self-Tuning Regulator (STR) structure of adaptive control framework. The control structure is shown in Figure 2, where an estimation of the plant parameters is used to update the controller parameters online.

The system has two loops, the first is a unity feedback loop, and the second loop includes a parameter estimation that is provided to the control law design algorithm. For system identification, the following nth order process is used:

$$\frac{y(k)}{u(k)} = \frac{b_1 z^{-1} + b_2 z^{-2} + \cdots + b_n z^{-n}}{1 + a_1 z^{-1} + a_2 z^{-2} + \cdots + a_n z^{-n}}, \tag{2}$$

which is executed online with the difference equation:

$$\begin{aligned}
y(k) = {} & -a_1 y(k-1) - a_2 y(k-2) - \cdots - a_n y(k-n) \\
& + b_1 u(k-1) + b_2 u(k-2) + \cdots + b_n u(k-n).
\end{aligned} \tag{3}$$

The model parameters a's and b's are included in a parameter vector Θ of order $(2n * 1)$, and the following Recursive Least

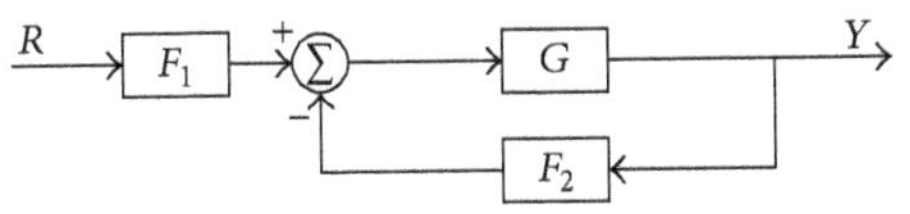

FIGURE 3: The Model Matching Controller structure.

Square (RLS) identification algorithm is used to identify the aircraft dynamics [13]:

$$\Theta(t) = \Theta(t-1) + K(t)\left[\phi(t) - x^t(t)\,\Theta(t-1)\right],$$
$$K(t) = \frac{P(t-1)\;x(t)}{\gamma + x^t(t)\;P(t-1)\;x(t)}, \tag{4}$$
$$P(t) = \frac{\left[I - K(t)\;x^t(t)\right]P(t-1)}{\gamma}.$$

A forgetting factor γ has been included in the algorithm for improved identification, which allows the estimation process to credit more recently obtained data.

3.1. The MMC Real-Time Algorithm. The proposed MMC structure is shown in Figure 3, where F_2 is the main feedback control transfer function and F_1 is updated online to satisfy the model matching condition. Consider G and F_2 as follows:

$$G(z) = \frac{B(z)}{A(z)},$$
$$F_2(z) = \frac{\beta(z)}{\alpha(z)}. \tag{5}$$

Plant G is of order n, and controller F_2 is of order $(n-1)$:

$$G(z) = \frac{B(z)}{A(z)} = \frac{b_0 z^n + b_1 z^{n-1} + \cdots + b_{n-1} z + b_n}{z^n + a_1 z^{n-1} + \cdots + a_{n-1} z + a_n},$$
$$F_2(z) = \frac{\beta(z)}{\alpha(z)} = \frac{\beta_0 z^{n-1} + \beta_1 z^{n-2} + \cdots + \beta_{n-2} z + \beta_{n-1}}{\alpha_0 z^{n-1} + \alpha_1 z^{n-2} + \cdots + \alpha_{n-2} z + \alpha_{n-1}}. \tag{6}$$

F_1 is chosen such that the overall response will match a specified design model:

$$\frac{Y}{R} = F_1(z)\cdot\frac{\alpha(z)\,B(z)}{\alpha(z)\,A(z) + \beta(z)\,B(z)} = \frac{B_m(z)}{A_m(z)}, \tag{7}$$

where the desirable model is given by

$$\frac{Y}{R} = \frac{B_m(z)}{A_m(z)}. \tag{8}$$

The closed form Diophantine equation solution (DES) [14] is then implemented online with a choice of the closed loop characteristic polynomial as

$$D(z) = \alpha(z)\,A(z) + \beta(z)\,B(z),$$
$$D(z) = H(z)\,F(z) = H_1(z)\,B(z)\,F(z), \tag{9}$$

and with the designer specifying H_1 and F, and B is the process nominator polynomial as determined by the identification algorithm. H is of order n, and F is of order $n-1$. If B is of order m, then the H_1 polynomial is of order $n-m$. If B is a stable polynomial, m can equal n and H_1 must be equal to 1, and $\alpha(z)$ and $\beta(z)$ are given by the DES as

$$M = E^{-1}D, \tag{10}$$

where M is the vector of the α and β polynomials. F_1 transfer is then given by

$$F_1(z) = \frac{H_1(z)\,F(z)}{\alpha(z)}\cdot\frac{B_m(z)}{A_m(z)}, \tag{11}$$

and the design requirement is satisfied as

$$\frac{Y}{R} = \frac{H_1(z)\,F(z)}{\alpha(z)}\cdot\frac{B_m(z)}{A_m(z)}\cdot\frac{\alpha(z)\,B(z)}{H_1(z)\,B(z)\,F(z)} = \frac{B_m(z)}{A_m(z)}. \tag{12}$$

The controller transfer function and its execution difference equation are given by:

$$u = -\frac{\beta(z)}{\alpha(z)}y + R\cdot\frac{H_1(z)\,F(z)}{\alpha(z)}\cdot\frac{B_m(z)}{A_m(z)},$$
$$\alpha A_m u = H_1 F B_m R - \beta A_m Y. \tag{13}$$

If the identification algorithm happens to yield an unstable pole within the process polynomial B, a modified controller algorithm is used, starting with the expression for the Diophantine equation:

$$D(z) = \alpha(z)\,A(z) + \beta(z)\,B(z). \tag{14}$$

In the case of the unstable pole, we state that:

$$B(z) = B^-(z)\,B^+(z),$$
$$B_m(z) = B^-(z)\,B_1(z), \tag{15}$$

where B^+ is stable and B_1 is the only polynomial specified in the design stage

$$D(z) = H(z)\,F(z) = H_1(z)\,B^+(z)\,F(z). \tag{16}$$

The Diophantine equation is solved for $\alpha(z)$ and $\beta(z)$ with the new definition for $D(z)$, and the front-end transfer function is stated in (17) to satisfy the design requirement as in (18)

$$F_1(z) = \frac{H_1(z)\,F(z)}{\alpha(z)}\cdot\frac{B_1(z)}{A_m(z)}, \tag{17}$$

$$\frac{Y}{R} = \frac{H_1(z)\,F(z)}{\alpha(z)}\cdot\frac{B_1(z)}{A_m(z)}\cdot\frac{\alpha(z)\,B^-(z)\,B^+(z)}{H_1(z)\,B^+(z)\,F(z)}$$
$$= \frac{B_1(z)\,B^-(z)}{A_m(z)} = \frac{B_m(z)}{A_m(z)}. \tag{18}$$

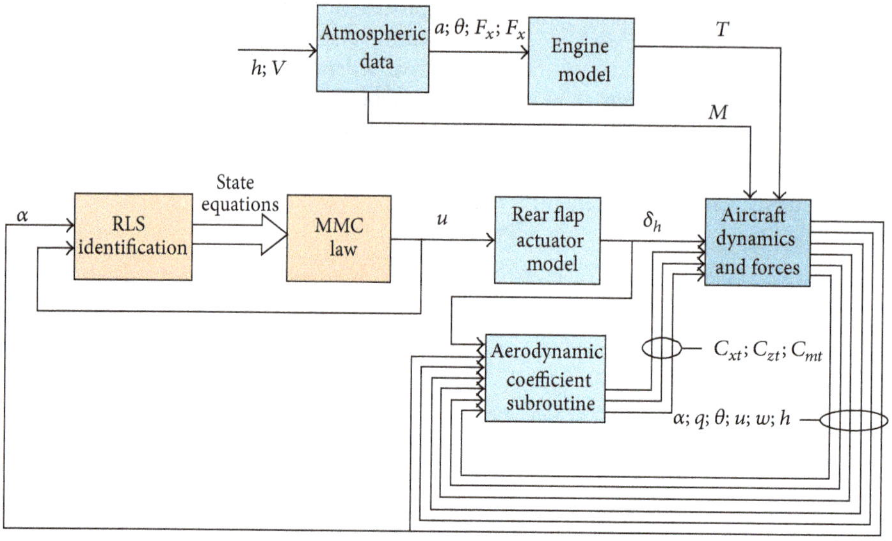

FIGURE 4: Complete aircraft system simulation layout.

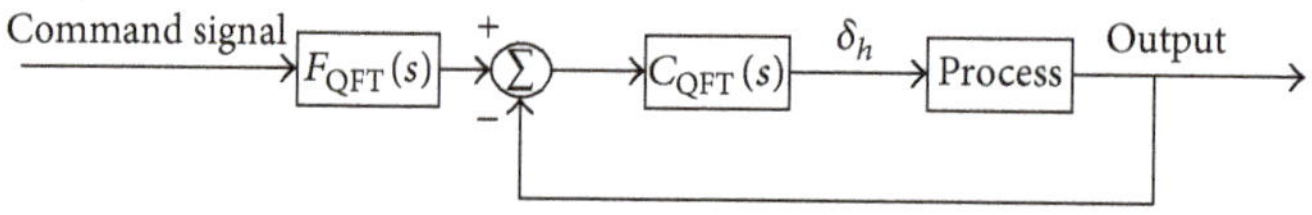

FIGURE 5: QFT controller configuration.

And the corresponding controller equations are then given by

$$u = -\frac{\beta(z)}{\alpha(z)}y + R \cdot \frac{H_1(z)F(z)}{\alpha(z)} \cdot \frac{B_1(z)}{A_m(z)},$$
$$\alpha A_m u = H_1 F B_1 R - \beta A_m Y. \quad (19)$$

The online model matching control algorithm incorporates both cases by checking the poles of $B(z)$ during each iteration and using the appropriate control method in each case.

4. Controller Assessment and Simulation Investigation Results

The investigation simulation system layout, including detailed representation of the aircraft dynamics is shown in Figure 4. The aircraft, earth atmosphere, and RLS data are as follows:

$$m = 9295, \quad I_Y = 75674, \quad S = 27.87,$$
$$\bar{c} = 3.45, \quad x_{cg} = 0.35, \quad g_c = 9.81, \quad R = 287$$
$$n = 2, \quad \gamma = 0.95, \quad f = 10000, \quad t_f = 0.02. \quad (20)$$

The aircraft is considered in a steady flight path prior to a command step in angle of attack. Initial conditions for each simulation are

$$u = 163, \quad w = 6.0, \quad q = 0.17,$$
$$\theta = 2.02, \quad \alpha = 4.0, \quad h = 9144. \quad (21)$$

The altitude and velocities were chosen based on other studies [13]. The orientation values were found by trial and error to provide a short initialization period for simulation purposes. These values exist prior to a 2-second initialization that stabilizes the aircraft to a +5 degree angle of attack from initial conditions. During the 1st second, the model for the elevator actuator is bypassed to allow the plane to move unrestricted by physical limitations. This is merely for the purpose of establishing a true set of states for the system and is necessary due to the simulation setting. In the next second, the real dynamics of the actuator are included, and the aircraft is steadied. Following this is time zero, or the beginning of the relevant simulation.

For comparison, a QFT controller was designed on the basis of a given F-16 aircraft model [6, 8]. The objective of the controller design is to select the two transfer functions in the control configuration shown in Figure 5. The first of these, $F_{QFT}(s)$, is an input transducer. The second, $C_{QFT}(s)$, is a front-end transfer function to the process whose output is the elevator deflection angle, δ_h, in the case of longitudinal flight control. The goal of both transfer functions is to change

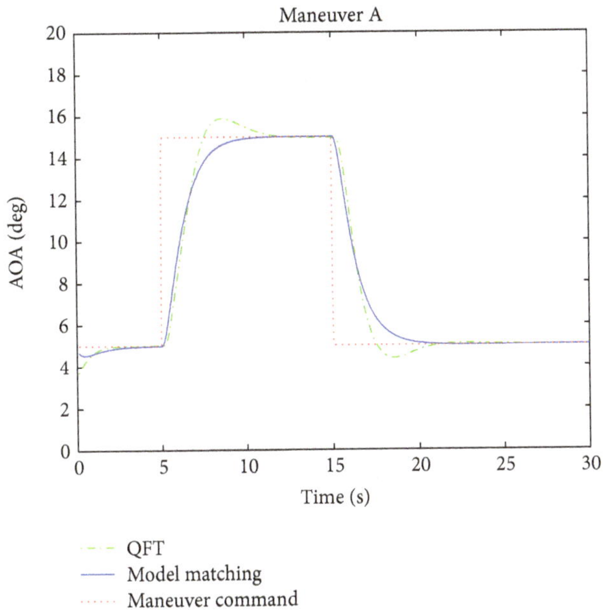

FIGURE 6: AOA response: maneuver A.

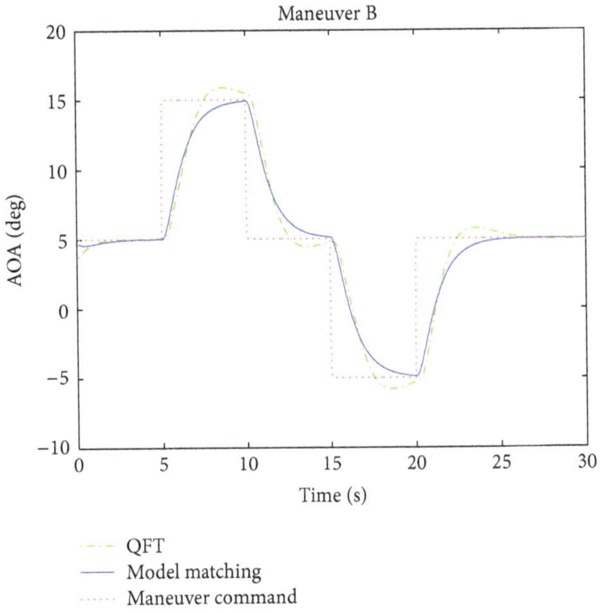

FIGURE 7: AOA Response: Maneuver B.

the elevator deflection angle such that the process output will track the command signal, the desired angle of attack. As shown in the figure, unity feedback is employed.

During the design, the plant is represented by a series of linear models. The linear models are used to determine the parameters for $F_{QFT}(s)$ and $C_{QFT}(s)$, and the following controller transfer functions have been obtained:

$$C_{QFT}(s) = \frac{-48,800\left(s^2 + 1.2(0.38)s + 0.38^2\right)(s + 2.1)}{s(s + 0.05)\left(s^2 + 1.4(20)s + 20^2\right)},$$
$$F_{QFT}(s) = \frac{(1.15)^2}{s^2 + 1.4(1.15)s + 1.15^2}. \quad (22)$$

The model matching control design was based on the following reference model and polynomials:

$$\frac{B_m(z)}{A_m(z)} = \frac{0.051z - 0.05}{z^2 - 1.93z + 0.931},$$
$$H_1(z) = z + 0.05, \quad (23)$$
$$F(z) = z.$$

The performance of the proposed MMC in comparison with the robust QFT design has been investigated under *four aircraft maneuvers* that have been synthesized to demonstrate the relative effectiveness of both techniques.

In the first *maneuver A*, the command reference is a +10 degree angle of attack step increase followed by a return to steady flight. This movement is consistent with a positive pitch angle and increase in altitude. And in *maneuver B*, the command reference involves a number of changes. First, a +10 degree step up in angle of attack is set, followed by a −10 degree return to level flight, then another −10 degree reduction to pitch the plane downward and followed by +10 degrees to level off. The *MMC* shows an outstanding response in maneuver A as shown in Figure 6. The response is just as fast as the baseline controller yet has no overshoot and promptly converges with the step command. Similar activity is found in the return from the step. Simulation of maneuver B shows the same features as shown in Figure 7, although there are periods where a more abrupt response by the baseline controller is noted. These are only found in a negative AOA command and are considered minor differences.

The third *maneuver C* investigated is the opposite of A. It is a −10 degree pushover followed by a command to level off and come out of a dive. And the fourth *maneuver D* is similar to C in its intent. It differs in that the initial pushover is 15 degrees, which is more aggressive, but held 5 seconds rather than 10. In these two maneuvers C and D, which denote instability using the QFT controller, are shown in Figures 8 and 9, and in both cases, the MMC was shown to recognize a difference between the response of the system at extreme pitch angles and that of the same system at level flight. The model matching adaptive controller performed well in every simulation and features a more consistent result in both a pull-up and pushover situations. In all simulations, it is found that the baseline controller overshoots the set point most of the time. This is not always a negative trait, but maneuvers C and D clearly demonstrate the manner in which this controller behaves differently when the aircraft is in a unique situation or orientation. The overshoot develops into instability in these cases. In contrast to the MMC response in every situation, the overshoot is unnecessary as the model determined online by the identification algorithm manipulates the control variable in a way that provides for a response that is both fast and dampened enough to eliminate overshoot.

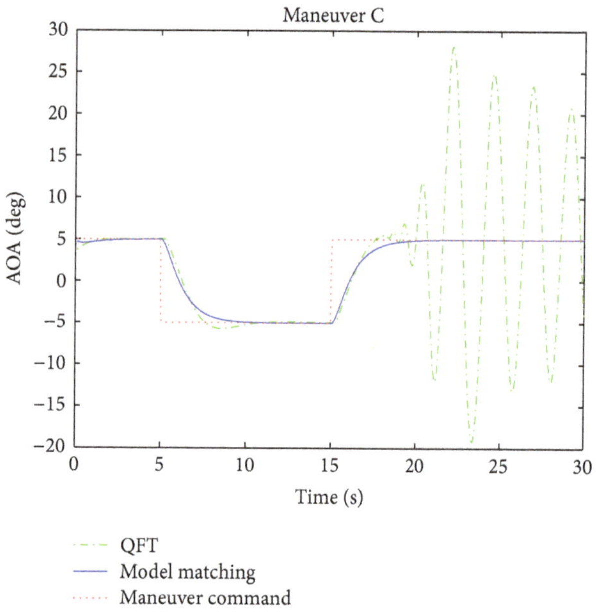

FIGURE 8: AOA response: maneuver C.

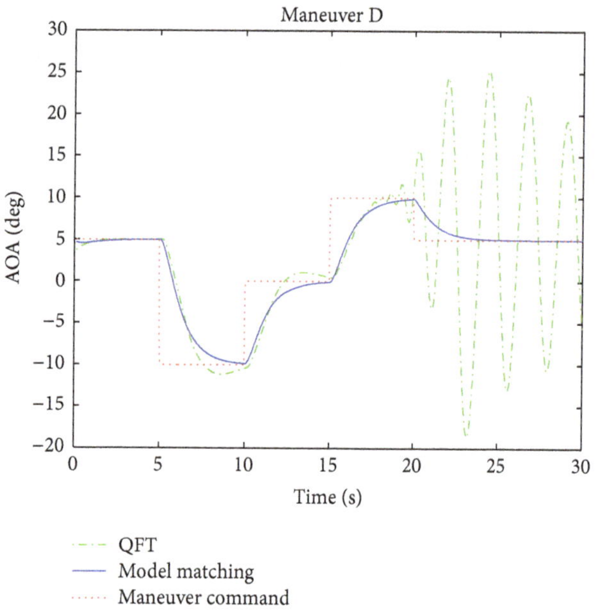

FIGURE 9: AOA response: maneuver D.

5. Conclusions

A development as well as investigation of a Model Matching Controller for a high performance (F-16) aircraft based on the STR framework has been presented. The proposed controller includes Recursive Least Squares identification algorithm and an MMC technique, both to be implemented in real time to produce adaptive systems that will track a reference signal optimally as the model undergoes various thirty second flight simulations. The MMC has been investigated and compared to a widely accepted and popular robust controller design, the QFT controller. This design was referred to as the baseline controller to assess the performance of the proposed controller. Simulations were performed for both controllers during four separate flight maneuvers. The baseline controller was found to be unstable in two of those maneuver demonstrations. Investigation results have shown that the proposed MMC can be a significant improvement to automatic flight control. While preserving the robustness of other control methods, stability of the aircraft can be maintained, giving a faster and more accurate response by using a control scheme that features adaptability. The ability to track a reference command well and remain stable throughout the flight envelope is necessary in design of a highly maneuverable aircraft. Challenges in automatic flight control are predominant over those in many systems due to the uncertainties that are involved in the aircraft itself, as well as its surroundings. Nonlinearities are found in the dynamics of the plane and the behavior of the actuators that control it. In addition, atmospheric conditions can always be given credit to the uncertainties in flight control. An aircraft's velocity, altitude, and orientation are all factors that decide how the plane will perform. Differences in these factors along with varying atmospheric conditions throughout the flight envelope can result in a less than optimum, or even unstable system. For the purpose of stability and control, the ability to cope with these different conditions cannot be compromised, and the proposed MMC has been shown to accomplish that.

Appendix

A. Six Degree of Freedom (DOF) F-16 Fighter Aircraft Dynamic Model

In developing a rigid body six DOF model of the F-16 aircraft, the aircraft flight mechanics can be aggregated from the object body (fuselage, wing, and stabilizers), engine, aerodynamics, gravity, atmosphere, and winds [1, 2, 13]. The dynamics of the aircraft are due to the force in the translational direction, which in effect creates the velocity. The velocity vector can be divided into three components based on three axes. Each has the effect of the moments on the other two axes, a gravitational effect developed based on the kinematic components and the force in each direction. Thus, the velocity vector components will constitute the three states of the dynamics. The kinematical equation shows the Euler angles of the motion. They are the effect of the moments and the attitude of the aircraft. In the x-axis, the angle is related to the moment "P," and the other two angles are due to the moments "Q" and "R." The z-axis component of the angle will have the effect of "Q" and "R" along with the angles in the x and y directions. The principles of flight dynamics involve the study of vehicle motions through air or space, where the aircraft experiences six degrees of freedom. These include three translational degrees which describe the movement of the center of mass, or the trajectory, and three attitude degrees that depict the orientation of the vehicle [2, 13].

A.1. Translational Dynamics.

$$\dot{u} = rv - qw - g_c \sin(\theta) + \frac{\bar{q}\, SC_{xt}}{m} + \frac{T}{m},$$
$$\dot{v} = pw - ru + g_c \cos(\theta)\sin(\varphi) + \frac{\bar{q}SC_{yt}}{m}, \qquad (A.1)$$
$$\dot{w} = qu - pv + g_c \cos(\theta)\cos(\varphi) + \frac{\bar{q}SC_{zt}}{m}.$$

A.2. Attitude Dynamics.

$$\dot{p} = \frac{(I_Y - I_Z)\, qr}{I_X} + \frac{I_{XZ}(\dot{r} + pq)}{I_X} + \frac{\bar{q}\, SbC_{lt}}{I_X},$$
$$\dot{q} = \frac{(I_Z - I_X)\, pr}{I_Y} + \frac{I_{XZ}(r^2 - p^2)}{I_Y} + \frac{\bar{q}\, S\bar{c}C_{mt}}{I_Y} - \frac{l_E r}{I_Y}, \qquad (A.2)$$
$$\dot{r} = \frac{(I_X - I_Y)\, pq}{I_Z} + \frac{I_{XZ}(\dot{p} - rq)}{I_Z} + \frac{\bar{q}\, SbC_{nt}}{I_Z} + \frac{l_E q}{I_Z}.$$

A.3. Kinematic Equations.

$$\dot{\varphi} = p + q\sin(\varphi)\tan(\theta) + r\cos(\varphi)\tan(\theta),$$
$$\dot{\theta} = q\cos(\varphi) - r\sin(\varphi), \qquad (A.3)$$
$$\dot{\psi} = q\frac{\sin(\varphi)}{\cos(\theta)} + r\frac{\cos(\varphi)}{\cos(\theta)}.$$

A.4. Navigation Equations.

$$\dot{p}_n = u\cos(\psi)\cos(\theta) + v(\cos(\psi)\sin(\theta)\sin(\varphi) - \sin(\psi)\cos(\varphi)) + w(\cos(\psi)\sin(\theta)\cos(\varphi) + \sin(\psi)\sin(\varphi)),$$
$$\dot{p}_e = u\sin(\psi)\cos(\theta) + v(\sin(\psi)\sin(\theta)\sin(\varphi) + \cos(\psi)\cos(\varphi)) + w(\sin(\psi)\sin(\theta)\cos(\varphi) - \cos(\psi)\sin(\varphi)), \qquad (A.4)$$
$$\dot{h} = u\sin(\theta) - v\cos(\theta)\sin(\varphi) - w\cos(\theta)\sin(\varphi).$$

The 6 degree of freedom aircraft dynamic model and navigation dynamics are given by the four sets of equations (A.1)–(A.4). In most cases, these 12 equations, which have very close relation to spacecraft and missile dynamics, joined by a robust depiction of geophysical and aerodynamic data, as well as thrust and control deflector subsystems, complete a thorough and accurate simulation in flight control.

References

[1] S. Kamalasadan and A. A. Ghandakly, "Multiple fuzzy reference model adaptive controller design for pitch-rate tracking," *IEEE Transactions on Instrumentation and Measurement*, vol. 56, no. 5, pp. 1797–1808, 2007.

[2] S. Kamalasadan and A. A. Ghandakly, "A neural network parallel adaptive controller for fighter aircraft pitch-rate tracking," *IEEE Transactions on Instrumentation and Measurement*, vol. 60, no. 1, pp. 258–267, 2011.

[3] S. E. Talole, A. Ghosh, and S. B. Phadke, "Proportional navigation guidance using predictive and time delay control," *Control Engineering Practice*, vol. 14, no. 12, pp. 1445–1453, 2006.

[4] M. Pachter, P. R. Chandler, and L. Smith, "Maneuvering flight control," *Journal of Guidance, Control, and Dynamics*, vol. 21, no. 3, pp. 368–374, 1998.

[5] R. Bhattacharya, G. J. Balas, M. A. Kaya, and A. Packard, "Nonlinear receding horizon control of an F-16 aircraft," *Journal of Guidance, Control, and Dynamics*, vol. 25, no. 5, pp. 924–931, 2002.

[6] T. Wagner and J. Valasek, "Digital autoland control laws using quantitative feedback theory and direct digital design," *Journal of Guidance, Control, and Dynamics*, vol. 30, no. 5, pp. 1399–1413, 2007.

[7] I. Fialho, G. J. Balas, A. K. Packard, J. Renfrow, and C. Mullaney, "Gain-scheduled lateral control of the F-14 aircraft during powered approach landing," *Journal of Guidance, Control, and Dynamics*, vol. 23, no. 3, pp. 450–458, 2000.

[8] S. A. Snell and P. W. Stout, "Robust longitudinal control design using dynamic inversion and quantitative feedback theory," *Journal of Guidance, Control, and Dynamics*, vol. 20, no. 5, pp. 933–940, 1997.

[9] C.-H. Lee, M.-G. Seo, M.-J. Tahk, J.-I. Lee, and B.-E. Jun, "Missile acceleration controller design using pi and time-delay adaptive feedback linearization methodology," *Proceedings of the Institution of Mechanical Engineers G*, vol. 226, no. 8, pp. 882–897, 2012.

[10] C. H. Lee, T. H. Kim, and M. J. Tahk, "Missile autopilot design for agile turn using time delay control with nonlinear observer," *International Journal of Aeronautical and Space Science and Technology*, vol. 12, no. 3, pp. 266–273, 2011.

[11] L. Bruyere, A. Tsourdos, and B. A. White, "Robust augmented lateral acceleration flight control design for a quasi-linear parameter-varying missile," *Proceedings of the Institution of Mechanical Engineers G*, vol. 219, no. 2, pp. 171–181, 2005.

[12] Q. Wang and R. F. Stengel, "Robust nonlinear control of a hypersonic aircraft," *Journal of Guidance, Control, and Dynamics*, vol. 23, no. 4, pp. 577–585, 2000.

[13] W. Gonsalves, *A multi-input multi-output self tuning regulator for nonlinear high performance aircraft control [M.S. thesis]*, ECE Department, California State University, Chico, Calif, USA, 2010.

[14] B. C. Kuo, *Automatic Control Systems*, Prentice Hall, 7th edition, 1995.

13

Noncrystalline Binder Based Composite Propellant

Mohamed Abdullah, F. Gholamian, and A. R. Zarei

Faculty of Chemistry, Malek Ashtar University, P.O. BOX 16705-3454, Tehran, Iran

Correspondence should be addressed to Mohamed Abdullah; mohamedazizam@gmail.com

Academic Editors: C. Meola and R. K. Sharma

This study reports on propellants based on cross-linked HTPE binder plasticized with butyl nitroxyethylnitramine (BuNENA) as energetic material and HP 4000D as noncrystalline prepolymer. This binder was conducted with solid loading in the 85%. The results showed an improvement in processability, mechanical properties and burning rate. In addition, its propellant delivers (about 6 seconds) higher performance (specific impulse) than the best existing composite solid rocket propellant. Thermal analyses have performed by (DSC, TGA). The thermal curves have showed a low glass transition temperature (T_g) of propellant samples, and there was no sign of binder polymer crystallization at low temperatures (−50°C). Due to its high molecular weight and unsymmetrical or random molecule distributions, the polyether (HP 4000D) has been enhanced the mechanical properties of propellants binder polymer over a large range of temperatures [−50, 50°C]. The propellants described in this paper have presented high volumetric specific impulse (>500 s·gr·cc^{-1}). These factors combined make BuNENA based composite propellant a potentially attractive alternative for a number of missions demanding composite solid propellants.

1. Introduction

Much research on composite solid propellants has been performed over the past few decades and much progress has been made, yet many of the fundamental processes are still unknown, and the development of new propellants remains highly empirical. Ways to enhance the performance of solid propellants for rocket and other applications continue to be explored experimentally, including the effects of various additives and the impact of fuel and oxidizer particle sizes on burning behavior.

In view of higher energy (I_{sp} > 264 s), composite propellants have been extensively used for rocket/missile applications and space missions. A higher specific impulse (I_{sp}) of composite propellants is obtained by incorporating a maximum possible amount of solids (oxidizer/metallic fuel) in the binder matrix and substituting the inert materials with energetic ones (energetic plasticizers). Present day applications demand propellants of superior mechanical properties in addition to higher energy content. Due to these contradictory requirements hydroxy-terminated polyether (HTPE) based propellants are plasticized with energetic plasticizers, such as BuNENA, bolster performance and mechanical properties [1].

HTPE with HP 4000D as Prepolymer is capable of taking up solids up to 85% and impart superior mechanical properties without compromising on high storage life, due to its random molecule distributions that prevent crystallization at low temperatures in addition, the presence of BuNENA (T_g = −86°C) [2] eliminates completely this phenomenon (crystallization of polyether) at operational temperature ranges [−50, 50°C] of solid rocket motors. Lately, much scholarly work has been done on the hydroxy-terminated polyether-based (HTPE) propellants instead of hydroxy-terminated polybutadiene-based (HTPB) propellants and has introduced BuNENA as energetic plasticizer in high-energy nitroester polyether (or polyester) (NEPE) propellants [3]. Plasticizer plays the essential role of complementary element to reduce the viscosity of the slurry and to improve the mechanical properties by lowering the T_g and the modulus of the binder. The use of BuNENA in composite propellants confers excellent properties, due to its characteristics such as, insensitive energetic material, low glass transition temperatures, and good thermal stability, so we are interested to use it in

TABLE 1: Materials used in composite propellant binder.

Reagent	Name	Company	Characteristics	Hydroxyl no. mg KOH/gm	Molecular weight g/mol
Prepolymer	HP 4000D	BASF	Water white liquid	28–30	4000
Curing agent	N-100	Bayer	Polyisocyanate	—	191
Stabilizer	Nitro diphenyl amine (2-NDPA)	Fluka	Orange crystal	—	214.22
Energetic plasticizer	n-Butyl-2-nitratoethyl-nitramine (BuNENA)	DYNO ASA, Norway	Slightly yellow liquid	—	207.18
Bonding agent	DYNAMAR Brand Curative (HX-878)	3M	Viscous amber liquid	—	316.44

TABLE 2: Composite propellant compositions.

Ingredients	Percentage by weight	
	Mixl	Mix2
Al (22 μm)	23.0	19.0
RDX (5 μm)	10.0	33.0
AP (200 μm)	36.4	23.1
AP (20 μm)	15.6	9.90
HP4000D	5.69	5.75
BuNENA	8.00	8.00
HX 878	0.20	0.20
2-NDPA	0.25	0.25
N100	0.86	0.80

our formulae to improve composite propellant compositions.

A number of studies have been carried out in the past on the formulation, processing, and improvement of mechanical properties and ballistic evaluation of HTPE based composite propellants, but with inert plasticizers and using conventional polyether that can be crystallized at low temperatures [3]. Other studies on double base and gun propellants have been published using BuNENA as energetic plasticizer [4–7]. However, detailed information on composite propellants with energetic plasticizers and new polyether like (HP-4000D) is not reported in the open literature. However, propellants based on cross-linked HTPE binders are being used as alternatives to HTPB compositions because they give a less severe response in slow cookoff tests for insensitive munitions (IM) compliance [8–10].

2. Experimental

2.1. Materials. Bimodal blends of AP were used, consisting of a medium sized (200 μm) fraction and a small sized (20 μm) fraction. This combination was recommended as offering an optimum AP particle size width distribution to give the best rheology and to improve propellant slurry processabillity, and also, RDX 5 microns and Al of 23 micron were used as filler in the propellant composition. Materials used for binder have been described in Table 1.

2.2. Propellant Processing and Characterization. The propellant ingredients were mixed in a 3-liter capacity sigma mixer for two batches (mixl, mix2) prepared according to Table 2. The processability of the slurry was monitored by measuring the end-of-mix viscosity (EOM) and viscosity build up for a period of 10 hours. Propellant slurry was cast into Teflon coated moulds, under vacuum, for evaluation the mechanical properties and strand burn rate. The mechanical properties of cured propellant samples were evaluated using dumb bells conforming to ASTM standards D-412-68 (Type-C) at a cross-head speed of 50 mm/minute at −40, 25, and 50°C. The cured propellant slabs were machined into strands of dimensions 175 × 5 × 5 mm. The strands were inhibited with coatings of phenolic epoxy resin or polyvinyl acetate paint. They were burned in a nitrogen pressurized Crawford-type bomb over a pressure range from 2 to 18 MPa.

3. Results and Discussion

3.1. Mechanical and Ballistic Properties. As seen in Table 3, propellant samples, mixl and mix2, exhibit a high density impulse (~500 g·s·cc^{-1}), good processabillity expressed by low viscosity of EOM (~4 K Poise), and reasonable pot life (8 hours). In addition, it has shown excellent mechanical properties especially at low temperatures (−40°C) when binder elongation at maximum stress reached to 65–70%. Thus, we have believed that the withstanding properties refer to kind of prepolymer used and to the presence of BuNENA in composition. The use of HP 4000D, as prepolymer, which is Ethylene-Oxide (EO) capped polypropylene glycol (PPG) polyols with low insaturation content, also when plasticized with BuNENA, which has low glass transition temperature (T_g = −86°C), has enhanced the binder elongation at low temperatures and prevented binder to crystallize (improved by thermal behavior next). When more RDX is used (mix2), the burning rate and pressure exponent decreased, respectively; see Table 3. In addition, mix2 showed increase in specific impulse (about 4 seconds).

3.2. DSC and TGA Analyses. In order to analyze the thermal behavior of propellant samples, differential scanning calorimeter (DSC) analysis was performed to determine glass transition temperature (T_g) and thermal decomposition (TA). Thermogravimetric analyses (TGA) were also carried

Table 3: Mechanical, rheological, and ballistic properties of propellant samples.

Composite propellants	Unit		Mix1			Mix2	
Density, ρ	$g\,cm^{-3}$		1.88			1.83	
Specific impulse (theoretical), 7/0.1 MPa	sec		266			270	
Density specific impulse ($\rho \cdot I_{sp}$)	$g \cdot cm^{-3}$ sec		500			494	
Burning rate at 7 MPa	$mm\,s^{-1}$		8.50			7.00	
Pressure exponent (n), (2–18) MPa	—		0.45			0.40	
Processability							
End of mixing (EOM) viscosity	K Poise		4.0			4.5	
Pot life (to 15 K poise)	hour		8			10	
Mechanical Properties T °C //50 mm min^{-1}	T °C	−40	25	50	−40	25	50
Max. tensile strength σ_{max}	$N\,mm^{-2}$	1.5	0.8	0.6	1.8	0.7	0.5
Elongation at ε_{max}	%	65	78	86	58	70	78
Elongation at break	%	70	86	91	63	76	81
E-modulus	$N\,mm^{-2}$	5.5	2.2	1.8	5.8	2.0	1.6
Glass transition DSC	°C		−75.03			−77.47	

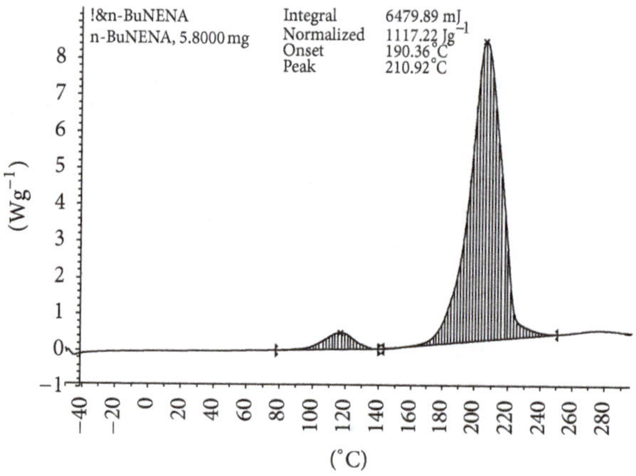

Figure 1: BuNENA DSC curve.

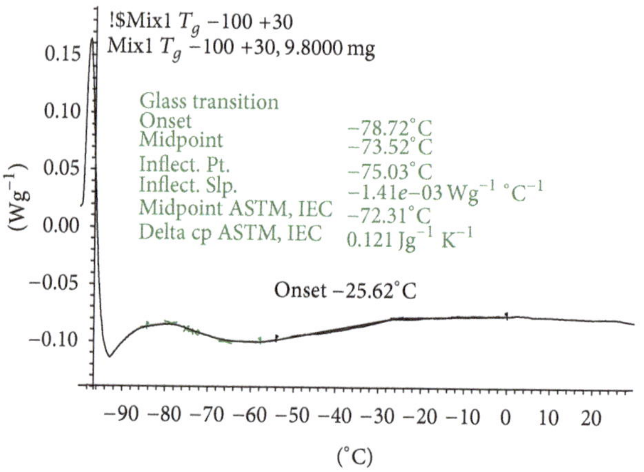

Figure 2: DSC T_g curve of propellant sample (mix1).

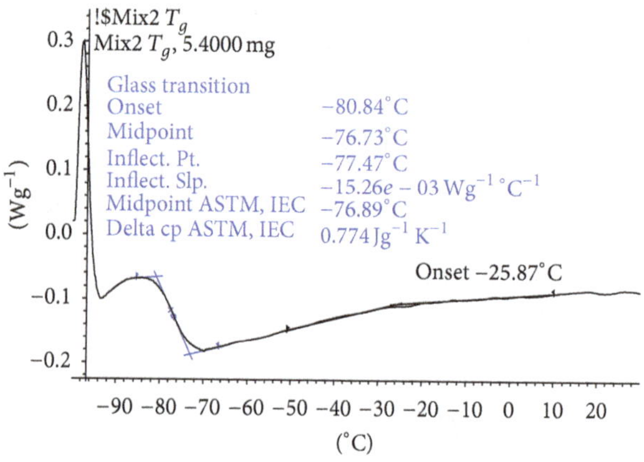

Figure 3: DSC T_g curve of propellant sample (mix2).

out. The thermal analyses were performed using Mettler TA4000 thermal analyzer equipped with a TA processor TC-11 and a DSC 30 measuring cell. An inert environment was maintained during all the analyses by using a flow of nitrogen of 40 cm^3 per min. Analyses were performed at a heating rate of 10°C per min in the temperature range from −100 to +30°C for T_g, 30 to 500°C for TA analysis, and 30 to 550°C for TGA. Figure 2 shows the DSC T_g curve of propellant sample mix1 (9.8 mg of sample), and Figure 3 is related to sample mix2 (5.4 mg of sample). As we have shown, T_g is 75.03°C corresponding to mix1 and 77.47°C to mix2. Figure 4 shows the DSC TA curve of (0.9 mg of sample) mix1 and Figure 5 shows the DSC TA curve (2.2 mg) of sample mix2. Figure 6 shows the thermogravimetric curve (4.3 mg) of sample mix1.

As can be seen from Figures 2 and 3, the glass transition temperatures for propellant samples mix1 and mix2 are very close. The T_g for these samples was around −76°C. On the other hand, the thermal decomposition observed in Figures 4 and 5 and several peaks, either exothermic or endothermic, are present during propellant thermal decomposition. Three endothermic peaks can be seen after 150°C from samples mix1 and mix2 in Figures 4 and 5, respectively. The exothermic peaks at around 188 and 193°C, with an onset around 160°C, can be assigned to the energetic plasticizer BuNENA. In propellant mix1, this peak can be seen at 199°C and it looks

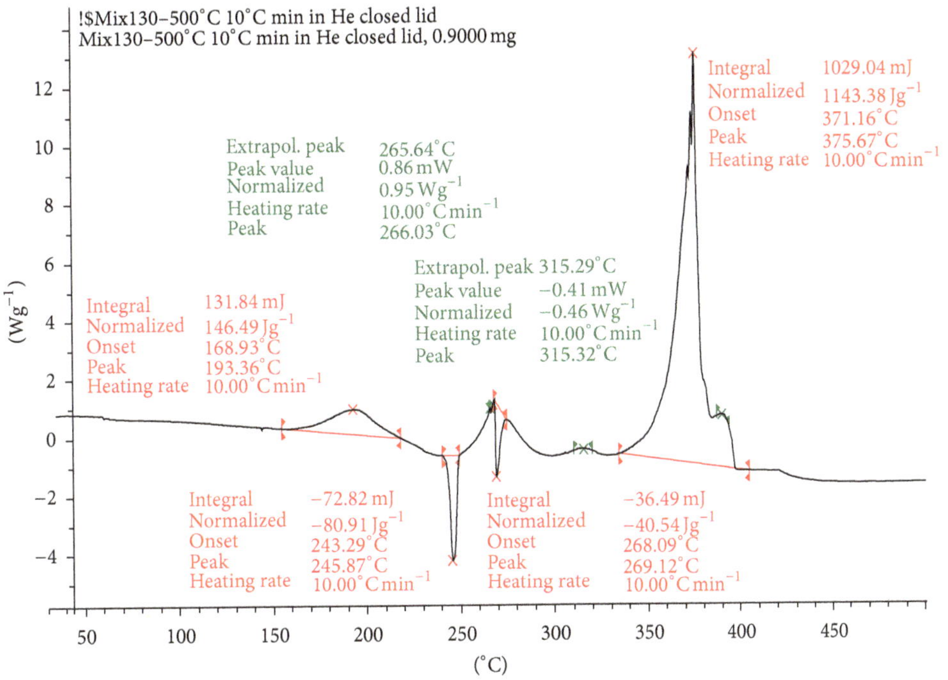
!$Mix130–500°C 10°C min in He closed lid
Mix130–500°C 10°C min in He closed lid, 0.9000 mg
Integral 1029.04 mJ
Normalized 1143.38 Jg−1
Onset 371.16°C
Peak 375.67°C
Heating rate 10.00°C min−1
Extrapol. peak 265.64°C
Peak value 0.86 mW
Normalized 0.95 Wg−1
Heating rate 10.00°C min−1
Peak 266.03°C
Extrapol. peak 315.29°C
Peak value −0.41 mW
Normalized −0.46 Wg−1
Heating rate 10.00°C min−1
Peak 315.32°C
Integral 131.84 mJ
Normalized 146.49 Jg−1
Onset 168.93°C
Peak 193.36°C
Heating rate 10.00°C min−1
Integral −72.82 mJ
Normalized −80.91 Jg−1
Onset 243.29°C
Peak 245.87°C
Heating rate 10.00°C min−1
Integral −36.49 mJ
Normalized −40.54 Jg−1
Onset 268.09°C
Peak 269.12°C
Heating rate 10.00°C min−1
(Wg−1)
(°C)

Figure 4: DSC TA curve of propellant sample (mix1).

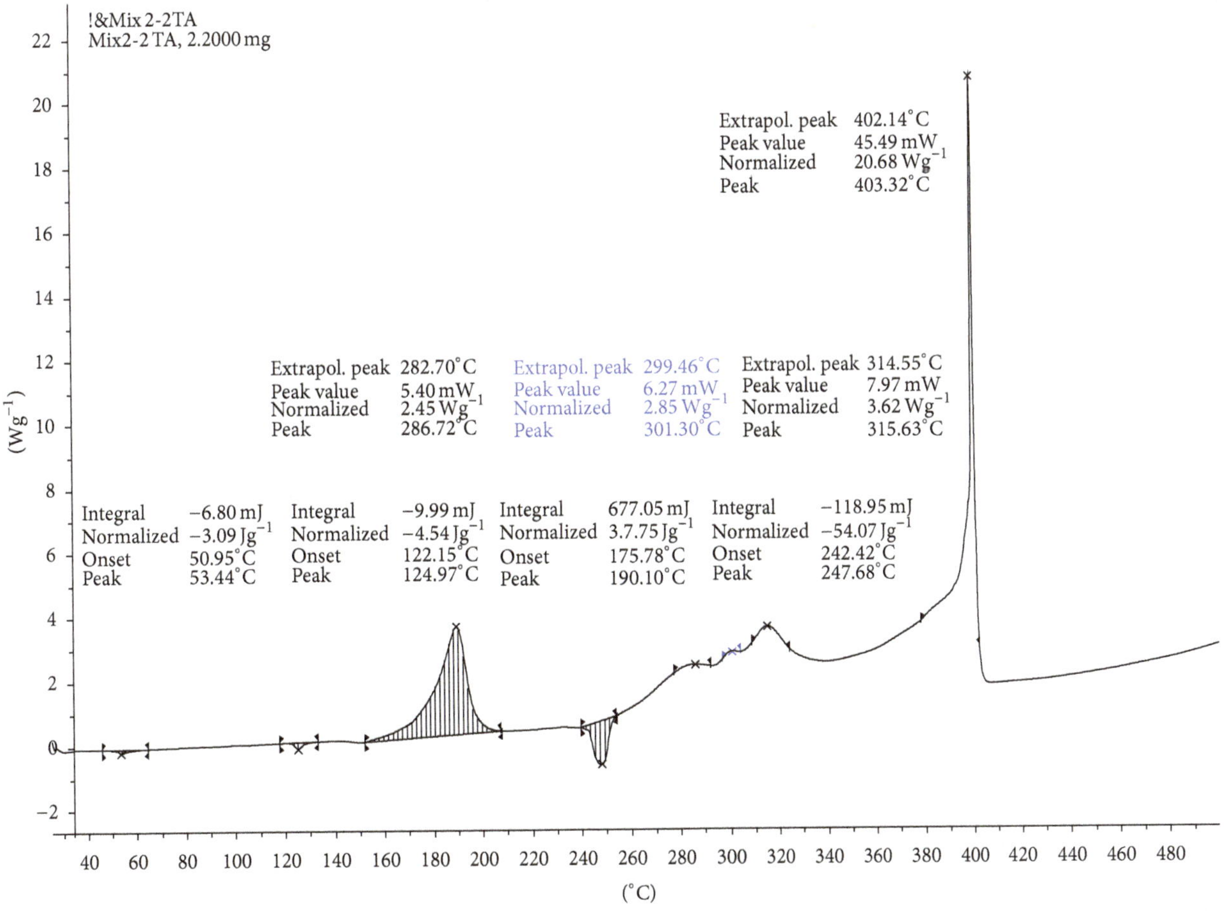
!&Mix 2-2TA
Mix2-2 TA, 2.2000 mg
Extrapol. peak 402.14°C
Peak value 45.49 mW
Normalized 20.68 Wg−1
Peak 403.32°C
Extrapol. peak 282.70°C
Peak value 5.40 mW
Normalized 2.45 Wg−1
Peak 286.72°C
Extrapol. peak 299.46°C
Peak value 6.27 mW
Normalized 2.85 Wg−1
Peak 301.30°C
Extrapol. peak 314.55°C
Peak value 7.97 mW
Normalized 3.62 Wg−1
Peak 315.63°C
Integral −6.80 mJ
Normalized −3.09 Jg−1
Onset 50.95°C
Peak 53.44°C
Integral −9.99 mJ
Normalized −4.54 Jg−1
Onset 122.15°C
Peak 124.97°C
Integral 677.05 mJ
Normalized 3.7.75 Jg−1
Onset 175.78°C
Peak 190.10°C
Integral −118.95 mJ
Normalized −54.07 Jg−1
Onset 242.42°C
Peak 247.68°C
(Wg−1)
(°C)

Figure 5: DSC TA curve of propellant sample (mix2).

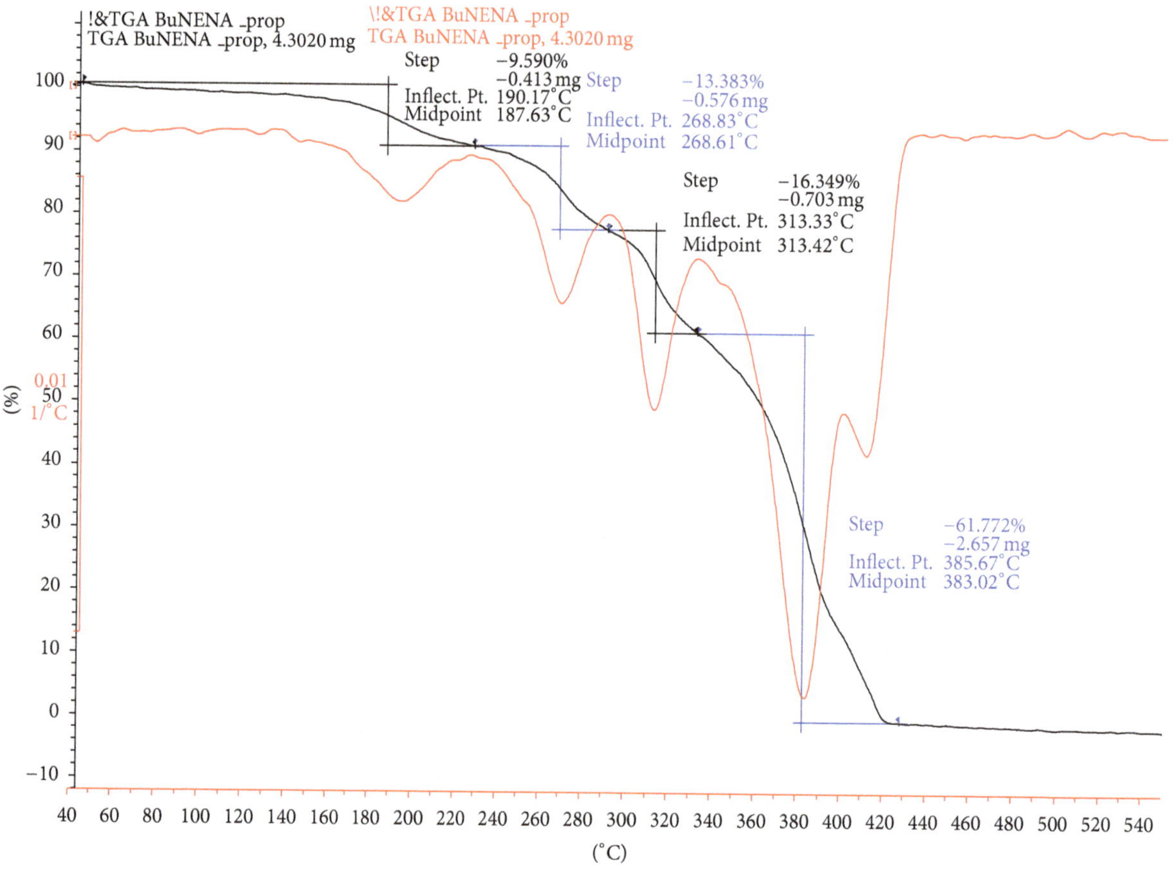

Figure 6: TGA curve of propellant sample (mix1).

more intense than in the other propellant mix2, as can be seen in Figure 5. In fact, in mix1 the heat of decomposition was 1143 J g^{-1} and the BuNENA heat of decomposition peak for the pure product was observed to be 1117 J g^{-1} at 210°C (Figure 1). Figure 6 shows the thermo gravimetric for 4.3 mg of sample Bu NENA prop. Several peaks are present during propellant thermal decomposition. The peak at around 190°C with weight loss 9.6% refers to Bu NENA, and around 268°C with weight loss 13.88%, refers to RDX. Two peaks at 313.33, and 385.65°C with weight losses 16.4, and 61.77% can be assigned to the AP decompositions on two stages.

4. Conclusions

Composite solid propellant, based on HP 4000D as prepolymer, BuNENA as energetic plasticizer, and Al/AP/RDX as fillers, clearly demonstrated that the specific impulse I_{sp} could be increased (about 6 seconds) as compared to the best existing type of composite propellant. The results are very promising, especially to the mechanical properties at low temperatures, as well as to the polymer rheology expressed by processability (good castability, low viscosity, and relatively long pot life). The paper found, however, that the difunctional end-capped propylene prepolymer block, that is made by and commercially available from BASF Corporation under the trade name HP-4000D, which is a high performance difunctional prepolymer, having an average molecular weight of 4000, is the best solution to reduce the tendency of the polyether to crystallize. The low temperature strain values indicate that the energetic composition has excellent elastomeric properties over a broad range of temperatures, as well as very high strain capability and tensile strength. These properties exceed even HTPB propellants in some instances, while providing higher oxygen content than HTPB. In the search for replacement of inert plasticizers for composite propellants, BuNENA is a new and very promising, interesting material with high performance and low vulnerability. On the other hand, RDX has been shown to be a very useful energetic ingredient in solid rocket propellant applications, due to its relative insensitivity to accidental energy stimuli. In addition, the absence of HCl in RDX combustion products makes it desirable on an environmental basis. Since the regression rate and burning behavior of a solid propellant can be greatly influenced by processes occurring in the burning surface and subsurface regions, so pressure exponent can be reduced. To improve the specific impulse of cross-linked HTPE propellant plasticized with BuNENA, RDX must be introduced in propellant composition (at least 10%), and when reaches 33%, the specific impulse increases to 270 s. The presence of RDX in composition formula has an effect on specific impulse and pressure exponent (n). In addition, more than 500 g·cm^{-3} sec of density impulse are expected.

References

[1] M. E. Sitzmann, N. J. Trivedi, and B. Patrick, "Investigation of an N-Butyl-N-(2-Nitroxyethyl)nitramine (BuNENA) process: identification of process intermediates, by-products and reaction pathways," *Propellants, Explosives, Pyrotechnics*, vol. 31, no. 2, pp. 124–130, 2006.

[2] N. Wingborg and C. Eldsäter, "2,2-Dinitro-1,3-bis-nitrooxy-propane (NPN): a new energetic plasticizer," *Propellants, Explosives, Pyrotechnics*, vol. 27, no. 6, pp. 314–319, 2002.

[3] B. S. Min and Y. C. Park, "A study on the aliphatic energetic plasticizers containing nitrate ester and nitramine," *Journal of Industrial and Engineering Chemistry*, vol. 15, no. 4, pp. 595–601, 2009.

[4] K. P. C. Rao, A. K. Sikder, M. A. Kulkarni, M. M. Bhalerao, and B. R. Gandhe, "Studies on n-Butyl nitroxyethylnitramine (n-BuNENA): synthesis, characterization and propellant evaluations," *Propellants, Explosives, Pyrotechnics*, vol. 29, no. 2, pp. 93–98, 2004.

[5] P. A. Silver and N. F. Stanley, "BuNENA gun propellants," in *Proceedings of the JANNAF Propulsion Meeting*, vol. 2, p. 515, Anaheim, Calif, USA, September 1990.

[6] Alu, Q. Shen, X. Liao, and G. Bao, "Preliminary study of BuNENA gun propellant," in *Proceedings of the 27th International Annual Conference on ICT, (Energetic Materials)*, p. 51. 1, Karlsruhe, Germany, 2006.

[7] L. A. Fang, S. Q. Hua, L. Xin, and V. G. Ling, "Preliminary study of Bu NENA gun propellants," in *Proceedings of the 27th International Annual Conference of ICT*, p. 51/1, Karlsruhe, Germany, June 1996.

[8] NIMIC Newsletter, "Solid rocket propellant for improved IM response, part 2," *IM Propellant Examples*, vol. 1, pp. 2–4, 2003.

[9] D. Schmitt, P. Eyever, and P. Elsner, "Insensitive high performance energetic materials-applied research for optimized products," *Propellants, Explosives, Pyrotechnics*, vol. 22, no. 3, pp. 109–111, 1997.

[10] T. F. Comfort, L. G. Dillman, K. O. Hartman, M. G. Mangum, and R. M. Steckman, "Insensitive HTPE propellants," in *Proceedings of the JANNAF Propulsion Meeting*, vol. 3, p. 87, CPIA Publication 630, Tampa, Fla, USA, December 1995.

14

Lateral Control Implementation for an Unmanned Aerial Vehicle

R. Samar,[1] M. Zamurad Shah,[1] and M. Nzar[2]

[1] *Mohammad Ali Jinnah University, Islamabad 44000, Pakistan*
[2] *Centres of Excellence in Science & Applied Technologies (CESAT), Islamabad, Pakistan*

Correspondence should be addressed to R. Samar; raza.samar@gmail.com

Academic Editors: A. Desbiens, Z. Mazur, and K. Peng

This paper presents practical aspects of guidance and control design for UAV and its flight test results. The paper focuses on the lateral-directional control and guidance aspects. An introduction to the mission and guidance problem is given first. Waypoints for straight and turning flight paths are defined. Computation of various flight path parameters is discussed, including formulae for real-time calculation of down-range (distance travelled along the desired track), cross-track deviation, and heading error of the vehicle; these are then used in the lateral guidance algorithm. The same section also describes how to make various mission-related decisions online during flight, such as when to start turning and when a waypoint is achieved. The lateral guidance law is then presented, followed by the design of a robust multivariable H_∞ controller for roll control and stability augmentation. The controller uses the ailerons and rudder for control of roll angle and stabilization of yaw rate of the vehicle. The reference roll angle is generated by the nonlinear guidance law. The sensors available on-board the vehicle do not measure yaw rate; hence, a practical method of its estimation is proposed. The entire guidance and control scheme is implemented on the flight control computer of the actual aerial vehicle and taken to flight. Flight test results for different mission profiles are presented and discussed.

1. Introduction

Successful control system design for high performance UAVs requires efficient and effective techniques for the design of guidance and control algorithms that ensure satisfactory operation in the face of system uncertainties. In this paper, we develop and present one such technique. The UAV under consideration is shown in Figure 1. The cruising speed of said UAV is about 45 m/sec. The main tasks of the guidance and control system are to:

(i) fly the vehicle on the desired mission path, with minimum cross-track deviation,

(ii) make decisions about various flight events, such as to start/stop turning or achievement of waypoints,

(iii) provide robust stabilization and control during flight.

A lateral track control law for small UAVs has been discussed in [1]. This is based on a pure geometrical concept. The idea is to make the ratio of lateral deviation to lateral velocity equal to the ratio of longitudinal distance to longitudinal velocity. Simulation results indicate that the yaw rate command generated by the guidance law exhibits oscillations in the vehicle roll channel, this could be a problem for implementation on a real vehicle. A receding horizon based controller for UAV guidance is presented in [2]. The dynamics of the problem are linearized and simplified, and a quadratic cost function based on the model predictive control methodology is set up. The output prediction horizon and the control horizon need to be selected keeping in view the computational resources available, and the actuator and vehicle dynamical constraints have to be modelled carefully. The problem is challenging with regards to real-time implementation (specially for critical applications like high performance UAVs), and infeasibility conditions for solutions have to be guarded against. Mixed integer linear programming- (MILP-) based guidance for UAVs has also been considered [3, 4]; however, here the optimization program generates a sequence of waypoints (positions) and velocities for the vehicle to follow. In other words a mission plan is generated, and deviations from this plan need to be corrected through a lower level guidance algorithm.

FIGURE 1: A photograph of the test vehicle.

A conceptually different guidance scheme employing vector fields for curved path following has also been pursued, see, for example, [5, 6]. Here, a vector field of course (or heading) commands is generated which is a function of vehicle position relative to the desired track. The difference between the actual and commanded headings forms the heading error which is driven to zero using an appropriate control algorithm. The course vector field can in cases give rise to large and sudden heading changes which can tax the capability of the control system.

The contribution of this paper is to present a practical and integrated guidance and control approach and demonstrate its effectiveness by experimental testing. Formulae for basic flight path computations are derived that provide the requisite input parameters for the guidance algorithm to function. The guidance algorithm is a computationally inexpensive algorithm based on the work presented in [7] with some improvements for practical considerations. The output of the guidance algorithm drives the control system, and design of a robust multivariable controller that tracks the guidance commands is presented. The overall system is integrated in a seamless fashion and implemented and test flown on an actual vehicle.

This paper is organized as follows. Section 2 describes the mission plan on which we intend the UAV to fly, including definition of straight, turn, and loiter waypoints. Here, we also outline the guidance and control problem in specific terms. Section 3 develops the basic formulae used for flight path computation. These form the basis for taking real-time decisions during flight and for implementing the guidance law. The nonlinear guidance law is discussed in Section 4. Section 5 describes the design of a robust multivariable H_∞ controller and discusses robustness of the controller across the flight envelope. Flight test results are presented and discussed in Section 6; Section 7 concludes the paper.

2. The Mission Plan

The mission of the UAV consists of a number of waypoints which define the path the UAV is required to take. The mission may be planned on-line or off-line. Here, we assume that information regarding the waypoints through which the UAV is desired to navigate is available; our main objective here is to discuss the practical aspects of how to guide and control the aerial vehicle through the given mission. The entire mission can be represented by a series interconnection of straight line segments, arcs, and circles (for loiter).

Figure 2 shows on the left-hand side, part of a mission comprising of a straight line segment. In this case, the UAV is required to fly through the waypoint WP2. The right side of Figure 2 shows part of a mission with a turn. Here, the UAV is not required to go through the central waypoint (WP2) but rather fly on a circular arc AB close to WP2. Point A is where the turn starts, and B is where it ends. Similarly, loiter missions are also possible where the vehicle flies in a circular orbit around a given waypoint. Figure 3 shows part of a mission which involves loiter at WP2. The arcs AB and CD are used for transitioning from the straight flight segments to the circular loiter pattern. In this study, we shall assume that the mission data in terms of waypoints is available.

We now formulate the problem in more specific terms. The task of the lateral guidance and control system is defined as to

(i) compute the down-range covered from one waypoint to the next and the cross-track deviation from the designated track/trajectory,

(ii) compute the error in heading angle of the vehicle,

(iii) determine if a waypoint is achieved,

(iv) compute "turn start" and "turn stop" decision flags for the autopilot,

(v) solve the guidance equations to calculate the roll reference command for steering the vehicle back onto the desired path,

(vi) stabilize and control the lateral-directional dynamics of the vehicle across the flight envelope in the presence of disturbances.

All the above computations have to be done in real time. The available control inputs to the vehicle are the ailerons and the rudder. Note that the cross-track deviation and heading error are used in the guidance equations.

3. Flight Path Computations

Here, we present the basic flight path computations that are performed in real time to enable navigation of the vehicle through a sequence of straight and turning waypoints. Waypoints are defined as geographic positions in terms of, latitude and longitude. For any three consecutive waypoints WP1, WP2 and WP3, if the turn angle (defined as the difference of azimuths of WP1 and WP3 at WP2) at the central waypoint (WP2) is nearly 180°, then the path from WP1 to WP2 will be considered a straight line, and WP2 will be referred to as a *straight* waypoint. On the other hand, if the turn angle is significant (say greater than 5°), then WP2 will be referred to as a *turning* waypoint, and the path from WP1 to WP3 will consist of straight and curved parts (Figure 2). The flight path computations are worked out separately for straight and turning parts of the flight.

The algorithm presented computes deviations from the reference track in the form of cross-track and heading errors both for straight and curved paths. The cross-track and heading errors are used in the guidance logic to generate appropriate roll commands for the autopilot so that the

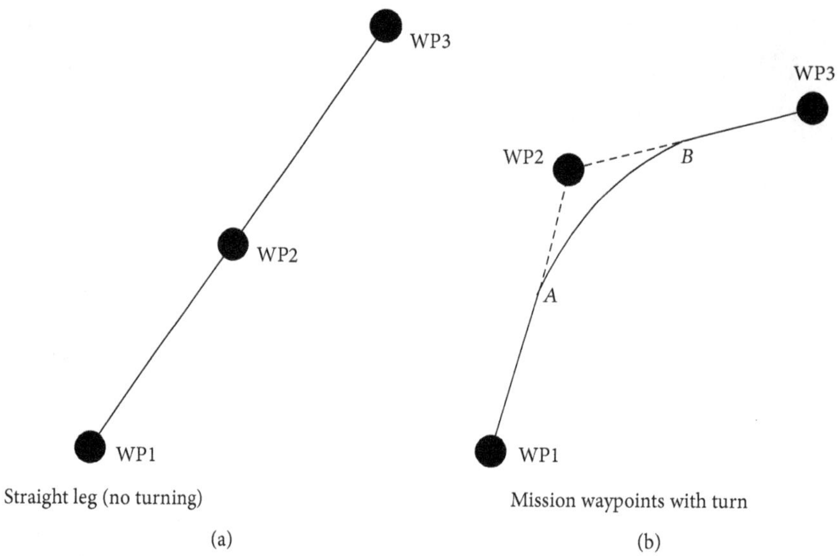

FIGURE 2: Waypoints for straight and turning flight.

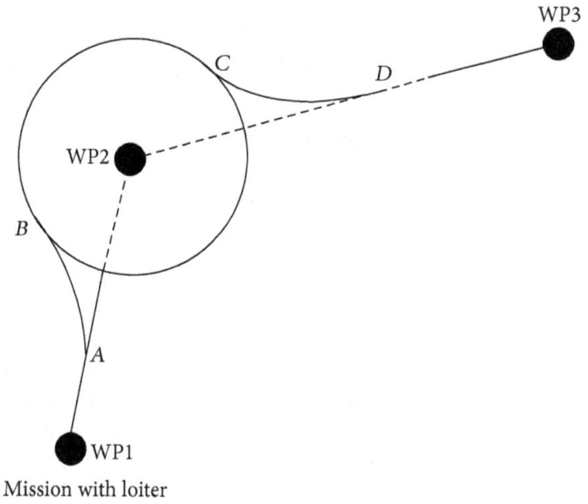

FIGURE 3: Mission with loiter about WP2.

vehicle is steered back onto the desired path. In addition to cross-track and heading errors, "waypoint achieved," "turn start," and "turn stop" decision flags are also computed for the autopilot.

Let WP1(ϕ_1, λ_1), WP2(ϕ_2, λ_2), and WP3(ϕ_3, λ_3) be three consecutive waypoints (ϕ and λ denote the latitude and longitude, resp.), and let $M(\phi_M, \lambda_M)$ be the current position of the vehicle. We define an Earth-Centred Earth-Fixed (ECEF) frame with its origin at the centre of mass of the earth, the z-axis directed northward along the polar axis, the x-axis in the equatorial plane and passing through the Greenwich Meridian, and the y-axis also in the equatorial plane and passing through 90° east longitude. The position vector of a point on the earth's surface with latitude ϕ and longitude λ in ECEF frame is given by

$$r_x = a\cos\phi_r\cos\lambda, \qquad r_y = a\cos\phi_r\sin\lambda, \qquad r_z = b\sin\phi_r, \tag{1}$$

where a is the earth's equatorial radius, b is the polar radius, and $\phi_r = \tan^{-1}((b/a)\tan\phi)$ is the reduced latitude of the point under consideration.

3.1. Straight Path Formulae

3.1.1. Distance Covered. Here, we consider the straight path from WP1 to WP2 (Figure 4). Let $\vec{r_1}$ and $\vec{r_2}$ be the position vectors of WP1 and WP2 in ECEF frame, respectively. The distance between WP1 and WP2 can be calculated as:

$$r_{12} = \frac{1}{2}\left(\left|\vec{r_1}\right| + \left|\vec{r_2}\right|\right)\cos^{-1}\left(\hat{r}_1 \cdot \hat{r}_2\right), \tag{2}$$

where $\hat{r}$ is a unit vector along $\vec{r}$.

3.1.2. Cross-Track Deviation. Cross-track error is the instantaneous normal displacement of the vehicle from the desired track. To compute the cross-track error, we first define a vector $\vec{r_N}$ normal to the trajectory plane (plane containing WP1, WP2, and the centre of the earth) as:

$$\vec{r_N} = \hat{r}_1 \times \hat{r}_2. \tag{3}$$

If $\vec{r_M}$ is the position vector of the vehicle in ECEF frame, then the cross-track error is

$$\text{Cross track} = -\vec{r_M} \cdot \hat{r}_N. \tag{4}$$

3.1.3. Heading Error. Let V_E and V_N denote the east and north velocity of the vehicle, respectively, then the velocity heading of the vehicle is given by $\psi_G = \tan^{-1}(V_E/V_N)$. The heading error is the difference between the actual velocity heading ψ_G and the desired heading ψ_R; that is, $\psi_E = \psi_G - \psi_R$.

If P is the projection of the vehicle's current position on the trajectory plane (Figure 4), then the desired heading angle (ψ_R) is defined as the angle between the trajectory plane and

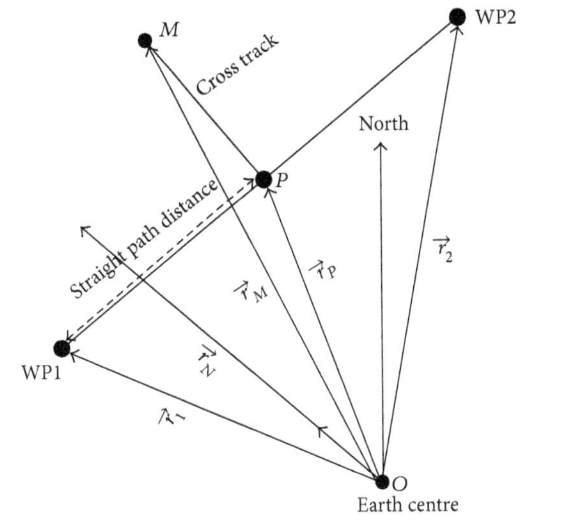

FIGURE 4: Straight flight path definitions.

the meridian plane passing through the projected point P. The position vector $\vec{r_P}$ of P in ECEF frame is given by

$$\vec{r_P} = \vec{r_M} - \left(\vec{r_M} \cdot \hat{r}_N\right)\hat{r}_N. \tag{5}$$

The desired heading angle ψ_R works out to be

$$\psi_R = \tan^{-1}\left[\frac{\left|\left(\hat{r}_P \times \hat{k}\right) \times \hat{r}_N\right|}{\left(\hat{r}_P \times \hat{k}\right) \cdot \hat{r}_N}\right], \tag{6}$$

where $\hat{k}$ is a unit vector along the earth's spin axis. The heading error ψ_E can now be readily computed.

3.1.4. Waypoint Achieved Criterion. The straight waypoint WP2 is achieved when the straight path distance covered exceeds r_{12}. The straight distance covered from WP1 to the current position of the vehicle is $(1/2)(|\vec{r_1}| + |\vec{r_P}|)\cos^{-1}(\hat{r}_1 \cdot \hat{r}_P)$.

3.2. Curved Path Formulae. For turning paths (Figure 5), we define another coordinate system (the East-North-Up or ENU frame) at the central waypoint WP2. The origin of the ENU frame is at WP2, the positive x-axis points towards east, the y-axis points towards north, and the z-axis points up along the local vertical. The position vector of a point on the earth's surface with latitude ϕ and longitude λ can be expressed in the ENU frame as:

$$x = R_N(\lambda - \lambda_2)\cos\phi_2, \qquad y = R_M(\phi - \phi_2), \qquad z = 0, \tag{7}$$

where R_N and R_M are the normal and meridian radii of curvature (of the earth's ellipsoidal model), respectively. These are given by

$$R_N = \frac{a}{\left(1 - e^2 \sin\phi_2\right)^{1/2}}, \qquad R_M = \frac{a\left(1 - e^2\right)}{\left(1 - e^2 \sin\phi_2\right)^{3/2}}, \tag{8}$$

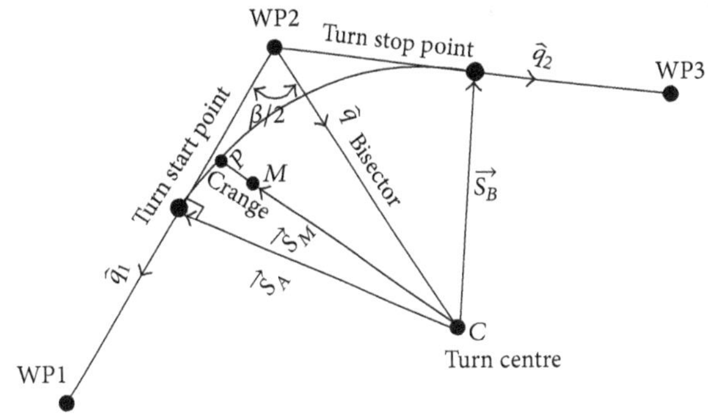

FIGURE 5: Flight path with turn.

where e denotes the eccentricity of the ellipsoidal earth:

$$e = \frac{\sqrt{a^2 - b^2}}{a}. \tag{9}$$

Let $\vec{q_1}$ and $\vec{q_2}$ be the position vectors of WP1 and WP3 in the ENU frame; that is,

$$\vec{q_1} = \left[R_N(\lambda_1 - \lambda_2)\cos\phi_2 \;\; R_M(\phi_1 - \phi_2) \;\; 0\right], \qquad \vec{q_2} = \left[R_N(\lambda_3 - \lambda_2)\cos\phi_2 \;\; R_M(\phi_3 - \phi_2) \;\; 0\right]. \tag{10}$$

The turn angle β ($\leq 180^\circ$) is given by

$$\beta = \cos^{-1}\left(\hat{q}_1 \cdot \hat{q}_2\right). \tag{11}$$

3.2.1. Turn Centre. The turn centre will be at a distance of $R/\sin(\beta/2)$ from WP2 on the line bisecting the turn angle β, R being the turn radius. The vector $\vec{q}$ along this line in the ENU frame is given by $\vec{q} = \hat{q}_1 + \hat{q}_2$. Thus, the position of the turn centre C in the ENU frame is

$$\vec{C} = \frac{R}{\sin(\beta/2)\,\hat{q}}. \tag{12}$$

The projection of the vehicle on the earth's surface (ϕ_M, λ_M) in the ENU frame is given as:

$$\vec{M} = \left[R_N(\lambda_M - \lambda_2)\cos\phi_2 \;\; R_M(\phi_M - \phi_2) \;\; 0\right], \tag{13}$$

and the position vector of the vehicle's projection w.r.t. the turn centre C is $\vec{S_M} = \vec{M} - \vec{C}$.

3.2.2. Turn Start Criterion. The turn start point is at a distance of $R/\tan(\beta/2)$ from WP2 on the line joining WP1 and WP2. Its position vector $\vec{A}$ in the ENU frame is $\vec{A} = R/\tan(\beta/2)\hat{q}_1$, and its position w.r.t. the turn centre C is $\vec{S_A} = \vec{A} - \vec{C}$. Turning is started when the cross-product $\vec{S_M} \times \vec{S_A}$ changes sign. During turn, the total distance (down-track) from WP1 to the current position is computed as:

$$\text{distance from WP1} = r_{12} - \frac{R}{\tan(\beta/2)} + R\cos^{-1}\left(\hat{S}_A \cdot \hat{S}_M\right). \tag{14}$$

3.2.3. Cross-Track Deviation. The cross track deviation during turning is given by $R - |\overrightarrow{S_M}|$.

3.2.4. Heading Error. During turn, the desired heading angle (ψ_R) is given by $\psi_R = \psi_s + \psi_T$, where ψ_s is the desired heading at the turn start point (computed using (6) at the turn start point) and ψ_T is the angular displacement from the turn start point to the current position:

$$\psi_T = \cos^{-1}\left(\widehat{S}_A \cdot \widehat{S}_M\right). \tag{15}$$

3.2.5. Waypoint Achieved Criterion. The waypoint WP2 is said to be achieved when the vehicle crosses the bisector line; that is, when the vector cross-product $\overrightarrow{S_M} \times \vec{C}$ changes sign.

3.2.6. Turn Stop Criterion. The turn stop point is at a distance of $R/\tan(\beta/2)$ from WP2 on the line joining WP2 and WP3. Its position vector $\vec{B}$ in the ENU frame is $\vec{B} = R/\tan(\beta/2)\widehat{q}_2$, and its position w.r.t. the turn centre C is $\vec{S_B} = \vec{B} - \vec{C}$. Turning is stopped when $\overrightarrow{S_M} \times \vec{S_B}$ changes sign.

4. Lateral Guidance Logic

The block diagram of the lateral-directional guidance and control system is shown in Figure 6. The outputs of the UAV that are measured by on-board sensors include the roll, pitch, and heading attitude angles and the (GPS) position and velocity of the vehicle. The roll and pitch angles are measured by a vertical gyro, and the heading angle is provided by a magnetic sensor. The outer loop is the lateral guidance loop that looks at the measured position and velocity of the vehicle, compares it with the desired mission path, and generates a roll angle command that acts as a reference for the inner loop to track. The inner loop has roll angle and yaw rate feedbacks; the yaw-rate is computed from the attitude angles as discussed in Section 5.1 below.

The lateral guidance scheme used is based on the work in [7]. The basic scheme is modified by introducing an adaptive tuning of the reference length to enhance the performance for large cross-track errors. Integral action is also added to improve tracking in the presence of disturbances. Here, we first introduce the basic scheme, followed by a discussion of the modifications.

4.1. The Basic Guidance Scheme. Figure 7 illustrates the guidance scheme. Let AB be the desired flight path of the vehicle, and let the instantaneous position of the vehicle be denoted by C, at a perpendicular distance y from the line AB. We define a point D at a distance L_1 from C such that the vehicle may follow a circular arc of radius R from C to the desired path at D. The centripetal acceleration a_L required for the vehicle to fly along this circular arc is given by

$$a_L = \frac{2V^2}{L_1}\sin\eta, \tag{16}$$

where V is the velocity of the vehicle and η is the angle between the velocity vector and the line CD. Since we shall employ a bank-to-turn scheme, the lateral acceleration command needs to be converted into a roll reference command. When the vehicle is banked at an angle ϕ, the lift vector L can be resolved into two components $L\cos\phi = mg$ and $L\sin\phi = (2mV^2/L_1)\sin\eta$, from which, we have

$$\phi = \tan^{-1}\left(\frac{2V^2}{gL_1}\sin\eta\right). \tag{17}$$

From Figure 7, we have $\eta = \eta_1 + \eta_2$, where the subangles are

$$\eta_1 = \sin^{-1}\left(\frac{y}{L_1}\right), \qquad \eta_2 = \psi_E, \tag{18}$$

where ψ_E is the heading error. Here, $y \leq L_1$ because the argument of the arcsin function should be less than or equal to 1. Now, we may write (17) as follows:

$$\phi_c = \tan^{-1}\left[\frac{2V^2}{gL_1}\sin\left\{\sin^{-1}\left(\frac{y}{L_1}\right) + \psi_E\right\}\right], \tag{19}$$

where the subscript in ϕ_c stands for roll angle *command.* Equation (19) defines the nonlinear lateral guidance law; it gives the roll angle which the vehicle should fly to get the lateral acceleration required to follow a circular arc back to its desired flight path.

4.2. Selection of L_1 Using Linear Analysis. A guideline for selection of the length L_1 is provided in [7] by linearizing (19) to yield a second-order system with damping and natural frequency given by

$$\zeta = \frac{1}{\sqrt{2}}, \qquad \omega_n = \frac{\sqrt{2}V}{L_1}, \tag{20}$$

where ζ and ω_n refer to the damping ratio and natural frequency of the standard second order system. It is stressed, however, that this is only a guideline as a very strong assumption of no inner loop dynamics ($a_L = -\ddot{y}$) is employed in its derivation.

4.3. Adaptive Adjustment of Length L_1. For system stability, we must have $y < L_1$ and $|\eta| < \pi/2$ [7]. For UAVs however, there is a real possibility of large cross-track errors developing during flight. This can be due to the runway not being in the direction of the first waypoint so that a cross-track deviation is necessarily generated in the initial phase of the flight or the GPS signal becoming masked for extended periods during flight. Furthermore, since lateral control is usually engaged sometime after take-off (to avoid large roll angles during take-off), the cross-track may build to a large value initially. In such cases, the condition $y < L_1$ may not be satisfied making ϕ_c undefined. Table 1 gives values of L_1 computed using (20) for different speeds typical of modern UAVs. The table shows that for small-to-medium UAVs, L_1 can vary from 200 m to 850 m depending upon the desired response time of the guidance algorithm (or ω_n). If a fast response is desired (high ω_n), a small L_1 will be selected, leading to large roll angles to bring the vehicle back on the desired path quickly, but with

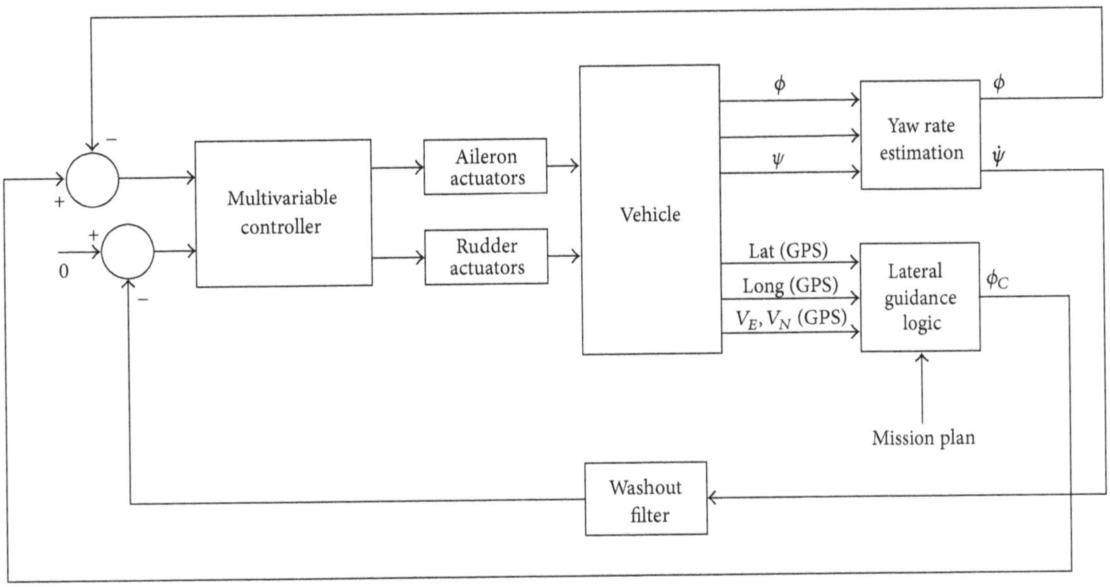

FIGURE 6: Block diagram of the overall guidance and control system.

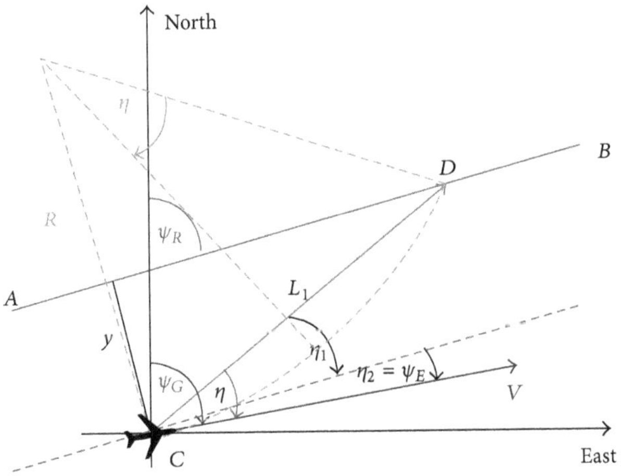

FIGURE 7: Nonlinear guidance logic setup.

TABLE 1: L_1 (m) for different flight speeds and ω_n.

Speed (m/s)	ω_n = 0.1 rad/s	ω_n = 0.2 rad/s	ω_n = 0.3 rad/s
40	565	280	190
50	705	350	235
60	850	425	285

TABLE 2: ϕ_c for different cross-track errors (ω_n = 0.2 rad/s).

Cross-track deviation (% of L_1)	ϕ_c (deg)
12.5	23
25	31
50	43
100	55

possibly high overshoots. A larger L_1 will make the system slower and reduce overshoot. This is discussed in [8] with reference to gain selection for a conventional proportional-derivative guidance law where the gains are scheduled with the magnitude of the cross-track error. It may be noted here that to cater for large cross-track errors (which are possible as discussed above), L_1 will have to be kept unduly large which may yield poor performance otherwise. On the other hand, if a relatively small L_1 is selected, the resulting roll angle might become too large and exceed the set limit. Table 2 shows the commanded roll angle ϕ_c for different cross-track errors for a fixed speed and L_1 chosen to yield an ω_n of 0.2 rad/sec. The last row of the table gives a ϕ_c of 55°, which may be too large for most vehicles, and thereafter for cross tracks larger than L_1, ϕ_c becomes undefined. A solution to this is to introduce an adaptive adjustment of L_1 with cross-track error [9]. We introduce a compensated reference length $\hat{L}_1$ to be used in (19) instead of a fixed L_1:

$$\hat{L}_1 = L_1 + k_1 |y|, \quad (21)$$

where $k_1 > 0$ is a gain to be tuned (a value of 1.5 for our application gave good results). Roll angle commands are computed using L_1 and $\hat{L}_1$ and compared (for $V = 50$ m/s and $\omega_n = 0.2$ rad/s) in Table 3. The table shows that for large cross-track errors, the roll angle command computed using $\hat{L}_1$ remains reasonably small and can be designed to lie within a desired range.

4.4. Addition of Integral Action. The roll angle command as given in (19) is modified by addition of an integrator for strict path tracking in the presence of constant (wind) disturbances. Addition of an integrator in the loop can pose problems such as wind-up leading to control surface saturation, for which the following actions are taken:

(i) integral action is activated only when the cross-range error becomes smaller than a set threshold y_{th},

TABLE 3: ϕ_c computed using L_1 and $\widehat{L}_1$.

Cross-track deviation (% of L_1)	ϕ_c (deg) using fixed L_1	ϕ_c (deg) using $\widehat{L}_1$
12.5	23	19
25	31	21
50	43	21
100	55	19
200	Undefined	14

(ii) the integrator state is limited to a predefined value; the integrator is switched off if its state exceeds the set limit.

With the addition of integral action and substitution of $\widehat{L}_1$ in place of L_1, (19) becomes

$$\phi_c = \tan^{-1}\left[\frac{2V^2}{g\widehat{L}_1}\sin\left\{\sin^{-1}\left(\frac{y}{\widehat{L}_1}\right)+\psi_E\right\}\right]+k_2\int \widehat{y}dt, \quad (22)$$

where the integral gain k_2 is tuned by simulation and

$$\widehat{y} = \begin{cases} y, & \text{if } y \le y_{\text{th}}, \\ 0, & \text{if } y > y_{\text{th}}. \end{cases} \quad (23)$$

5. Robust Multivariable Controller Design

5.1. The Plant Model. The lateral-directional control system is a multivariable control system as shown in Figure 6. The roll and pitch angles are sensed by the vertical gyro which has very fast dynamics as compared to the vehicle, and so this sensor is modelled as a simple gain in the feedback loop. The magnetic sensor (which senses the heading angle) dynamics are also neglected. The model of the UAV for controller design is taken as a linear approximation obtained at a cruise altitude of 2000 m and a speed of 42 m/sec. However, the flight envelope consists of an altitude range from 10 to 5000 m and a speed bracket from 35 to 60 m/sec. A number of linear models are available across the flight envelope to test the robustness of the control system at different operating conditions. Note that the plant has 2 inputs and 2 outputs. The inputs are the aileron and rudder deflection commands to the actuators. The outputs are the roll angle and the yaw-rate of the vehicle. The computation of the yaw-rate and its filtering is discussed below, but first, we look at computation of the heading angle.

5.1.1. Computation of the Heading Angle. The magnetic sensor installed on-board the vehicle measures the three components of the magnetic field of the earth in the body axes. This magnetic field vector is resolved into a local level frame using the pitch and roll measurements from the vertical gyro. We define the local level axis system ($X_lY_lZ_l$) in which the Z_l axis is upwards along the local vertical and the X_l and Y_l axes are in the local horizontal plane, with the Y_l axis directed along the projection of the longitudinal axis of the vehicle in the local level plane. The pitch and roll rotations θ and ϕ occur about the X_l and Y_l' axes (the Y_l' axis is obtained by rotating the Y_l axis by the pitch angle θ), respectively. The transformation matrix from the local level to the body frame is given by

$$T_{lb} = \begin{bmatrix} \cos\phi & \sin\phi\sin\theta & -\sin\phi\cos\theta \\ 0 & \cos\theta & \sin\theta \\ \sin\phi & -\sin\theta\cos\phi & \cos\phi\cos\theta \end{bmatrix}. \quad (24)$$

The magnetic field vector is transformed from the body axes to the local level axes by $M_l = T_{lb}^T M_b$, where M_b and M_l represent the magnetic field vector in the body and local level axes, respectively. The magnetic heading is now determined as $\tan^{-1}(M_{lx}/M_{ly})$, where M_{lx} and M_{ly} are the first two components of M_l. Note that this gives the magnetic heading of the vehicle, the true heading ψ (w.r.t. the true north) can be found by applying the declination correction to the magnetic heading [10].

5.1.2. Yaw-Rate Computation. Feedback of the yaw-rate is considered useful for providing damping to the Dutch-roll mode of the vehicle. The vertical gyro however does not provide measurement of angular rates; it only measures roll and pitch angles. The magnetic heading sensor measures the heading angle. Body angular rates are computed from attitude angles as follows.

We define a navigation axis system ($X_nY_nZ_n$) in which the Z_n axis is upwards along the local vertical, and the X_n and Y_n axes are in the local horizontal plane directed eastward and northward, respectively. The (true) heading, pitch, and roll angles are denoted by ψ, θ, and ϕ, respectively, and these rotations occur in this specific order to align the navigation axes with the body axes. The heading rotation occurs first about the Z_n axis, followed by the pitch rotation about the X_n' axis, finally followed by the roll rotation about the Y_n'' axis (note that the primed axes are the intermediate axes obtained while going from the navigation to the body frame through the ψ, θ, and ϕ rotation angles). Resolving the Euler angle rates $\dot{\psi}$, $\dot{\theta}$, and $\dot{\phi}$ into body axes, we can solve for the body axes rates. Denoting the roll, pitch, and yaw rates in the body axes by P, Q, and R, respectively, we have

$$\begin{aligned} P &= \dot{\psi}\sin\theta + \dot{\phi}, \\ Q &= \dot{\theta}\cos\phi - \dot{\psi}\cos\theta\sin\phi, \\ R &= \dot{\theta}\sin\phi + \dot{\psi}\cos\theta\cos\phi. \end{aligned} \quad (25)$$

It may be noted that the attitude sensors measure the body angles and not their rates. The derivatives of the attitude angles are computed by fitting a least squares line to n consecutive attitude measurements. We have chosen $n = 4$ here as we have seen it to yield good results; the derivative, thus, computed has enough noise smoothing and an acceptably small time delay. However, other choices for n can be made depending on the application and on the sampling time of the attitude sensor measurements. If, for example, the pitch angle measurements from the vertical gyro are denoted by θ_0, θ_1, θ_2, and θ_3, where θ_0 corresponds to the current measurement and θ_3 corresponds to the measurement taken three samples

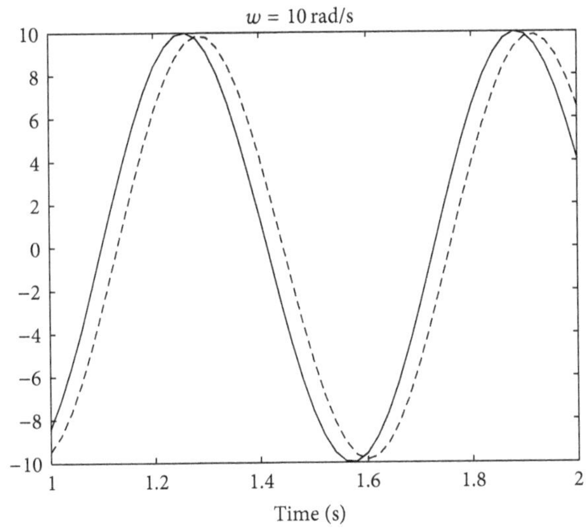

FIGURE 8: Delay introduced because of the 4-point derivative approximation.

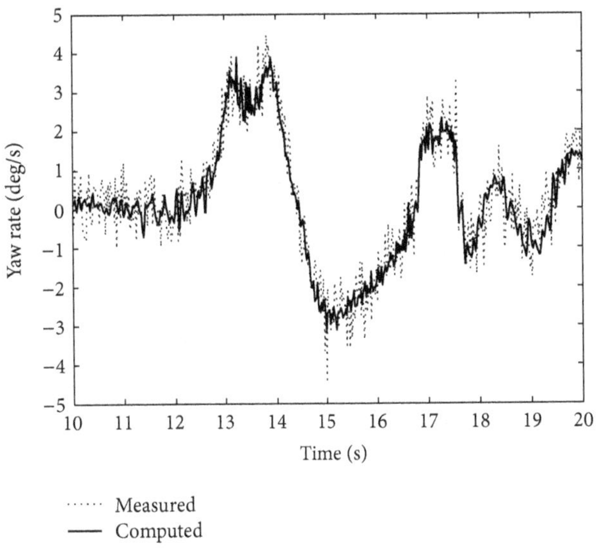

FIGURE 9: Measured and computed yaw rates during test flight.

previously, then the slope of the least squares line for these points approximates the derivative [11]:

$$\dot{\theta} \approx \frac{3\theta_0 + \theta_1 - \theta_2 - 3\theta_3}{10\Delta t}. \quad (26)$$

Here, Δt is the time interval at which the angle θ is sampled (20 msec for our case). This approximation of the derivative by slope of a 4-point least squares line gives a good compromise between sensitivity to measurement noise and the delay introduced into the estimation of $\dot{\theta}$. Figure 8 shows the delay between the actual and computed derivatives for a pure sinusoid of frequency 10 rad/sec, which is about 25 msec. This method of estimating body rates and using them for feedback control is demonstrated to work well for normal flight maneuvers. A test flight was carried out in which a rate sensor was specially installed on the vehicle for validation purposes only; this sensor is not available for feedback control otherwise. The comparison between the measured and computed yaw-rates shown in Figure 9 indicates the validity of the approximation. It is, thus, concluded that the approximation (26) can be used for Euler angle rate estimation from attitude measurements, and thereafter (25) can be employed for transforming the Euler angle rates into body-axes angular rates which are used for feedback control.

5.1.3. The Washout Filter. Yaw-rate feedback is typically useful during transients when the yaw-rate fluctuates or exhibits appreciable changes with time. The rudder is normally employed to suppress the yaw-rate oscillations to provide damping to the Dutch-roll mode of the vehicle. Thus, during steady turns when there is a nearly constant yaw-rate, it is not desirable for the rudder to act to oppose the turning motion. The washout filter is a high-pass filter that is used to feedback the yaw-rate at higher frequencies, while effectively breaking the feedback at dc to enable steady turning of the vehicle. The washout filter is chosen as $W_w = s/(s+1)$; it may be thought of as a frequency-dependent weight on the yaw-rate feedback, with a small gain at low frequencies and unity gain at high frequencies. The 3 dB point occurs at 1 rad/sec, so that frequencies above 1 rad/sec will be available for feedback.

5.2. H_∞ Loop-Shaping Design Procedure. For controller design, we shall use the H_∞ loop shaping design procedure proposed in [12]. The procedure is intuitive in that it is based on the multivariable generalization of classical loop-shaping ideas. The open-loop plant, once given the desired loop-shape, is robustly stabilized against coprime factor uncertainty. The resulting controller has been shown to enjoy some favourable properties, such as no pole-zero cancellation occurs in the closed-loop system (except for a certain special class of plants), see [13]. In addition, the controllers, thus, designed have been successful in various applications; examples are those described in [14–17].

In practical design applications, the performance specifications are first translated into the frequency domain, and the open-loop plant's singular value frequency response is given the desired shape. This is achieved by augmentation of the nominal plant model G by pre- and/or postcompensators (or weighting functions) W_1 and W_2, respectively. The shaped plant $G_s = W_2 G W_1$ is then robustly stabilized against coprime factor uncertainty, and the controller K, thus, obtained is cascaded with the weights to obtain the final controller $W_1 K W_2$. It can be shown that the controller does not significantly alter the specified loop-shape provided a sufficiently small value of the cost γ is achieved; for details, refer to [17].

We now outline a design procedure for designing robust controllers based on open-loop shaping and robust stabilization of the normalized coprime factors of the plant. The procedure consists of the following main steps.

(1) Plot the singular value frequency response of the open-loop plant $G(s)$. Based on this, select a precompensator $\widehat{W}_1$ and/or a postcompensator W_2 to give the plant a desired open-loop shape. Form the product $W_2(s)G(s)\widehat{W}_1(s)$.

(2) Align the singular values of $W_2G\widehat{W}_1$ at the desired bandwidth. The align gain K_a is the approximate real inverse of the system at the specified frequency. The cross-over (and hence the bandwidth) is thus adjusted to approximately the align frequency. An additional constant diagonal matrix K_g may sometimes be used in front of the align gain to exercise control over actuator usage. It is chosen so that the various actuator rate limits are not exceeded whilst following references or rejecting disturbances. The precompensator can now be written as $W_1 = \widehat{W}_1K_aK_g$; see Figure 10. Build the shaped plant $G_s(s) = W_2(s)G(s)W_1(s)$, and calculate the optimal cost γ_{opt} [12]. A high value (typically > 10) of γ_{opt} indicates that the specified loop-shape is inconsistent with robust stability; in such a case, the weights W_1 and W_2 should be modified.

(3) Use a slightly suboptimal value of γ, and compute the corresponding controller.

(4) Cascade the controller with the weights W_1 and W_2. Controller order reduction may be performed if desired.

(5) Form the closed-loop, and check the appropriate performance and robustness measures against the given specification.

5.3. Controller Design. Here, we will describe the design of the lateral controller, using the design procedure listed in Section 5.2. The block diagram of the closed-loop system is shown in Figure 11.

(1) Singular values of the open-loop plant $G(s)$ are plotted, and these indicate the need for boosting the low frequency gain for good tracking and disturbance rejection for the roll control channel at these frequencies. The low frequency gain is boosted by introducing integral action in the roll control loop; we choose to place a pole at $s = -0.01$ instead of using a pure integrator (because of the wind-up problems associated with integrators, and also because we do not want to introduce an unstable pole into the shaped plant G_s). $\widehat{W}_1$ in Figure 10 is chosen as $\begin{bmatrix} 28(s+2.5)/(s+0.01) & 0 \\ 0 & 40 \end{bmatrix}$ and W_2 as $\begin{bmatrix} 14(s+2)/(s+70) & 0 \\ 0 & 50/(s+50) \end{bmatrix}$. Since we are aiming for a closed-loop bandwidth of approximately 10–15 rad/sec, therefore zeros are introduced at −2.5 and −2 in $\widehat{W}_1$ and W_2 to reduce the roll-off at the cross-over frequencies. The poles at −70 and −50 in W_2 are placed to provide adequate roll-off at higher frequencies. Since $\widehat{W}_{11}$ (the upper left element of $\widehat{W}_1$) approximates an integrator, it will be implemented in its conditioned form (see [18, 19]). Gains are selected in $\widehat{W}_1$ and W_2 to adjust the cross-over frequency to around 10 rad/sec.

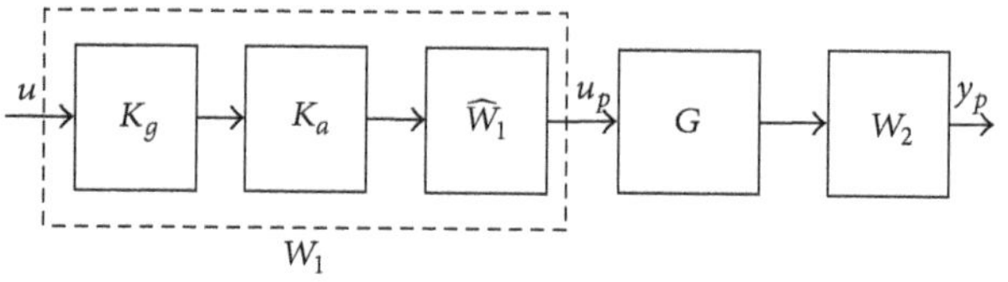

Figure 10: The shaped plant and the controller.

(2) The align matrix K_a is chosen to be the identity matrix for robustness reasons (aligning the nominal plant and designing the controller based on this may not hold good with other models of the plant across the flight envelope) and since both singular values are close enough at cross-over. The matrix K_g is also kept as the identity matrix I_2. The precompensator is therefore $W_1 = \widehat{W}_1K_aK_g = \widehat{W}_1$. The shaped plant $G_s = W_2GW_1$ is now formed, and its singular values are shown in Figure 12. The low frequency gain in the roll channel is boosted, as indicated by the larger singular value. The washout filter in the yaw-rate channel means that the low frequency gain in this channel remains low, as indicated by the smaller singular value of the shaped plant.

(3) An optimal γ of 1.925 is computed; the controller is obtained for a slightly suboptimal γ of 2.02.

(4) The controller is reduced using optimal Hankel-norm approximation and cascaded with the weights W_1 and W_2 to form the complete controller W_1KW_2. This reduced controller is discretized using the bilinear (Tustin's) approximation for implementation.

The singular values of the sensitivity function are shown in Figure 13. One of the singular values, the one that remains flat at low frequencies, corresponds to yaw-rate and indicates no disturbance rejection on that output. The other singular value shows adequate disturbance rejection on the roll angle channel for frequencies less than 5 rad/sec. A number of linearized models are taken along the periphery of the flight envelope. Step responses for these plant models in the envelope using the designed controller are given in Figure 14. The controller performs acceptably well throughout the envelope, and the robustness of the design is illustrated.

5.4. Robustness Analysis. A more formal robustness analysis of the controller is now performed. Note that the controller design is based on a normalized coprime factor representation of the plant: $G_s = W_2GW_1 = \widetilde{M}^{-1}\widetilde{N}$, where the transfer functions $\widetilde{M}$ and $\widetilde{N}$ represent the normalized coprime factors of the (shaped) plant. The controller is designed to maximize robust stability in the face of coprime factor perturbations. The perturbations (or uncertainties) in the plant are represented as perturbations on the normalized coprime factors of the plant. The perturbed plant G_{s_Δ} is given by

$$G_{s_\Delta} = \left(\widetilde{M} + \Delta_{\widetilde{M}}\right)^{-1}\left(\widetilde{N} + \Delta_{\widetilde{N}}\right), \qquad (27)$$

where $\Delta_{\widetilde{M}}$, $\Delta_{\widetilde{N}}$ are stable unknown transfer functions representing uncertainty in the plant model. Robust stability to

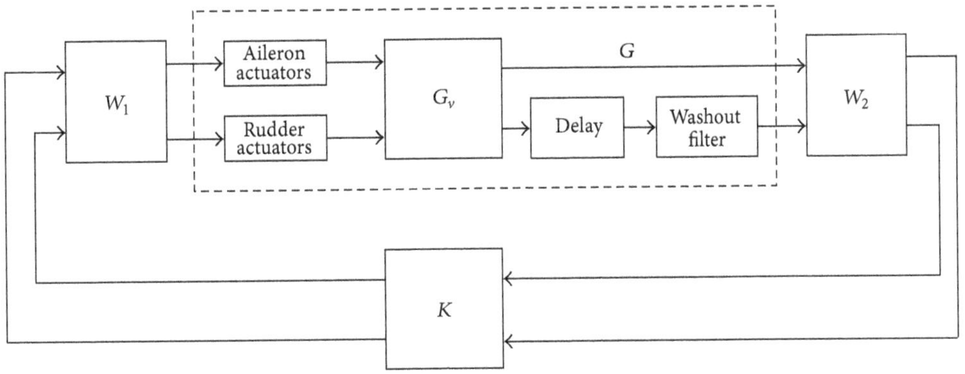

FIGURE 11: Plant, controller, and weighting functions.

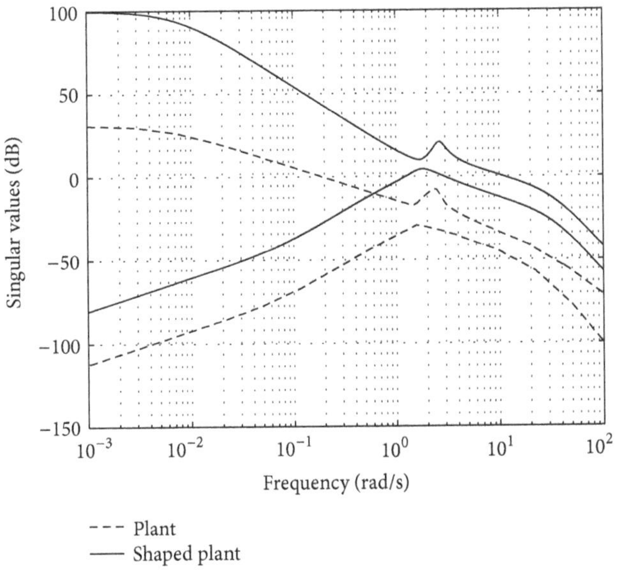

FIGURE 12: Singular values of the original and shaped plants.

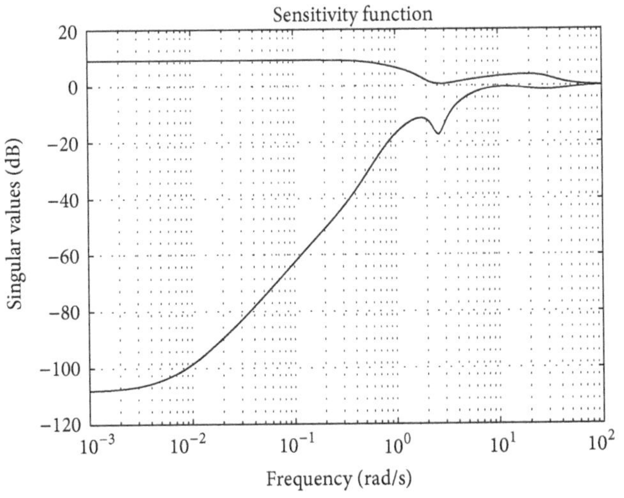

FIGURE 13: Sensitivity function $(I + GW_1KW_2)^{-1}$.

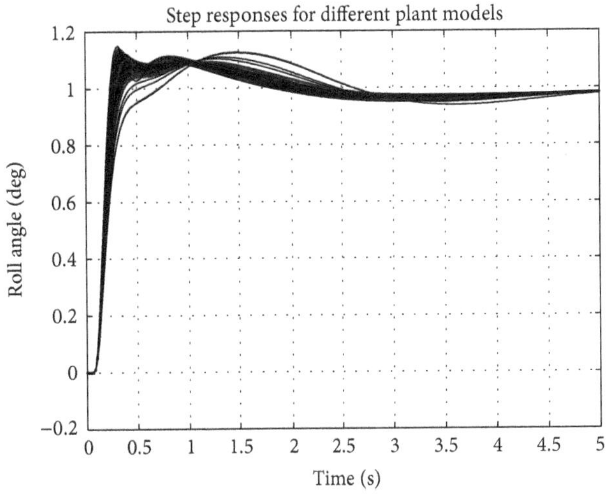

FIGURE 14: Step responses for different plant models across the flight envelope.

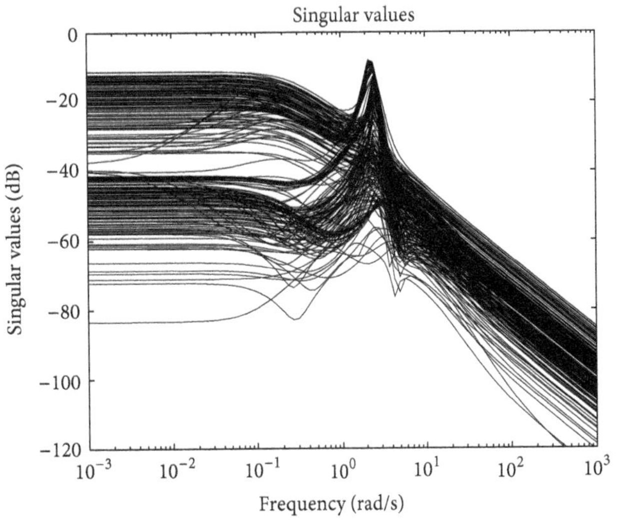

FIGURE 15: Singular values of $(\Delta_{\widetilde{M}} \;\; \Delta_{\widetilde{N}})$ for different plants in the envelope.

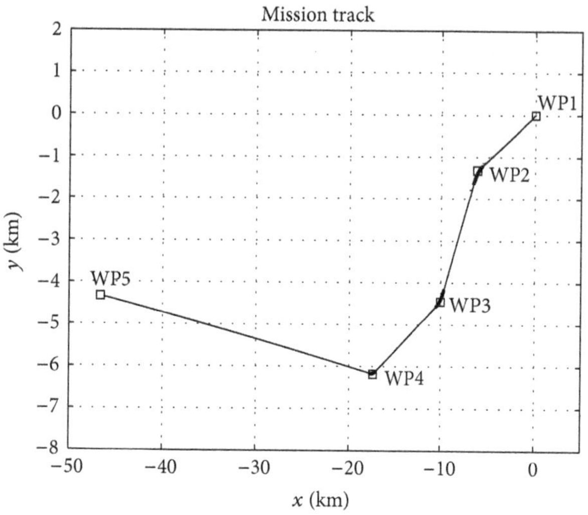

FIGURE 16: Part of Mission 1 and the path flown by the vehicle.

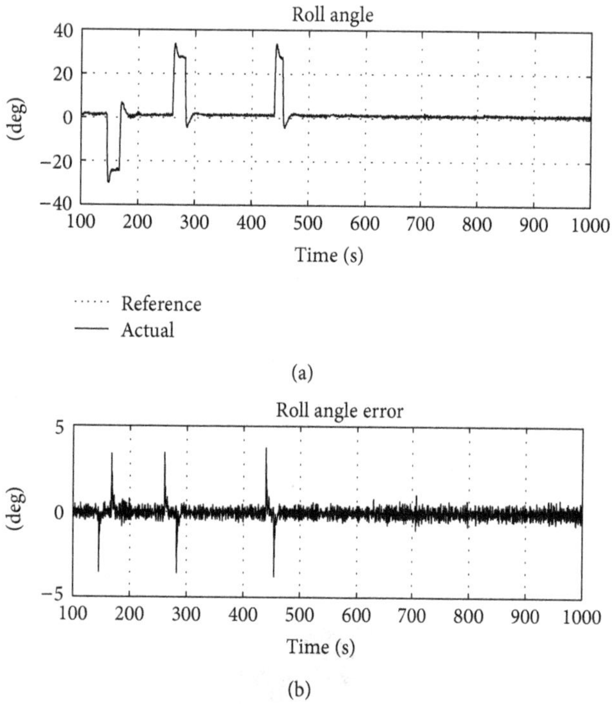

FIGURE 17: The roll angle and the roll angle error.

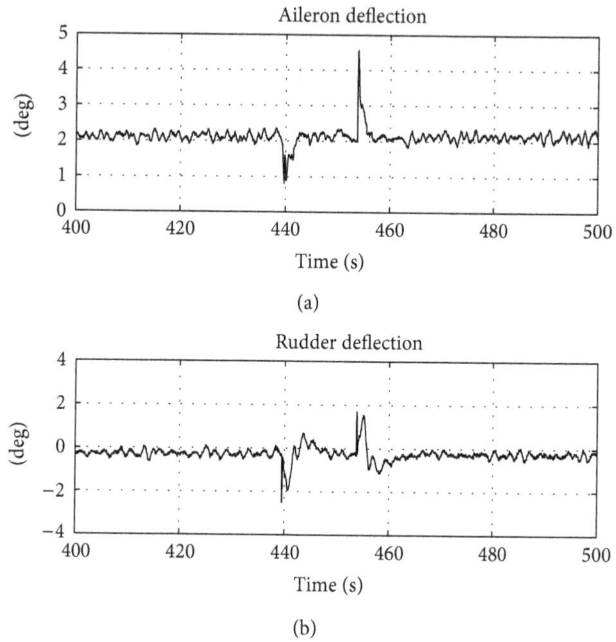

FIGURE 18: The actuator (aileron and rudder) deflections.

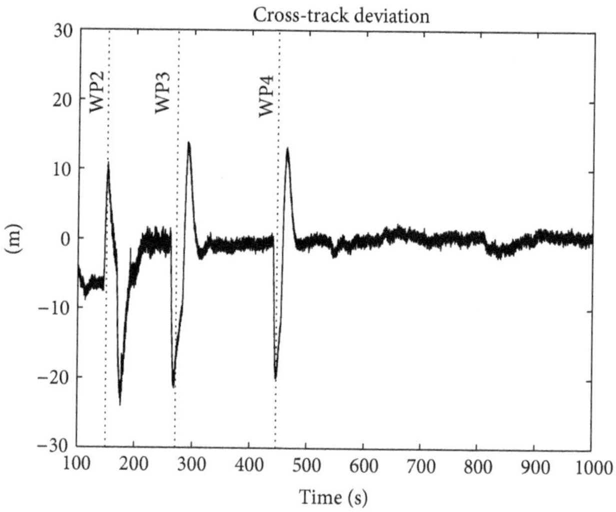

FIGURE 19: Cross-track deviation of the vehicle.

coprime factor perturbations is maximized by a controller K which stabilizes the nominal plant G_s and minimizes

$$\gamma = \left\| \begin{bmatrix} K \\ I \end{bmatrix} (I - G_s K)^{-1} \widetilde{M}^{-1} \right\|_\infty . \tag{28}$$

The perturbed closed-loop system is then guaranteed to remain stable in the face of all $\Delta_{\widetilde{M}}, \Delta_{\widetilde{N}}$ such that [20]

$$\left\| [\Delta_{\widetilde{M}} \ \ \Delta_{\widetilde{N}}] \right\|_\infty < \gamma^{-1}. \tag{29}$$

Robust stability across the flight envelope is tested by obtaining normalized coprime factorizations of all plants in the envelope and then taking their differences from the nominal plant to get uncertainty transfer functions $[\Delta_{\widetilde{M}_i} \ \ \Delta_{\widetilde{N}_i}]$, where the subscript i refers to the i"th" plant in the envelope. Singular values of these uncertainty transfer functions are plotted in Figure 15; the maximum of these across all frequencies equals 0.42. The cost γ for our controller is 2.02, which indicates that (29) is satisfied, thus ensuring robust stability for the family of plants in the envelope; that is,

$$\left\| [\Delta_{\widetilde{M}_i} \ \ \Delta_{\widetilde{N}_i}] \right\|_\infty = 0.42 < 2.02^{-1}. \tag{30}$$

It is seen that a single controller is robust enough to stabilize all plants in the flight envelope; also robust performance is

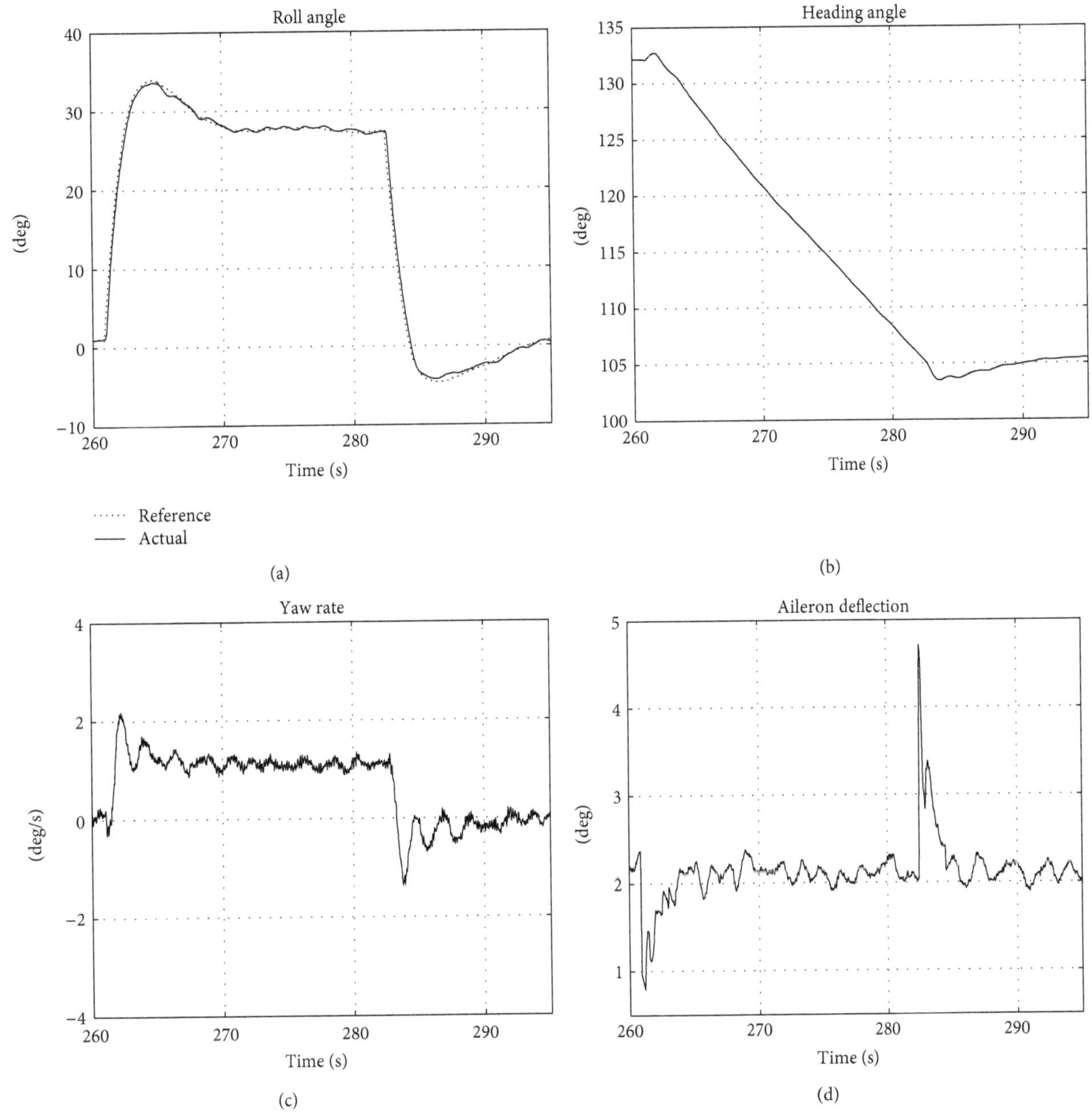

FIGURE 20: The turn performance.

checked for and is found to be adequate by performing step response analysis of the family of plants (Figure 14).

6. Flight Test Results

The complete guidance and control scheme is implemented on a real-time embedded computer and flown on the actual UAV for various missions. Part of the two typical missions and their flight results are presented here. Figure 16 shows part of the first mission in which waypoints are indicated by small squares, and the actual path flown by the vehicle is shown in solid line. The thick portions of the line indicate that the vehicle is in a state of turn; that is, the turn start criterion is satisfied (see Section 3.2). As soon as the turn stop criterion is met, the line thickness is changed back to normal. The figure shows very good path following by the vehicle during flight. Figure 17 shows the performance of the multivariable roll/yaw-rate controller during flight. In the top part of the figure the commanded and actual roll angles are plotted in dotted and solid lines, respectively. The two lines are almost on top of one another, thus demonstrating the tracking performance of the roll controller. The lower part of the figure shows the roll error which peaks to a maximum value of around 3 degrees at times when there is a sudden change in the roll reference; the error in the steady-state is almost zero. Figure 18 shows the actuator (aileron and rudder) deflections

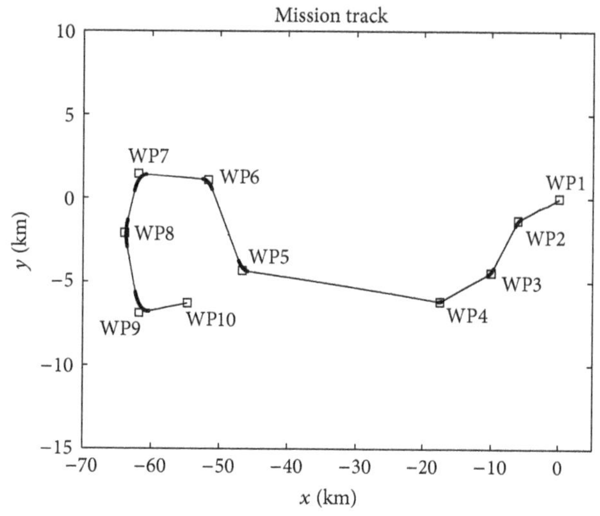

FIGURE 21: Part of Mission 2 and the path flown by the vehicle.

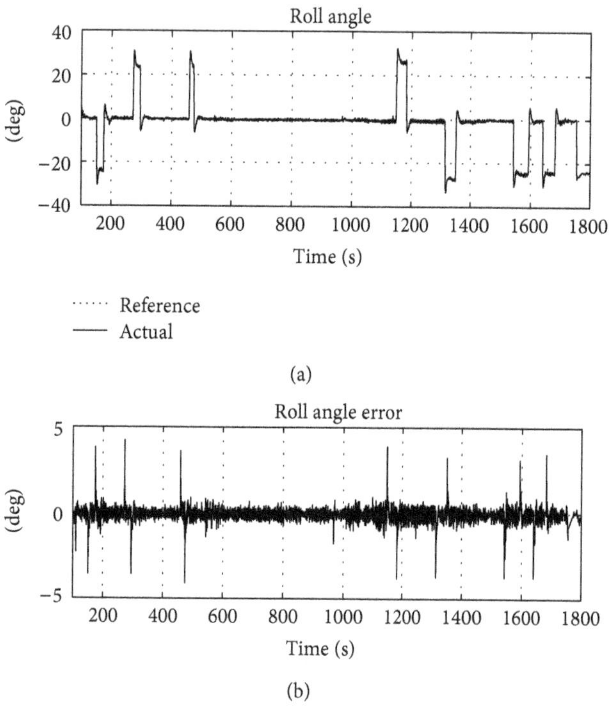

FIGURE 22: The roll angle and the roll angle error for Mission 2.

recorded during the same flight. The figure shows a 100 second zoomed view of the flight data. The deflections are small, and the aileron is deflected to about 5 degrees, whereas the rudder deflection stays within 2 degrees during a turn. The multivariable controller engages both actuators to provide good roll angle tracking and stability augmentation. Figure 19 shows the cross-track deviation of the vehicle from the desired path and thus illustrates the tracking performance of the guidance and control system. The performance is excellent with the off-track deviation controlled to within a few meters. The larger 10–20 m errors arise only when a turn is initiated or when one ends, since the reference path

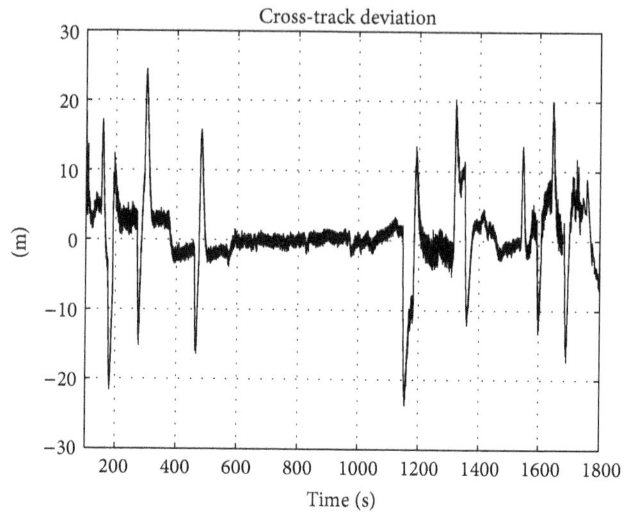

FIGURE 23: Cross-track deviation of the vehicle for Mission 2.

to be followed suddenly changes at these instants. Figure 20 shows a zoomed view of different variables of interest related to the turn performance of the vehicle. The bank-to-turn scheme is illustrated as the rolling of the vehicle causes the heading angle to change continuously. The yaw-rate of the vehicle is also shown (which is fed back to the controller through the washout filter). The washout filter blocks the yaw-rate feedback during steady turns and only feeds through the higher frequency transients for stabilization of the Dutch-roll mode of the vehicle.

Figure 21 shows part of the second mission with the waypoints indicated by small boxes and also the actual path flown by the vehicle. The line width is increased to indicate turnings. The top part of Figure 22 shows the commanded and actual roll angles, the commanded angle being generated by the guidance (22). The roll angle tracking provided by the multivariable controller is quite good. The bottom half of the figure shows the roll error which again is seen to be very small and peaks only during transients. The cross-track deviation is shown in Figure 23.

7. Conclusion

A practical methodology for the design of lateral-directional guidance and control systems for high performance UAVs is presented. The methodology has been developed keeping in view the practical aspects of the problem and implementation on a real-time platform. Flight path formulae are derived both for straight and turning flight. A nonlinear lateral guidance law is presented that uses the angular error and the cross-track deviation of the vehicle and generates roll angle commands to guide the vehicle back onto the desired path. An adaptive adjustment in the basic law is suggested to cater for large deviations from the nominal track as may arise in case of GPS outages. A robust multivariable controller is designed for the lateral-directional control of the vehicle. The controller uses measurements of the roll angle and yaw-rate of the vehicle and optimally actuates the ailerons and the rudder to provide roll control and Dutch-roll damping. A practical method of computing the heading angle from the magnetometer and then estimating the yaw-rate is presented. The robustness analysis of the controller is performed, and it is seen to perform well throughout the flight envelope of the vehicle. Finally, the flight test results are presented and discussed; these demonstrate the overall performance of the guidance and control system.

References

[1] M. Niculescu, "Lateral track control law for Aerosonde UAV," in *Proceedings of the 39th AIAA Aerospace Sciences Meeting and Exhibit*, Reno, Nev, USA, January 2001.

[2] T. Keviczky and G. J. Balas, "Software-enabled receding horizon control for autonomous UAV guidance," in *Proceedings of the AIAA Guidance, Navigation, and Control Conference*, pp. 527–552, San Francisco, Calif, USA, August 2005.

[3] B. Mettler, M. Valenti, T. Schouwenaars et al., "Autonomous UAV guidance build-up: flight-test demonstration and evaluation plan," in *Proceedings of the AIAA Guidance, Navigation, and Control Conference and Exhibit*, Austin, Tex, USA, August 2003.

[4] T. Schouwenaars, M. Valenti, E. Feron, and J. How, "Implementation and flight test results of MILP-based UAV guidance," in *Proceedings of the IEEE Aerospace Conference*, March 2005.

[5] S. R. Griffiths, "Vector field approach for curved path following for miniature aerial vehicles," in *Proceedings of the AIAA Guidance, Navigation, and Control Conference*, pp. 3375–3389, Keystone, Colo, USA, August 2006.

[6] W. J. Pisano, D. A. Lawrence, and P. C. Gray, "Autonomous UAV control using a 3-sensor autopilot," in *Proceedings of the AIAA InfoTech at Aerospace Conference*, pp. 423–436, Rohnert Park, Calif, USA, May 2007.

[7] S. Park, J. Deyst, and J. P. How, "A new nonlinear guidance logic for trajectory tracking," in *Proceedings of the AIAA Guidance, Navigation, and Control Conference*, pp. 941–956, Providence, RI, USA, August 2004.

[8] R. Samar, S. Ahmed, and F. Aftab, "Lateral control with improved performance for UAVs," in *Proceedings of the 17th IFAC Symposium on Automatic Control in Aerospace (ACA '07)*, pp. 37–42, Toulouse, France, June 2007.

[9] R. Samar, S. Ahmed, and M. Nzar, "Lateral guidance & control design for an unmanned aerial vehicle," in *Proceedings of the 17th World Congress, International Federation of Automatic Control (IFAC '08)*, Seoul, South Korea, July 2008.

[10] J. M. Quinn, R. J. Coleman, M. R. Peck, and S. E. Lauber, "The joint US/UK, 1990 epoch world magnetic model," Tech. Rep. 304, Naval Oceanographic Office, October 1991.

[11] J. H. Mathews, *Numerical Methods for Mathematics, Science, and Engineering*, Prentice-Hall, Englewood Cliffs, NJ, USA, 2nd edition, 1992.

[12] D. McFarlane and K. Glover, "A loop-shaping design procedure using H∞ synthesis," *IEEE Transactions on Automatic Control*, vol. 37, no. 6, pp. 759–769, 1992.

[13] M. C. Tsai, E. J. M. Geddes, and I. Postlethwaite, "Pole-zero cancellations and closed-loop properties of an H∞ mixed sensitivity design problem," *Automatica*, vol. 28, no. 3, pp. 519–530, 1992.

[14] R. Samar, G. Murad, I. Postlethwaite, and D. W. Gu, "A discrete time H∞ observer-based controller & its application to a glass tube production process," *European Journal of Control*, vol. 2, no. 2, pp. 112–125, 1996.

[15] A. J. Smerlas, D. J. Walker, I. Postlethwaite, M. E. Strange, J. Howitt, and A. W. Gubbels, "Evaluating H∞ controllers on the NRC Bell 205 fly-by-wire helicopter," *Control Engineering Practice*, vol. 9, no. 1, pp. 1–10, 2001.

[16] I. Postlethwaite, E. Prempain, E. Turkoglu, M. C. Turner, K. Ellis, and A. W. Gubbels, "Design and flight testing of various H∞ controllers for the Bell 205 helicopter," *Control Engineering Practice*, vol. 13, no. 3, pp. 383–398, 2005.

[17] S. Skogestad and I. Postlethwaite, *Multivariable Feedback Control Analysis and Design*, John Wiley & Sons, West Sussex, UK, 2nd edition, 2005.

[18] R. Hanus, M. Kinnaert, and J. L. Henrotte, "Conditioning technique, a general anti-windup and bumpless transfer method," *Automatica*, vol. 23, no. 6, pp. 729–739, 1987.

[19] R. Hanus and M. Kinnaert, "Control of constrained multivariable systems using the conditioning technique," in *Proceedings of the American Control Conference*, pp. 1712–1718, Pittsburgh, Penn, USA, June 1989.

[20] D. C. McFarlane and K. Glover, *Robust Controller Design Using Normalized Coprime Factor Plant Descriptions*, vol. 138 of *Lecture Notes in Control and Information Sciences*, Springer, Berlin, Germany, 1990.

15

Approximate Solutions to Nonlinear Optimal Control Problems in Astrodynamics

Francesco Topputo and Franco Bernelli-Zazzera

Department of Aerospace Science and Technology, Politecnico di Milano, Via La Masa 34, 20156 Milano, Italy

Correspondence should be addressed to Francesco Topputo; francesco.topputo@polimi.it

Academic Editors: H. Baoyin, C. Bigelow, and D. Yu

A method to solve nonlinear optimal control problems is proposed in this work. The method implements an approximating sequence of time-varying linear quadratic regulators that converge to the solution of the original, nonlinear problem. Each subproblem is solved by manipulating the state transition matrix of the state-costate dynamics. Hard, soft, and mixed boundary conditions are handled. The presented method is a modified version of an algorithm known as "approximating sequence of Riccati equations." Sample problems in astrodynamics are treated to show the effectiveness of the method, whose limitations are also discussed.

1. Introduction

Optimal control problems are solved with indirect or direct methods. Indirect methods stem from the calculus of variations [1, 2]; direct methods use a nonlinear programming optimization [3, 4]. Both methods require the solution of a complex set of equations (Euler-Lagrange differential equations or Karush-Kuhn-Tucker algebraic equations) for which iterative numerical methods are used. These iterative procedures implement some form of Newton's method to find the zeros of a nonlinear function. They are initiated by providing an initial guess solution. Guessing an appropriate initial solution is not trivial and requires a deep knowledge of the problem at hand. In indirect methods, the initial value of the Lagrange multiplier has to be provided, whose lack of physical meaning makes it difficult to formulate a good guess. In direct methods, the initial trajectory and control have to be guessed at discrete points over the whole time interval.

This paper presents an approximate method to solve nonlinear optimal control problems. This is a modification of the method known as "approximating sequence of Riccati equations" (ASRE) [5, 6]. It transforms the nonlinear dynamics and objective function into a pseudolinear and quadratic-like structure, respectively, by using state- and control-dependent functions. At each iteration, these functions are evaluated by using the solutions at the previous iteration, and therefore, a series of time-varying linear quadratic regulators is treated. This sequence is solved with a state transition matrix approach, where three different final conditions are handled: final state fully specified, final state not specified, and final state not completely specified. These define hard, soft, and mixed constrained problems, respectively.

The main feature of the presented method is that it does not require guessing any initial solution or Lagrange multiplier. In fact, iterations start by evaluating the state- and control-dependent functions using the initial condition and zero control, respectively. The way the dynamics and objective function are factorized recalls the state-dependent Riccati equations (SDRE) method [7–9]. These two methods possess some similarities, although the way they solve the optimal control problem is different. As the method is approximated, suboptimal solutions are derived. These could be used as first guess solutions for either indirect or direct methods.

2. The Nonlinear Optimal Control Problem

The optimal control problem requires that, given a set of n first-order differential equations

$$\dot{\mathbf{x}} = \mathbf{f}(\mathbf{x}, \mathbf{u}, t), \tag{1}$$

the m control functions $\mathbf{u}(t)$ must be determined within initial, final time t_i, t_f, such that the performance index

$$J = \varphi\left(\mathbf{x}\left(t_f\right), t_f\right) + \int_{t_i}^{t_f} L(\mathbf{x}, \mathbf{u}, t)\, dt \tag{2}$$

is minimized while satisfying $n + q$ two-point conditions

$$\mathbf{x}(t_i) = \mathbf{x}_i, \qquad \psi\left(\mathbf{x}\left(t_f\right), t_f\right) = 0. \tag{3}$$

The problem consists in finding a solution that represents a stationary point of the augmented performance index

$$\begin{aligned}\bar{J} &= \varphi\left(\mathbf{x}\left(t_f\right), t_f\right) + \boldsymbol{\nu}^T \psi\left(\mathbf{x}\left(t_f\right), \mathbf{u}\left(t_f\right), t_f\right) \\ &+ \int_{t_i}^{t_f} \left[L(\mathbf{x}, \mathbf{u}, t) + \boldsymbol{\lambda}^T (\mathbf{f}(\mathbf{x}, \mathbf{u}, t) - \dot{\mathbf{x}})\right] dt,\end{aligned} \tag{4}$$

where $\boldsymbol{\lambda}$ is the vector of costate and $\boldsymbol{\nu}$ is the multiplier of the boundary condition. The necessary conditions for optimality, also referred to as Euler-Lagrange equations, are

$$\dot{\mathbf{x}} = \frac{\partial H}{\partial \boldsymbol{\lambda}}, \qquad \dot{\boldsymbol{\lambda}} = -\frac{\partial H}{\partial \mathbf{x}}, \qquad \frac{\partial H}{\partial \mathbf{u}} = 0, \tag{5}$$

where H, the Hamiltonian, is

$$H(\mathbf{x}, \boldsymbol{\lambda}, \mathbf{u}, t) = L(\mathbf{x}, \mathbf{u}, t) + \boldsymbol{\lambda}^T \mathbf{f}(\mathbf{x}, \mathbf{u}, t). \tag{6}$$

The differential-algebraic system (5) must be solved together with the final boundary conditions (3) and the transversality conditions

$$\boldsymbol{\lambda}\left(t_f\right) = \left[\frac{\partial \varphi}{\partial \mathbf{x}} + \left(\frac{\partial \psi}{\partial \mathbf{x}}\right)^T \boldsymbol{\nu}\right]_{t=t_f}, \tag{7}$$

which define a differential-algebraic parametric two-point boundary value problem whose solution supplies $\boldsymbol{\nu}$ and the functions $\mathbf{x}(t)$, $\boldsymbol{\lambda}(t)$, $\mathbf{u}(t)$, $t \in [t_i, t_f]$.

3. The Approximating Sequence of Riccati Equations

Let the controlled dynamics (1) be rewritten in the form

$$\dot{\mathbf{x}} = A(\mathbf{x}, t)\, \mathbf{x} + B(\mathbf{x}, \mathbf{u}, t)\, \mathbf{u}, \tag{8}$$

and let the objective function (2) be rearranged as

$$\begin{aligned}J &= \frac{1}{2}\mathbf{x}^T\left(t_f\right) S\left(\mathbf{x}\left(t_f\right), t_f\right) \mathbf{x}\left(t_f\right) \\ &+ \frac{1}{2}\int_{t_i}^{t_f} \left[\mathbf{x}^T Q(\mathbf{x}, t)\, \mathbf{x} + \mathbf{u}^T R(\mathbf{x}, \mathbf{u}, t)\, \mathbf{u}\right] dt,\end{aligned} \tag{9}$$

where the operators $A, B, S, Q,$ and R have appropriate dimensions. The nonlinear dynamics (8) and the performance index (9) define an optimal control problem. The initial state, $\mathbf{x}_i$, is assumed to be given, while the final condition (ψ in (3)) can assume three different forms (see Section 4). The problem is formulated as an approximating sequence of Riccati equations. This method reduces problem (8)-(9) to a series of time-varying linear quadratic regulators that are defined by evaluating the state- and control-dependent matrices using the solution of the previous iteration (the first iteration considers the initial condition and zero control).

The initial step consists in solving *problem 0*, which is defined as follows:

$$\begin{aligned}\dot{\mathbf{x}}^{(0)} &= A(\mathbf{x}_i, t)\, \mathbf{x}^{(0)} + B(\mathbf{x}_i, \mathbf{0}, t)\, \mathbf{u}^{(0)}, \\ J &= \frac{1}{2}\mathbf{x}^{(0)T}\left(t_f\right) S\left(\mathbf{x}_i, t_f\right) \mathbf{x}^{(0)}\left(t_f\right) \\ &+ \frac{1}{2}\int_{t_i}^{t_f} \left[\mathbf{x}^{(0)T} Q(\mathbf{x}_i, t)\, \mathbf{x}^{(0)} + \mathbf{u}^{(0)T} R(\mathbf{x}_i, \mathbf{0}, t)\, \mathbf{u}^{(0)}\right] dt.\end{aligned} \tag{10}$$

Problem 0 is a standard time-varying linear quadratic regulator (TVLQR), as the arguments of A, B, S, Q, and R are all given except for the time. This problem is solved to yield $\mathbf{x}^{(0)}(t)$ and $\mathbf{u}^{(0)}(t)$, $t \in [t_i, t_f]$, where the superscript denotes the problem that the solution refers to.

At a generic, subsequent iteration, problem k has to be solved. This is defined as follows:

$$\begin{aligned}\dot{\mathbf{x}}^{(k)} &= A\left(\mathbf{x}^{(k-1)}(t), t\right) \mathbf{x}^{(k)} + B\left(\mathbf{x}^{(k-1)}(t), \mathbf{u}^{(k-1)}(t), t\right) \mathbf{u}^{(k)}, \\ J &= \frac{1}{2}\mathbf{x}^{(k)T}\left(t_f\right) S\left(\mathbf{x}^{(k-1)}\left(t_f\right), t_f\right) \mathbf{x}^{(k)}\left(t_f\right) \\ &+ \frac{1}{2}\int_{t_i}^{t_f} \left[\mathbf{x}^{(k)T} Q\left(\mathbf{x}^{(k-1)}(t), t\right) \mathbf{x}^{(k)} \right. \\ &\left. + \mathbf{u}^{(k)T} R\left(\mathbf{x}^{(k-1)}(t), \mathbf{u}^{(k-1)}(t), t\right) \mathbf{u}^{(k)}\right] dt.\end{aligned} \tag{11}$$

Problem k is again a TVLQR; note that $\mathbf{x}^{(k-1)}$ and $\mathbf{u}^{(k-1)}$ are the solutions of problem $k-1$ achieved at previous iteration. Solving problem k yields $\mathbf{x}^{(k)}(t)$ and $\mathbf{u}^{(k)}(t)$, $t \in [t_i, t_f]$.

Iterations continue until a certain convergence criterion is satisfied. In the present implementation of the algorithm, the convergence is reached when

$$\begin{aligned}&\left\|\mathbf{x}^{(k)} - \mathbf{x}^{(k-1)}\right\|_\infty \\ &= \max_{t \in [t_i, t_f]} \left\{\left|x_j^{(k)}(t) - x_j^{(k-1)}(t)\right|, j = 1, \ldots, n\right\} \le \varepsilon,\end{aligned} \tag{12}$$

where ε is a prescribed tolerance. That is, iterations terminate when the difference between each component of the state, evaluated for all times, changes by less than ε between two consecutive iterations.

4. Solution of the Time-Varying Linear Quadratic Regulator by the State Transition Matrix

With the approach sketched in Section 3, a fully nonlinear optimal control problem is reduced to a sequence of time-varying linear quadratic regulators. These can be solved a number of times to achieve an approximate solution of the original, nonlinear problem. This is done by exploiting the structure of the problem as well as its state transition matrix. This scheme differs from that implemented in [5, 6], and, in part, is described in [1].

Suppose that the following dynamics are given:

$$\dot{\mathbf{x}} = A(t)\mathbf{x} + B(t)\mathbf{u}, \quad (13)$$

together with the quadratic objective function

$$J = \frac{1}{2}\mathbf{x}^T(t_f)S(t_f)\mathbf{x}(t_f) + \frac{1}{2}\int_{t_i}^{t_f}\left[\mathbf{x}^T Q(t)\mathbf{x} + \mathbf{u}^T R(t)\mathbf{u}\right]dt, \quad (14)$$

where Q, R, and S are positive semidefinite and positive definite time-varying matrices with appropriate dimensions, respectively. The Hamiltonian of this problem is

$$H = \frac{1}{2}\left[\mathbf{x}^T Q(t)\mathbf{x} + \mathbf{u}^T R(t)\mathbf{u}\right] + \boldsymbol{\lambda}^T\left[A(t)\mathbf{x} + B(t)\mathbf{u}\right], \quad (15)$$

and the optimality conditions (5) read

$$\dot{\mathbf{x}} = A(t)\mathbf{x} + B(t)\mathbf{u}, \quad (16)$$

$$\dot{\boldsymbol{\lambda}} = -Q(t)\mathbf{x} - A^T(t)\boldsymbol{\lambda}, \quad (17)$$

$$0 = R(t)\mathbf{u} + B^T(t)\boldsymbol{\lambda}. \quad (18)$$

From (18), it is possible to get

$$\mathbf{u} = -R^{-1}(t)B^T(t)\boldsymbol{\lambda}, \quad (19)$$

which can be substituted into (16)-(17) to yield

$$\begin{aligned}\dot{\mathbf{x}} &= A(t)\mathbf{x} - B(t)R^{-1}(t)B^T(t)\boldsymbol{\lambda},\\ \dot{\boldsymbol{\lambda}} &= -Q(t)\mathbf{x} - A^T(t)\boldsymbol{\lambda}.\end{aligned} \quad (20)$$

In a compact form, (20) can be arranged as

$$\begin{bmatrix}\dot{\mathbf{x}}\\ \dot{\boldsymbol{\lambda}}\end{bmatrix} = \begin{bmatrix}A(t) & -B(t)R^{-1}(t)B^T(t)\\ -Q(t) & -A^T(t)\end{bmatrix}\begin{bmatrix}\mathbf{x}\\ \boldsymbol{\lambda}\end{bmatrix}. \quad (21)$$

Since (21) is a system of linear differential equations, the exact solution can be written as

$$\mathbf{x}(t) = \phi_{xx}(t_i,t)\mathbf{x}_i + \phi_{x\lambda}(t_i,t)\boldsymbol{\lambda}_i, \quad (22)$$

$$\boldsymbol{\lambda}(t) = \phi_{\lambda x}(t_i,t)\mathbf{x}_i + \phi_{\lambda\lambda}(t_i,t)\boldsymbol{\lambda}_i, \quad (23)$$

where the functions ϕ_{xx}, $\phi_{x\lambda}$, $\phi_{\lambda x}$, and $\phi_{\lambda\lambda}$ are the components of the state transition matrix, which can be found by integrating the following dynamics:

$$\begin{bmatrix}\dot{\phi}_{xx} & \dot{\phi}_{x\lambda}\\ \dot{\phi}_{\lambda x} & \dot{\phi}_{\lambda\lambda}\end{bmatrix} = \begin{bmatrix}A(t) & -B(t)R^{-1}(t)B^T(t)\\ -Q(t) & -A^T(t)\end{bmatrix}\begin{bmatrix}\phi_{xx} & \phi_{x\lambda}\\ \phi_{\lambda x} & \phi_{\lambda\lambda}\end{bmatrix}, \quad (24)$$

with the initial conditions

$$\begin{aligned}\phi_{xx}(t_i,t_i) &= \phi_{\lambda\lambda}(t_i,t_i) = I_{n\times n},\\ \phi_{x\lambda}(t_i,t_i) &= \phi_{\lambda x}(t_i,t_i) = 0_{n\times n}.\end{aligned} \quad (25)$$

If both $\mathbf{x}_i$ and $\boldsymbol{\lambda}_i$ were given, it would be possible to compute $\mathbf{x}(t)$ and $\boldsymbol{\lambda}(t)$ through (22)-(23), and therefore the optimal control function $\mathbf{u}(t)$ with (19). The initial condition is assumed to be given, whereas the computation of $\boldsymbol{\lambda}_i$ depends on the final condition, which, in the present algorithm, can be defined in three different ways.

4.1. Hard Constrained Problem. In a hard constrained problem (HCP), the value of the final state is fully specified, $\mathbf{x}(t_f) = \mathbf{x}_f$, and therefore, (14) does not account for S. The value of $\boldsymbol{\lambda}_i$ can be found by writing (22) at final time

$$\mathbf{x}_f = \phi_{xx}(t_i,t_f)\mathbf{x}_i + \phi_{x\lambda}(t_i,t_f)\boldsymbol{\lambda}_i \quad (26)$$

and by solving for $\boldsymbol{\lambda}_i$; that is,

$$\boldsymbol{\lambda}_i(\mathbf{x}_i,\mathbf{x}_f,t_i,t_f) = \phi_{x\lambda}^{-1}(t_i,t_f)\left[\mathbf{x}_f - \phi_{xx}(t_i,t_f)\mathbf{x}_i\right]. \quad (27)$$

4.2. Soft Constrained Problem. In a soft constrained problem (SCP), the final state is not specified, and thus S in (14) is an $n\times n$ positive definite matrix. The transversality condition (7) sets a relation between the state and costate at final time

$$\boldsymbol{\lambda}(t_f) = S(t_f)\mathbf{x}(t_f), \quad (28)$$

which can be used to find $\boldsymbol{\lambda}_i$. This is done by writing (22)-(23) at final time and using (28)

$$\begin{aligned}\mathbf{x}(t_f) &= \phi_{xx}(t_i,t_f)\mathbf{x}_i + \phi_{x\lambda}(t_i,t_f)\boldsymbol{\lambda}_i,\\ S(t_f)\mathbf{x}(t_f) &= \phi_{\lambda x}(t_i,t_f)\mathbf{x}_i + \phi_{\lambda\lambda}(t_i,t_f)\boldsymbol{\lambda}_i.\end{aligned} \quad (29)$$

Equations (29) represent a linear algebraic system of $2n$ equations in the $2n$ unknowns $\{\mathbf{x}(t_f), \boldsymbol{\lambda}_i\}$. The system can be solved by substitution to yield

$$\begin{aligned}\boldsymbol{\lambda}_i(\mathbf{x}_i,t_i,t_f) &= \left[\phi_{\lambda\lambda}(t_i,t_f) - S(t_f)\phi_{x\lambda}(t_i,t_f)\right]^{-1}\\ &\quad\times\left[S(t_f)\phi_{xx}(t_i,t_f) - \phi_{\lambda x}(t_i,t_f)\right]\mathbf{x}_i.\end{aligned} \quad (30)$$

4.3. Mixed Constrained Problem. In a mixed constrained problem (MCP), some components of the final state are specified and some are not. Without any loss of generality, let the state be decomposed as $\mathbf{x} = (\mathbf{y},\mathbf{z})$, where $\mathbf{y}$ are the p known components at final time, $\mathbf{y}(t_f) = \mathbf{y}_f$, and $\mathbf{z}$ are remaining $n-p$ elements. The costate is decomposed accordingly as $\boldsymbol{\lambda} = (\boldsymbol{\xi},\boldsymbol{\eta})$. With this formalism, S in (14) is $(n-p)\times(n-p)$, and it is pre- and postmultiplied by $\mathbf{z}(t_f)$. The transversality condition (7) is $\boldsymbol{\eta}(t_f) = S(t_f)\mathbf{z}(t_f)$.

The MCP is solved by partitioning the state transition matrix in a suitable form such that, at final time, (22)-(23) read

$$\begin{bmatrix} \mathbf{y}(t_f) \\ \mathbf{z}(t_f) \end{bmatrix} = \begin{bmatrix} \phi_{yy} & \phi_{yz} \\ \phi_{zy} & \phi_{zz} \end{bmatrix} \begin{bmatrix} \mathbf{y}_i \\ \mathbf{z}_i \end{bmatrix} + \begin{bmatrix} \phi_{y\xi} & \phi_{y\eta} \\ \phi_{z\xi} & \phi_{z\eta} \end{bmatrix} \begin{bmatrix} \boldsymbol{\xi}_i \\ \boldsymbol{\eta}_i \end{bmatrix}, \quad (31)$$

$$\begin{bmatrix} \boldsymbol{\xi}(t_f) \\ \boldsymbol{\eta}(t_f) \end{bmatrix} = \begin{bmatrix} \phi_{\xi y} & \phi_{\xi z} \\ \phi_{\eta y} & \phi_{\eta z} \end{bmatrix} \begin{bmatrix} \mathbf{y}_i \\ \mathbf{z}_i \end{bmatrix} + \begin{bmatrix} \phi_{\xi\xi} & \phi_{\xi\eta} \\ \phi_{\eta\xi} & \phi_{\eta\eta} \end{bmatrix} \begin{bmatrix} \boldsymbol{\xi}_i \\ \boldsymbol{\eta}_i \end{bmatrix}, \quad (32)$$

where the dependence of the state transition matrix components on t_i, t_f is omitted for brevity. From the first row of (31), it is possible to get

$$\boldsymbol{\xi}_i = \phi_{y\xi}^{-1} \left[\mathbf{y}_f - \phi_{yy}\mathbf{y}_i - \phi_{yz}\mathbf{z}_i \right] - \phi_{y\xi}^{-1}\phi_{y\eta}\boldsymbol{\eta}_i, \quad (33)$$

which can be substituted in the second row of (31) to yield

$$\begin{aligned} \mathbf{z}(t_f) &= \left[\phi_{zy} - \phi_{z\xi}\phi_{y\xi}^{-1}\phi_{yy} \right] \mathbf{y}_i + \left[\phi_{zz} - \phi_{z\xi}\phi_{y\xi}^{-1}\phi_{yz} \right] \mathbf{z}_i \\ &\quad + \phi_{z\xi}\phi_{y\xi}^{-1}\mathbf{y}_f + \left[\phi_{z\eta} - \phi_{z\xi}\phi_{y\xi}^{-1}\phi_{y\eta} \right] \boldsymbol{\eta}_i. \end{aligned} \quad (34)$$

Equations (33)-(34), together with the transversality condition $\boldsymbol{\eta}(t_f) = S(t_f)\mathbf{z}(t_f)$, can be substituted in the second row of (32) to compute the component of the initial costate

$$\boldsymbol{\eta}_i(\mathbf{x}_i, \mathbf{y}_f, t_i, t_f) = \left[\widetilde{\phi}_{\eta\eta} \right]^{-1} \mathbf{w}(\mathbf{x}_i, \mathbf{y}_f, t_i, t_f), \quad (35)$$

where

$$\begin{aligned} \widetilde{\phi}_{\eta\eta} &= \phi_{\eta\eta} - \phi_{\eta\xi}\phi_{y\xi}^{-1}\phi_{y\eta} - S\left(\phi_{z\eta} - \phi_{z\xi}\phi_{y\xi}^{-1}\phi_{y\eta} \right), \\ \mathbf{w}&(\mathbf{x}_i, \mathbf{y}_f, t_i, t_f) \\ &= \left[S\left(\phi_{zy} - \phi_{z\xi}\phi_{y\xi}^{-1}\phi_{yy} \right) - \phi_{\eta y} + \phi_{\eta\xi}\phi_{y\xi}^{-1}\phi_{yy} \right] \mathbf{y}_i \\ &\quad + \left[S\left(\phi_{zz} - \phi_{z\xi}\phi_{y\xi}^{-1}\phi_{yz} \right) + \phi_{\eta z} + \phi_{\eta\xi}\phi_{y\xi}^{-1}\phi_{yz} \right] \mathbf{z}_i \\ &\quad + \left[S\left(\phi_{z\xi}\phi_{y\xi}^{-1} \right) - \phi_{\eta\xi}\phi_{y\xi}^{-1} \right] \mathbf{y}_f. \end{aligned} \quad (36)$$

Once $\boldsymbol{\eta}_i$ is know, the remaining part of the initial costate, $\boldsymbol{\xi}_i$, is computed through (33), and therefore, the full initial costate is obtained as a function of the initial condition, given final condition, initial and final time; that is, $\boldsymbol{\lambda}_i(\mathbf{x}_i, \mathbf{y}_f, t_i, t_f) = (\boldsymbol{\xi}_i(\mathbf{x}_i, \mathbf{y}_f, t_i, t_f), \boldsymbol{\eta}_i(\mathbf{x}_i, \mathbf{y}_f, t_i, t_f))$.

5. Numerical Examples

Two simple problems with nonlinear dynamics are considered to apply the developed algorithm. These correspond to the controlled relative spacecraft motion and to the controlled two-body dynamics for low-thrust transfers.

5.1. Low-Thrust Rendezvous. This problem is taken from the literature where a solution is available, for comparison's sake [10, 11]. Consider the planar, relative motion of two particles in a central gravity field expressed in a rotating frame with normalized units: the length unit is equal to the orbital radius, the time unit is such that the orbital period is 2π, and the gravitational parameter is equal to 1. In these dynamics, the state, $\mathbf{x} = (x_1, x_2, x_3, x_4)$, represents the radial, tangential displacements (x_1, x_2) and the radial, tangential velocity deviations (x_3, x_4), respectively. The control, $\mathbf{u} = (u_1, u_2)$, is made up by the radial and tangential accelerations, respectively.

The equations of motion are

$$\begin{aligned} \dot{x}_1 &= x_3, \\ \dot{x}_2 &= x_4, \\ \dot{x}_3 &= 2x_4 - (1 + x_1)\left(\frac{1}{r^3} - 1 \right) + u_1, \\ \dot{x}_4 &= -2x_3 - x_2\left(\frac{1}{r^3} - 1 \right) + u_2, \end{aligned} \quad (37)$$

with $r = \sqrt{(x_1+1)^2 + x_2^2}$. The initial condition is $\mathbf{x}_i = (0.2, 0.2, 0.1, 0.1)$. Two different problems are solved to test the algorithm in both hard and soft constrained conditions.

Hard Constrained Rendezvous. The HCP consists in minimizing

$$J = \frac{1}{2}\int_{t_i}^{t_f} \mathbf{u}^T\mathbf{u}\, dt \quad (38)$$

with the final, given condition $\mathbf{x}_f = (0, 0, 0, 0)$ and $t_i = 0$, $t_f = 1$.

Soft Constrained Rendezvous. The SCP considers the following objective function:

$$J = \frac{1}{2}\mathbf{x}^T(t_f)\, S\mathbf{x}(t_f) + \frac{1}{2}\int_{t_i}^{t_f} \mathbf{u}^T\mathbf{u}\, dt, \quad (39)$$

with $S = \operatorname{diag}(25, 15, 10, 10)$, $t_i = 0$ and $t_f = 1$ ($\mathbf{x}_f$ is free).

The differential equations (37) are factorized into the form of (8) as

$$\begin{bmatrix} \dot{x}_1 \\ \dot{x}_2 \\ \dot{x}_3 \\ \dot{x}_4 \end{bmatrix} = \underbrace{\begin{bmatrix} 0 & 0 & 1 & 0 \\ 0 & 0 & 0 & 1 \\ f(x_1,x_2)\left(1 + \dfrac{1}{x_1}\right) & 0 & 0 & 2 \\ 0 & f(x_1,x_2) & -2 & 0 \end{bmatrix}}_{A(\mathbf{x})} \times \begin{bmatrix} x_1 \\ x_2 \\ x_3 \\ x_4 \end{bmatrix} + \underbrace{\begin{bmatrix} 0 & 0 \\ 0 & 0 \\ 1 & 0 \\ 0 & 1 \end{bmatrix}}_{B} \begin{bmatrix} u_1 \\ u_2 \end{bmatrix}, \quad (40)$$

with $f(x_1, x_2) = -1/[(x_1+1)^2 + x_2^2]^{3/2} + 1$. Thus, the problem is put into the pseudo-LQR form (8)-(9) by defining $A(\mathbf{x})$ and B as in (40) and by setting $Q = 0_{4\times4}$ and $R = I_{2\times2}$.

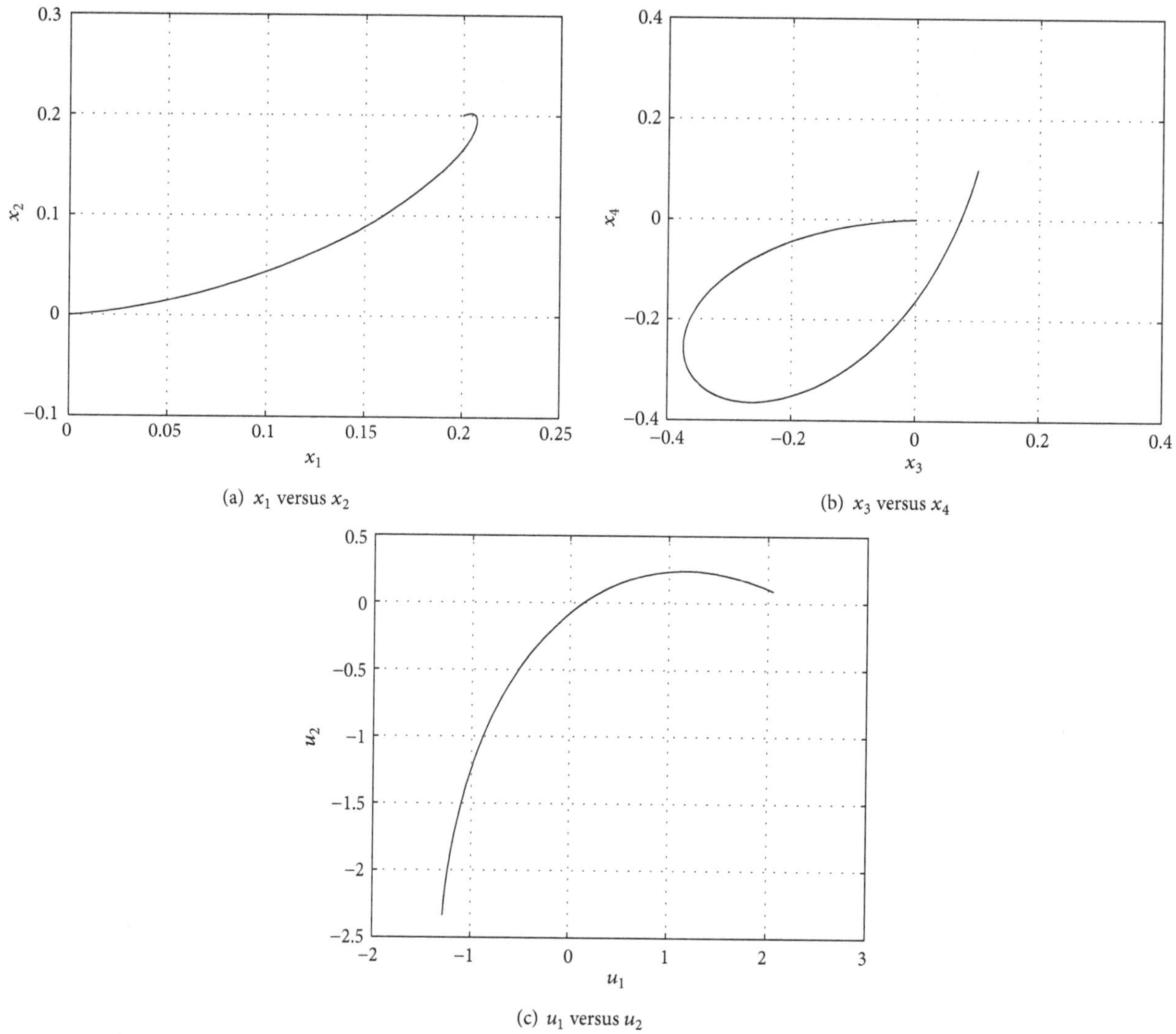

FIGURE 1: Hard constrained rendezvous.

The two problems have been solved with the developed method. Table 1 reports the details of the HCP and SCP, whose solutions are shown in Figures 1 and 2, respectively. In Table 1, J is the objective function at the final iteration, "Iter" is the number of iterations, and the "CPU time" is the computational time (this refers to an Intel Core 2 Duo 2 GHz with 4 GB RAM running Mac OS X 10.6). The termination tolerance ε in (12) is 10^{-9}. The optimal solutions found replicate those already known in the literature [10, 11], indicating the effectiveness of the developed method.

5.2. Low-Thrust Orbital Transfer. In this problem, the controlled, planar Keplerian motion of a spacecraft in polar coordinates is studied. The dynamics are written in scaled coordinates, where the length unit corresponds to the radius of the initial orbit, the time unit is such that its period is 2π, and the gravitational parameter is 1. The state, $\mathbf{x} = (x_1, x_2, x_3, x_4)$, is made up by the radial distance from the attractor (x_1), the phase angle (x_2), the radial velocity (x_3), and the transversal velocity (x_4), whereas the control, $\mathbf{u} = (u_1, u_2)$, corresponds to the radial and transversal accelerations, respectively [12, 13]. The equations of motion are

$$\begin{aligned}
\dot{x}_1 &= x_3, \\
\dot{x}_2 &= x_4, \\
\dot{x}_3 &= x_1 x_4^2 - \frac{1}{x_1^2} + u_1, \\
\dot{x}_4 &= -\frac{2x_3x_4}{x_1} + \frac{u_2}{x_1},
\end{aligned} \tag{41}$$

and the objective function is

$$J = \frac{1}{2}\int_{t_i}^{t_f} \mathbf{u}^T \mathbf{u}\, dt, \tag{42}$$

with $t_i = 0$ and $t_f = \pi$. The initial state corresponds to the conditions at the initial orbit; that is, $\mathbf{x}_i = (1, 0, 0, 1)$. Two different HCPs are solved, which correspond to the final states

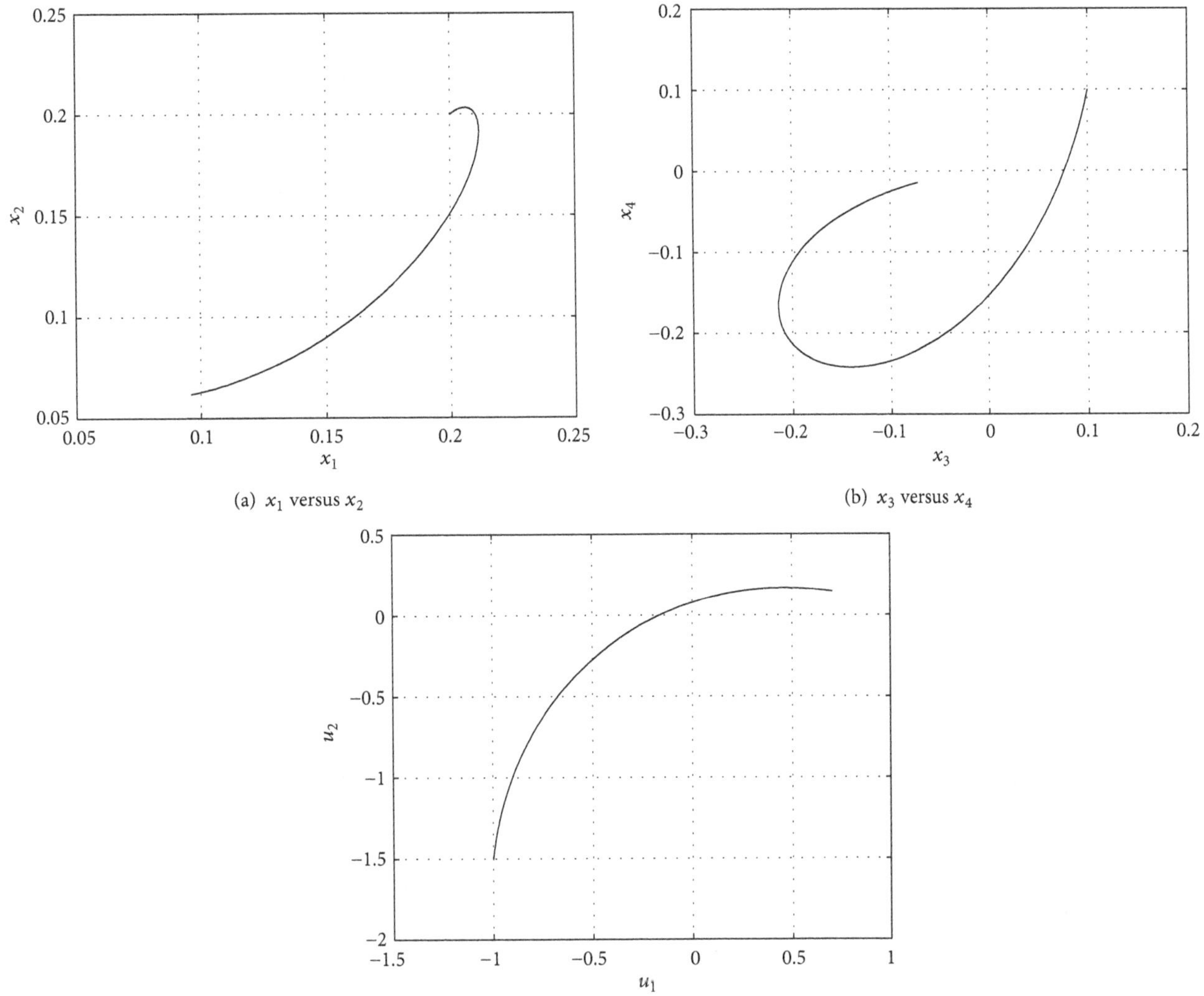

(a) x_1 versus x_2

(b) x_3 versus x_4

(c) u_1 versus u_2

FIGURE 2: Soft constrained rendezvous.

$\mathbf{x}_f = (1.52, \pi, 0, 1.52^{-3/2})$ and $\mathbf{x}_f = (1.52, 1.5\pi, 0, 1.52^{-3/2})$, respectively. This setup mimics an Earth-Mars low-thrust transfer. The dynamics (41) and the objective function (42) are put in the form (8)-(9) by defining $Q = 0_{4\times4}$, $R = I_{2\times2}$, and

$$A(\mathbf{x}) = \begin{bmatrix} 0 & 0 & 1 & 0 \\ 0 & 0 & 0 & 1 \\ x - \dfrac{1}{x_1^3} & 0 & 0 & x_1 x_4 \\ 0 & 0 & -\dfrac{2x_4}{x_1} & 0 \end{bmatrix},$$

$$B(\mathbf{x}) = \begin{bmatrix} 0 & 0 \\ 0 & 0 \\ 1 & 0 \\ 0 & \dfrac{1}{x_1} \end{bmatrix}. \qquad (43)$$

TABLE 1: Rendezvous solutions details.

Problem	J	Iter	CPU time (s)
HCP	0.9586	5	0.375
SCP	0.5660	6	0.426

TABLE 2: Earth-Mars transfer details.

Problem	J	Iter	CPU time (s)
$x_{2,f} = \pi$	0.5298	22	5.425
$x_{2,f} = 1.5\pi$	4.8665	123	41.831

The two HCPs have been solved with the developed method. The solutions' details are reported in Table 2, whose columns have the same meaning as in Table 1. It can be seen that more iterations and an increased computational burden are required to solve this problem. The solution with $x_{2,f} = 1.5\pi$ is reported in Figure 3.

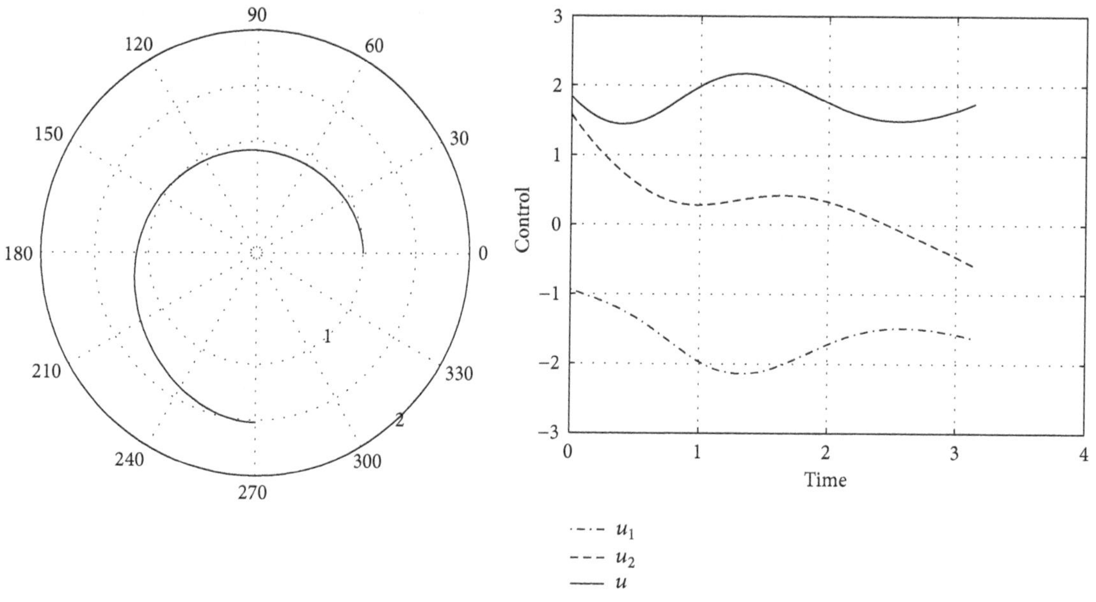

(a) Transfer trajectory

(b) Control profile

FIGURE 3: Orbital transfer with $x_{2,f} = 1.5\pi$.

6. Conclusion

In this paper, an approximated method to solve nonlinear optimal control problems has been presented, with applications to sample cases in astrodynamics. With this method, the nonlinear dynamics and objective function are factorized in a pseudolinear and quadratic-like forms, which are similar to those used in the state-dependent Riccati equation approach. Once in this form, a number of time-varying linear quadratic regulator problems are solved. A state transition matrix approach is used to deal with the time-varying linear quadratic regulators. The results show the effectiveness of the method, which can be used to either have suboptimal solutions or to provide initial solutions to more accurate optimizers.

References

[1] A. E. Bryson and Y. C. Ho, *Applied Optimal Control*, John Wiley & Sons, New York, NY, USA, 1975.

[2] L. S. Pontryagin, V. G. Boltyanskii, R. V. Gamkrelidze, and E. F. Mishchenko, *The Mathematical Theory of Optimal Processes*, John Wiley & Sons, New York, NY, USA, 1962.

[3] J. T. Betts, *Practical Methods for Optimal Control and Estimation Using Nonlinear Programming*, SIAM, Philadelphia, Pa, USA, 2010.

[4] B. Conway, "Spacecraft trajecory optimization using direct transcription and nonlinear programming," in *Spacecraft Trajectory Optimization*, pp. 37–78, Cambridge University Press, Cambridge, UK, 2010.

[5] T. Çimen and S. P. Banks, "Global optimal feedback control for general nonlinear systems with nonquadratic performance criteria," *Systems and Control Letters*, vol. 53, no. 5, pp. 327–346, 2004.

[6] T. Çimen and S. P. Banks, "Nonlinear optimal tracking control with application to super-tankers for autopilot design," *Automatica*, vol. 40, no. 11, pp. 1845–1863, 2004.

[7] C. P. Mracek and J. R. Cloutier, "Control designs for the nonlinear benchmark problem via the state-dependent Riccati equation method," *International Journal of Robust and Nonlinear Control*, vol. 8, no. 4-5, pp. 401–433, 1998.

[8] J. D. Pearson, "Approximation methods in optimal control," *Journal of Electronics and Control*, vol. 13, pp. 453–469, 1962.

[9] A. Wernli and G. Cook, "Suboptimal control for the nonlinear quadratic regulator problem," *Automatica*, vol. 11, no. 1, pp. 75–84, 1975.

[10] C. Park, V. Guibout, and D. J. Scheeres, "Solving optimal continuous thrust rendezvous problems with generating functions," *Journal of Guidance, Control, and Dynamics*, vol. 29, no. 2, pp. 321–331, 2006.

[11] C. Park and D. J. Scheeres, "Determination of optimal feedback terminal controllers for general boundary conditions using generating functions," *Automatica*, vol. 42, no. 5, pp. 869–875, 2006.

[12] A. Owis, F. Topputo, and F. Bernelli-Zazzera, "Radially accelerated optimal feedback orbits in central gravity field with linear drag," *Celestial Mechanics and Dynamical Astronomy*, vol. 103, no. 1, pp. 1–16, 2009.

[13] F. Topputo, A. H. Owis, and F. Bernelli-Zazzera, "Analytical solution of optimal feedback control for radially accelerated orbits," *Journal of Guidance, Control, and Dynamics*, vol. 31, no. 5, pp. 1352–1359, 2008.

16

Cooperative Object Manipulation by a Space Robot with Flexible Appendages

P. Zarafshan and S. Ali A. Moosavian

Center of Excellence in Robotics and Control, Advanced Robotics and Automated Systems Laboratory, Department of Mechanical Engineering, K. N. Toosi University of Technology, P.O. Box 19395-1999, Tehran, Iran

Correspondence should be addressed to P. Zarafshan; payam.zarafshan@gmail.com

Academic Editors: R. V. Kruzelecky and I. Taymaz

Modelling and control of rigid-flexible multibody systems is studied in this paper. As a specified application, a space robotic system with flexible appendages during a cooperative object manipulation task is considered. This robotic system necessitates delicate force exertion by several end-effectors to move an object along a desired path. During such maneuvers, flexible appendages like solar panels may get stimulated and vibrate. This vibrating motion will cause some oscillatory disturbing forces on the spacecraft, which in turn produces error in the motion of the end-effectors of the cooperative manipulating arms. In addition, vibration control of these flexible members to protect them from fracture is another challenging problem in an object manipulation task for the stated systems. Therefore, the multiple impedance control algorithm is extended to perform an object manipulation task by such complicated rigid-flexible multibody systems. This extension in the control algorithm considers the modification term which compensates the disturbing forces due to vibrating motion of flexible appendages. Finally, a space free-flying robotic system which contains two 2-DOF planar cooperative manipulators, appended with two highly flexible solar panels, is simulated. Obtained results reveal the merits of the developed model-based controller which will be discussed.

1. Introduction

Robotic manipulators are widely used in unsafe, costly, and repetitive boring tasks. Most available robotic manipulators are designed such that they can provide essential stiffness for end-effector to reach its desired position without flexible deformations [1]. This stiffness is usually attained by massive links. Consequently, design and use of weighty rigid manipulators may be deficient in energy consumption and the speed of operation. In particular, for space and on-orbit applications minimum weight design does not allow using such stiff manipulators. On the other hand, even assuming rigid manipulators, existence of flexible components on space robotic systems such as solar panels, necessitates considering their effect. The required settling time for vibration of such parts may delay the operation and so conflicts with increasing time efficiency of the system. This conflict of high speed and high accuracy during any operation makes these robots a disputative research problem [2–5].

Robotic systems with flexible components include continuous dynamic systems that are simplified by using a finite number of rigid degrees of freedom and a limited number of modes. This leads to a set of ordinary and partial differential equations that are usually nonlinear and coupled. Precise solution of these systems in most cases is almost impossible [6]. In studying these systems, if we ignore the flexibility effects in mathematic model, two types of error will be produced. The first one is related to the actuator torques, and the second one corresponds to the position of end-effectors. The position/orientation of end-effectors for precise tasks should not experience any vibration even with small amplitude. Therefore, to achieve high accuracy, we must begin with more precise mathematic models [7–11]. To study the dynamics of a rigid-flexible multibody space system, an inertia frame is used as a universal reference frame. Moreover, an intermediate reference frame is attached to each flexible or rigid body which is usually called floating frame. The motion relative to this intermediate frame for flexible parts occurs because of the body deformation only. This selection simplifies the calculations of internal forces since the magnitude of the stress and strain does not vary under the rigid body motion. To develop dynamics model of such systems, various approaches have been used, including the Lagrange method [9, 12], the Hamilton principle [13], the Newton-Euler equations [14], the virtual work principle [15], and Kane method [16].

Controller design for multibody systems with flexible members requires the development of proper dynamics model of such systems. Such models are also required to be as concise as possible for implementation of model-based control algorithms. In most researches on dynamics analysis, the modelling approach introduces an accumulation in the dynamics of rigid-flexible multibody systems, [17, 18]. This is done while the modelling approach does not affect their non-model based controller design, whereas, using an accumulated dynamic model to control these complicated systems by a model-based control algorithm will become as a challenging problem. On the other hand, control of flexible multibody systems is currently an attractive research subject because of its application in flexible manipulators and the articulated space structures [19–22]. This depends on determining the actuator torques that can produce the desired motion of such a complicated multibody system. In other words, the inverse dynamics is part of controller design, though control can be directly applied on a physical system without using a numerical model [23, 24]. In fact, operational problems with robotic manipulators in space due to structural flexibility lead to subsequent difficulties with their position control and have been widely studied [25, 26]. However, force interaction with the environment makes a more challenging problem than position control of such systems which is the main focus of this paper. The object manipulation operation by the rigid-flexible multibody systems is a problem that was less proceeded in the researches. Of course, various control algorithms were presented for the object manipulation task in the rigid robotic systems in which each of them has the advantages. These algorithms can be cited including multiple impedance control (MIC) [27], augmented object control (AOM) [28], and non-model-based impedance control (NMIC) [29]. Although, each of these algorithms were shown their capabilities for rigid systems, but their performance must be studied for rigid-flexible multibody systems.

As stated above, in this paper, a space robotic system with flexible appendages is considered to perform cooperative object manipulation task. It necessitates delicate force exertion by two or more end-effectors to move the object along a desired path. First, we must extract the simpler sets of dynamics equations which can be used for model-based controllers. To this end, the system dynamics is virtually partitioned into two rigid and flexible portions, and a convenient model is developed for control purposes of rigid-flexible multibody systems. Next, based on the genetic algorithm approach using MATLAB/GATOOL, appropriate trajectories are designed to study the stimulation effects of the flexible appendages. Then, an object manipulation operation is studied by extended multiple impedance control algorithm to suppress the vibration of the flexible members. Finally, using a comprehensive simulation routine, obtained results of the implementations of this controller on the rigid-flexible multibody system for the designed minimum time trajectory (CASE-I) and a circular path (CASE-II) will be discussed.

2. Dynamics Modelling

2.1. Modelling of Rigid Components. Space free-flying robots (SFFRs) are space systems that include an actuated spacecraft equipped with few manipulators. Distinct from fixed-based manipulators, the spacecraft (base) of an SFFR responds to dynamic reaction forces due to the arms motion. Unlike long-reach space manipulators, SFFRs are suggested to be comparable to a human body and an astronaut so are usually investigated under the assumption of rigid elements. The motion equations of a space robot with rigid components which were described by [30] can be extended to consider flexible elements as

$$\mathbf{H}(\boldsymbol{\beta}_0,\boldsymbol{\theta})\ddot{\mathbf{q}} + \mathbf{C}_1(\boldsymbol{\beta}_0,\dot{\boldsymbol{\beta}}_0,\boldsymbol{\theta},\dot{\boldsymbol{\theta}})\dot{\mathbf{q}} + \mathbf{C}_2(\boldsymbol{\beta}_0,\dot{\boldsymbol{\beta}}_0,\boldsymbol{\theta},\dot{\boldsymbol{\theta}}) = \mathbf{Q}(\boldsymbol{\beta}_0,\boldsymbol{\theta}) + \mathbf{Q}_{\text{flex.}}(\boldsymbol{\beta}_0), \quad (1)$$

where $\mathbf{Q}_{\text{flex.}}$ and $\mathbf{Q}$ are the resultant forces/torques applied on the main body of the space robot due to vibrating motion of the flexible solar panels and the generalized forces, respectively. Also, $\mathbf{C}_1$ and $\mathbf{C}_2$ are the vector of centrifugal and the Coriolis terms, and the mass matrixes $\mathbf{H}$ can be partitioned as

$$\begin{aligned}\mathbf{H}_{ij} &= M_{\text{sys}}\frac{\partial \mathbf{R}_{C_0}}{\partial q_i}\cdot\frac{\partial \mathbf{R}_{C_0}}{\partial q_j} + \frac{{}^0\partial\omega_0}{\partial\dot{q}_i}\cdot\mathbf{I}_0\cdot\frac{{}^0\partial\omega_0}{\partial\dot{q}_j} \\ &+ \sum_{m=1}^{n}\sum_{k=1}^{N_m}\left(m_k^{(m)}\frac{\partial \mathbf{r}_{C_k}^{(m)}}{\partial q_i}\cdot\frac{\partial \mathbf{r}_{C_k}^{(m)}}{\partial q_j} + \frac{{}^k\partial\omega_k^{(m)}}{\partial\dot{q}_i}\cdot\mathbf{I}_k^{(m)}\cdot\frac{{}^k\partial\omega_k^{(m)}}{\partial\dot{q}_j}\right) \\ &+ \left(\sum_{m=1}^{n}\sum_{k=1}^{N_m}m_k^{(m)}\frac{\partial \mathbf{r}_{C_k}^{(m)}}{\partial q_i}\right)\cdot\frac{\partial \mathbf{R}_{C_0}}{\partial q_j} \\ &+ \left(\sum_{m=1}^{n}\sum_{k=1}^{N_m}m_k^{(m)}\frac{\partial \mathbf{r}_{C_k}^{(m)}}{\partial q_j}\right)\cdot\frac{\partial \mathbf{R}_{C_0}}{\partial q_i},\end{aligned} \quad (2a)$$

$$\mathbf{C}_{1ij} = M_{\text{sys}}\frac{\partial \mathbf{R}_{C_0}}{\partial q_i}\cdot\left(\sum_{s=1}^{N}\frac{\partial^2 \mathbf{R}_{C_0}}{\partial q_s\partial q_j}\dot{q}_s\right)$$

$$+\frac{{}^{0}\partial\omega_0}{\partial\dot{q}_i}\mathbf{I}_0\frac{{}^{0}\partial\omega_0}{\partial q_j}+\omega_0\mathbf{I}_0\frac{{}^{0}\partial^2\omega_0}{\partial\dot{q}_i\partial q_j}$$

$$+\sum_{m=1}^{n}\sum_{k=1}^{N_m}\left(m_k^{(m)}\frac{\partial\mathbf{r}_{C_k}^{(m)}}{\partial q_i}\cdot\left(\sum_{s=1}^{N}\frac{\partial^2\mathbf{r}_{C_k}^{(m)}}{\partial q_s\partial q_j}\dot{q}_s\right)\right.$$

$$\left.+\frac{{}^{k}\partial\omega_k^{(m)}}{\partial\dot{q}_i}\mathbf{I}_k^{(m)}\frac{{}^{k}\partial\omega_k^{(m)}}{\partial q_j}+\omega_k^{(m)}\mathbf{I}_k^{(m)}\frac{{}^{k}\partial^2\omega_k^{(m)}}{\partial\dot{q}_i\partial q_j}\right)$$

$$+\left(\sum_{s=1}^{N}\frac{\partial^2\mathbf{R}_{C_0}}{\partial q_s\partial q_i}\dot{q}_s\right)\cdot\sum_{m=1}^{n}\sum_{k=1}^{N_m}\left(m_k^{(m)}\frac{\partial\mathbf{r}_{C_k}^{(m)}}{\partial q_j}\right)$$

$$+\frac{\partial\mathbf{R}_{C_0}}{\partial q_i}\cdot\sum_{m=1}^{n}\sum_{k=1}^{N_m}\left(m_k^{(m)}\sum_{s=1}^{N}\frac{\partial^2\mathbf{r}_{C_k}^{(m)}}{\partial q_s\partial q_j}\dot{q}_s\right), \tag{2b}$$

$$\mathbf{C}_{2i}=-\left(\omega_0\mathbf{I}_0\frac{{}^{0}\partial\omega_0}{\partial q_i}+\sum_{m=1}^{n}\sum_{k=1}^{N_m}\omega_k^{(m)}\mathbf{I}_k^{(m)}\frac{{}^{k}\partial\omega_k^{(m)}}{\partial q_i}\right), \tag{2c}$$

where ω_0 and $\mathbf{I}_0$ are angular velocity and moment of inertia matrix of the base, M_{sys} is the total mass of the rigid subsystem, $\omega_k^{(m)}$ is the angular velocity of the kth link of the (m)th manipulator, $\mathbf{r}_{C_k}^{(m)}$ is the position vector of mass center of this link with mass of $m_k^{(m)}$ and inertia matrix of $\mathbf{I}_k^{(m)}$, and $\mathbf{R}_{C_0}$ is the inertial position vector of mass center of the base. Array $\mathbf{q}$ describes the rigid subsystem variable state that is defined as

$$\mathbf{q}=\left\{\mathbf{R}_{C_0}^T,\boldsymbol{\beta}_0^T,\boldsymbol{\theta}^T\right\}^T, \tag{3}$$

where $\boldsymbol{\beta}_0$ is a set of Euler angles that determine the orientation of the base and $\boldsymbol{\theta}$ describes joint angle of the links that is defined as

$$\boldsymbol{\theta}^T=\left\{\theta_1^{(1)},\ldots,\theta_{N_1}^{(1)},\ldots,\theta_1^{(m)},\ldots,\theta_{N_m}^{(m)},\ldots,\theta_1^{(n)},\ldots,\theta_{N_n}^{(n)}\right\}^T, \tag{4}$$

where n is the number of manipulators, N_m is the number of links of the (m)th manipulator, and $\theta_{N_m}^{(m)}$ is the joint angle of the (m)th of manipulator N_m. Also, $\mathbf{Q}$ is vector of generalized forces and torques that are fully described by [31] with determining all dynamics and kinematics parameters. Thus, (1) presents the motion's equation of a rigid robot in space, that is, a microgravity environment.

2.2. Modelling of Flexible Components. As mentioned before, based on real systems, it is assumed that a SFFR with rigid components is appended by flexible appendages, for example, the solar panels to supply the required electrical energy. In this section, dynamics of the flexible panels is developed by using a floating frame reference. It will be shown that the equations of motion of such systems can be written in terms of a set of inertia shape integrals in addition to the mass of the body, the inertia matrix, the generalized forces and the stiffness matrix. These inertia shape integrals depend on the assumed displacement field that appear in the nonlinear terms. They signify the inertia coupling between the reference motion and the elastic deformation of the body. It will be shown that the deformable body inertia matrix depends on the elastic deformation of the body, and hence, it is an implicit function of time. Here, the configuration of each flexible body in the multibody system is recognized by using two sets of coordinates, that is, reference and elastic coordinates. Reference coordinates describe the location and orientation of a selected body reference. On the other hand, elastic coordinates explain the body deformation with respect to the body reference. In order to avoid the computational difficulties of infinite-dimensional spaces, these coordinates are introduced by using classical approximation techniques such as Rayleigh-Ritz methods [31]. Thus, the global position of an arbitrary point on the flexible body is defined by using a coupled set of reference and elastic coordinates. Also, the kinetic energy of the flexible body is developed, and the inertia coupling between the reference motion and the elastic deformation is recognized. The kinetic energy as the virtual work of the forces acting on the body is written in terms of the coupled sets of reference and elastic coordinates [31]. Then, the motion's equations of the flexible members can be obtained as [4, 5]

$$\mathbf{M}_f^{\{i\}}\ddot{\overline{\mathbf{q}}}^{\{i\}}+\mathbf{K}^{\{i\}}\overline{\mathbf{q}}^{\{i\}}=\mathbf{Q}_e^{\{i\}}+\mathbf{Q}_v^{\{i\}},\quad \{i\}=\{1,2,\ldots,n_h\}, \tag{5}$$

where n_b is the total number of the flexible bodies in the multibody system. Also, $\mathbf{Q}_v^{\{i\}}$ and $\mathbf{Q}_e^{\{i\}}$ are correspondingly a quadratic velocity vector which contains all gyroscopic and the Coriolis components and the vector of generalized forces associated with the $\{i\}$th body. Moreover, $\mathbf{M}_f^{\{i\}}$ and $\mathbf{K}^{\{i\}}$ are, respectively, recognized as the symmetric mass matrix and the symmetric positive definite stiffness matrix of the body $\{i\}$. It is recommended that $\overline{\mathbf{q}}^{\{i\}}$ is the vector of reference and elastic coordinates of the flexible body. This equation can be written in a partitioned matrix form as

$$\begin{bmatrix}\mathbf{m}_{rr}^{\{i\}} & \mathbf{m}_{rf}^{\{i\}}\\ \mathbf{m}_{fr}^{\{i\}} & \mathbf{m}_{ff}^{\{i\}}\end{bmatrix}\begin{bmatrix}\ddot{\overline{\mathbf{q}}}_r^{\{i\}}\\ \ddot{\overline{\mathbf{q}}}_f^{\{i\}}\end{bmatrix}+\begin{bmatrix}0 & 0\\ 0 & \mathbf{k}_{ff}^{\{i\}}\end{bmatrix}\begin{bmatrix}\overline{\mathbf{q}}_r^{\{i\}}\\ \overline{\mathbf{q}}_f^{\{i\}}\end{bmatrix}=\begin{bmatrix}\left(\mathbf{Q}_e^{\{i\}}\right)_r\\ \left(\mathbf{Q}_e^{\{i\}}\right)_f\end{bmatrix}+\begin{bmatrix}\left(\mathbf{Q}_v^{\{i\}}\right)_r\\ \left(\mathbf{Q}_v^{\{i\}}\right)_f\end{bmatrix}, \tag{6}$$

where "r" and "f", respectively, refer to rigid and flexible coordinates of the flexible members. These dynamics parameters are fully described by [4, 5].

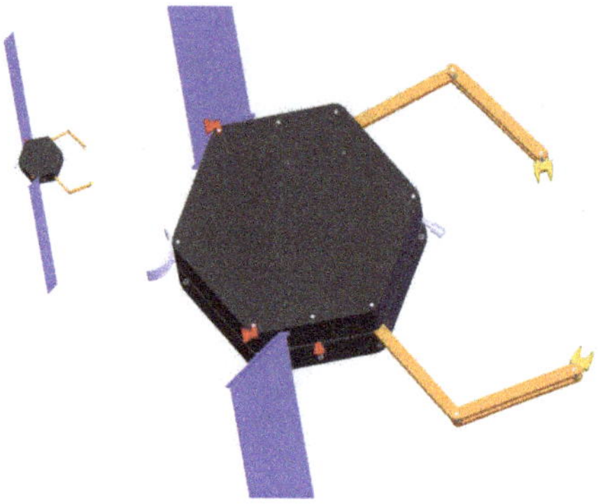

FIGURE 1: The considered space robotic system.

2.3. Complete Dynamics Equations of Motion. It should be noted that, in the developed modelling approach, the equations of motion for the $\{i\}$th flexible body are separated from those equations of the main rigid body system of the assumed space robot in Figure 1. This can be desirably used in designing the controller through a simulation study. In fact, after we solve the equations of motion for the rigid body system at each time step, the acceleration terms are considered as inputs for those equations of the flexible part. Then, we solve the equations of motion for the flexible subsystem considering these inputs. The obtained constraint forces are attained as the outputs which will be exerted on the mobile base of the space robot as estimated disturbance forces, [4, 5].

Considering (6), the generalized forces due to the stimulation of the flexible members which are applied on the rigid subsystem as the modification term or the constraint force or $\mathbf{Q}_{\text{flex.}}(\boldsymbol{\beta}_0)$ can be achieved as

$$\mathbf{Q}_{\text{flex.}}(\boldsymbol{\beta}_0) = \sum_{\{i\}=\{1\}}^{\{n_b\}} \mathbf{J}_f^{\{i\}^T} \left(\mathbf{Q}_e^{\{i\}}\right)_r, \quad (7)$$

where $\mathbf{J}_f^{\{i\}}$ is the Jacobian matrix of the floating frame of each flexible body related to the inertial frame of the main body. As detailed previously, this dynamics modelling approach combines the Lagrange and the Newton-Euler methods. To use this approach, the computation procedure at each time step includes the following calculations. First, the motion's equations of the rigid subsystem or (1) are solved. Then, the acceleration, velocity, and position terms of the rigid subsystem, that is, $\ddot{\mathbf{q}}$, $\dot{\mathbf{q}}$, $\mathbf{q}$ in the previous formulation, are calculated. Then, the rigid components of the acceleration, velocity, and position terms of each flexile body, that is, $\ddot{\overline{\mathbf{q}}}_r^{\{i\}}$, $\dot{\overline{\mathbf{q}}}_r^{\{i\}}$, $\overline{\mathbf{q}}_r^{\{i\}}$ in the previous formulation, are computed. After that, they were inserted into the motion equations of the flexible members as input terms. The relationship between these two sets of variables is established by the kinematics constraints between the origin of the floating frame which is attached to the flexible member and the reference frame of the rigid subsystem. Considering these inputs and by solving the second row of (6), the flexible components of the acceleration, velocity, and position terms of each flexile body, that is, $\ddot{\overline{\mathbf{q}}}_f^{\{i\}}$, $\dot{\overline{\mathbf{q}}}_f^{\{i\}}$, $\overline{\mathbf{q}}_f^{\{i\}}$ in the previous formulation, are calculated. Using these values and substituting into the first row of (6), the constraint forces or $\mathbf{Q}_e^{\{i\}}$ are computed. Also, these results are applied to the equations of the rigid subsystem as the produced forces from the incitement of the flexible member, that is, $\mathbf{Q}_{\text{flex.}}(\boldsymbol{\beta}_0)$ by using (7). As being clear, this dynamics modelling approach of a rigid multibody system with flexible members increases the computations of dynamics analysis, and it is useful in model-based control algorithms. For instance, the inverse of a mass matrix of the accumulated rigid-flexible multibody system requires much more calculations than inversing two matrices of the rigid and flexible members of the multibody system separately. Next, we study the designated path which it is applied in the simulations of the object manipulation control algorithm.

3. Time Optimal Trajectory Planning

The minimum time trajectory (MTT) can be considered as a useful strategy for mobile robotic systems to move on a given path between the two points during a minimum time. This task is performed by maximum available force/torque capacity of the actuators. This causes the fast dynamics of the multibody system to be stimulated, and its effects can be studied. So, we consider an object manipulation operation along a straight path, and then, the various scenarios for MTT would be presented to perform this task. Considering the mobile base and a desired path length, several solutions exist. The first scenario is that the two manipulators of the robot perform this operation by the base movement. The second one is that the base remains stationary and performs the task using its manipulators whereas the desired path to follow is within the fixed work space of the manipulators. The third scenario is achieved by combining the two, which this one is considered in all of simulations [32].

Designing procedure of the MTT includes expressing the considered path using a path variable, then computation of the velocity and acceleration of the system variables in terms of the path variable and its derivatives [33]. So, replacing these in the dynamic equations and after some simplifications, the dynamic equations are obtained in terms of the path variable. Therefore, considering the actuators bound, the equation of the desired MTT is obtained. In this procedure, it is assumed that the path of each end-effector in its work space is specified. Thus, the path parameter "s" is selected on the given path. Then, relation between the joint variables $\mathbf{q}$ and the path parameter can be stated. By differentiating this equation and solving that, $\dot{\mathbf{q}}$, $\ddot{\mathbf{q}}$ are obtained according to the path parameter. Then, replacing these parameters in the motion's equations yields

$$\overline{\mathbf{C}}_1(s)\ddot{s} + \overline{\mathbf{C}}_2(s,\dot{s}) = \boldsymbol{\tau}, \quad (8)$$

where $\overline{\mathbf{C}}_1(s)$ and $\overline{\mathbf{C}}_2(s,\dot{s})$ are the mass matrix and the quadratic velocity vector of the robotic system which are explained as the path parameter, respectively. We should note that for the movement on the given path in minimum time, the acceleration $\ddot{s}$ is determined by considering the extreme bounds of all actuators as

$$\boldsymbol{\tau}_{\min} \leq \overline{\mathbf{C}}_1\ddot{s} + \overline{\mathbf{C}}_2 \leq \boldsymbol{\tau}_{\max}. \tag{9}$$

Thus

$$f_i(s,\dot{s}) \leq \ddot{s} \leq g_i(s,\dot{s}), \tag{10}$$

where $f_i(s,\dot{s})$ is the lower acceleration and $g_i(s,\dot{s})$ is the higher acceleration. We should not that $\ddot{s}$ is a scalar quantity and should be determined by considering the capacity of the weakest actuators. Therefore,

$$\begin{aligned} \ddot{s} &= f = \max_{i=1,2} f_i(s,\dot{s}), \\ \ddot{s} &= g = \min_{i=1,2} g_i(s,\dot{s}). \end{aligned} \tag{11}$$

Considering the limitation of actuators of the assumed space robot in the following case studies as [27],

$$|\tau_i| \leq 7\,\mathrm{N}\cdot\mathrm{m}, \quad \text{for } i = 1,2,3,4. \tag{12}$$

Using genetic algorithm based on GATOOL toolbox of MATLAB for one manipulator [24], the $\overline{\mathbf{C}}_1(s)$ and $\overline{\mathbf{C}}_2(s,\dot{s})$ functions are optimized based on the assumed limitations or (12). Therefore, the acceleration is obtained. As shown in Figure 2, the designed trajectory and its velocity are achieved according to the obtained acceleration. It should be noted that the movement amplitude is 10 m and the optimum time in this maneuver is 5.8026 sec for this designed trajectory.

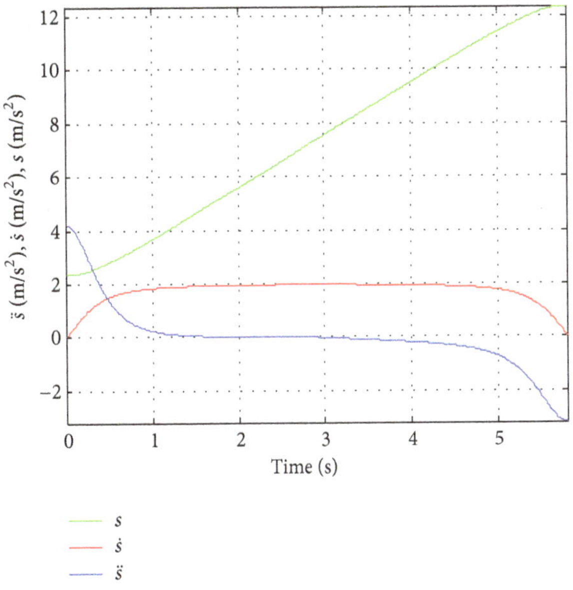

FIGURE 2: The position, velocity, and acceleration of the path parameter or s for the designed MTT.

4. Controller Design

There are several main issues that make the control problem of a rigid-flexible multibody system more complicated than control of a rigid system [34, 35]. First, the number of degrees of freedom is much larger than the number of actuators. A flexible body has an infinite number of degrees of freedom. As an example, the body can be discretized into a finite number of degree of freedom using various techniques such as finite element method or modal analysis, but the number of actuators is still generally much less than number of degrees of freedom, which may make the controller incapable to achieve an exact performance. At best, the controller can follow a trajectory that minimizes the error between the desired and the actual trajectories. The second issue is related to wave propagation delays. An action at one end of a flexible beam takes time to propagate to its tip. The third one is reversal action. This effect can be observed in a rotating flexible beam [36]. When a torque is applied to the beam in one direction, its tip position initially moves in the opposite direction. On the other hand, there are two main requirements for a rigid-flexible multibody system controller. These are fast and precise responses in following the desired trajectory. These two requirements are usually in conflict. The faster controller is the less accurate and vice versa. Several types of control laws have been proposed each offering benefits under some conditions. So, often more than one type of control law is used in these systems in order to improve the performance.

As studied in some researches, the multiple impedance control (MIC) law has the better performance rather than other object manipulation control algorithm [27]. So, the MIC law is introduced here. On the other hand, the rigid system assumption is considered for the MIC algorithm, whereas this assumption must be changed to control severe vibrations of flexible members in a rigid-flexible system and also to attain a successful object manipulation operation. Consequently, the MIC law is extended to compensate the disturbing forces due to the vibration of the flexible appendages. The modification term is placed on the control forces where the flexible members are installed on the affiliated part. In this assumed space robot, this means that the compensator term must be added on the control force of the robot, because the flexible appendages are located on this part. Also, to perform an object manipulation task in a planar maneuver, we should consider the following constraint in grasping the object:

$$(X_A - X_B)^2 + (Y_A - Y_B)^2 = l^2, \tag{13}$$

where X_A and Y_A are the end-effectors position of the left manipulator, X_B and Y_B are the end-effectors position of the right manipulator, and l is the object length.

Considering the formulation of MIC law, the end-effectors forces are

$$\mathbf{F}_e = \{\mathbf{F}_A^x \ \mathbf{F}_A^y \ \mathbf{F}_B^x \ \mathbf{F}_B^y\}^T. \quad (14)$$

Therefore, the equations of motion for the object are

$$F_A^x + F_B^x = m\ddot{X}_o,$$
$$F_A^y + F_B^y = m\ddot{Y}_o, \quad (15)$$
$$\frac{(F_B^x - F_A^x)\,l}{2C_0} + \frac{(F_B^y - F_A^y)\,l}{2S_0} = I\ddot{\theta}_o,$$

where S_0 and C_0 stand for the SIN and COS functions of the object orientation or θ_o, respectively. Thus, the grasp matrix is obtained as

$$\mathbf{G} = \begin{bmatrix} 1 & 0 & 1 & 0 \\ 0 & 1 & 0 & 1 \\ -\frac{l}{2C_0} & -\frac{l}{2S_0} & \frac{l}{2C_0} & \frac{l}{2S_0} \end{bmatrix}. \quad (16)$$

For the object motion, we have

$$\mathbf{M}\ddot{\mathbf{X}}_o = \mathbf{G}\mathbf{F}_e + \mathbf{F}_o + \mathbf{F}_c, \quad (17)$$

where $\mathbf{F}_c$ is the force applied on the object due to contact with environment, $\mathbf{F}_o$ is the vector of other external forces applied on the object, and $\mathbf{M}$ is the mass matrix of the object. Choosing the impedance law for the object motion as

$$\mathbf{M}_{\text{des}}\ddot{\mathbf{e}}_o + \mathbf{K}_d\dot{\mathbf{e}}_o + \mathbf{K}_p\mathbf{e}_o = -\mathbf{F}_c, \quad (18)$$

where $\mathbf{e}_o = \mathbf{X}_{o_{\text{des}}} - \mathbf{X}_o$ is the tracking error of object variables and $\mathbf{M}_{\text{des}}$, $\mathbf{K}_d$, $\mathbf{K}_p$ are the gain matrices for the proposed controller. Then, the desired exerted forces from end-effectors to move the object are obtained as:

$$\mathbf{F}_{e_{\text{req}}} = \mathbf{G}^{\#}\left\{\mathbf{M}\mathbf{M}_{\text{des}}^{-1}\left(\mathbf{M}_{\text{des}}\ddot{\mathbf{X}}_{o_{\text{des}}} + \mathbf{K}_d\dot{\mathbf{e}}_o + \mathbf{K}_p\mathbf{e}_o + \mathbf{F}_c\right) + \mathbf{F}_w + (\mathbf{F}_c + \mathbf{F}_o)\right\}, \quad (19)$$

where $\mathbf{G}^{\#}$ is the pseudoinverse of $\mathbf{G}$ as

$$\mathbf{G}^{\#} = \mathbf{W}^{-1}\mathbf{G}^T\left(\mathbf{G}\mathbf{W}^{-1}\mathbf{G}^T\right)^{-1}, \quad (20)$$

where it is weighted by a task weighting matrix $\mathbf{W}$ [27]. Thus, the force that is applied on the object by the (i)th end-effectors is directly obtained from $\mathbf{F}_{e_{\text{req}}}$ as

$$\widetilde{\mathbf{Q}}_f^{(i)} = \mathbf{F}_{e_{\text{req}}}. \quad (21)$$

Next, to compute the required force for motion control, if the equations of motion of the space robotic system or (1) can be written in the task space as

$$\widetilde{\mathbf{H}}^{(i)}\left(\mathbf{q}^{(i)}\right)\ddot{\widetilde{\mathbf{X}}}^{(i)} + \widetilde{\mathbf{C}}^{(i)}\left(\mathbf{q}^{(i)}, \dot{\mathbf{q}}^{(i)}\right) = \widetilde{\mathbf{Q}}^{(i)} + \widetilde{\mathbf{Q}}_{\text{flex.}}^{(i)}, \quad (22)$$

where (i) indicates the (i)th manipulator and $\widetilde{\mathbf{X}}^{(i)}$ is the output coordinate as:

$$\widetilde{\mathbf{X}} = \left[X_{C_0} \ Y_{C_0} \ \delta_{C_0} \ X_{E,A} \ Y_{E,A} \ X_{E,B} \ Y_{E,B} \ \delta_{\text{Ant}} \ \delta_{\text{Cam}}\right]^T, \quad (23a)$$

$$\widetilde{\mathbf{H}}^{(i)} = \mathbf{J}_c^{(i)^{-T}}\mathbf{H}^{(i)}\mathbf{J}_c^{(i)^{-1}},$$
$$\widetilde{\mathbf{C}}^{(i)} = \mathbf{J}_c^{(i)^{-T}}\left[\mathbf{C}_1^{(i)} + \mathbf{C}_2^{(i)}\right] - \widetilde{\mathbf{H}}^{(i)}\dot{\mathbf{J}}_c^{(i)}\dot{\mathbf{q}}^{(i)}, \quad (23b)$$
$$\widetilde{\mathbf{Q}}^{(i)} = \mathbf{J}_c^{(i)^{-T}}\mathbf{Q}^{(i)},$$
$$\widetilde{\mathbf{Q}}_{\text{flex.}}^{(i)} = \mathbf{J}_c^{(i)^{-T}}\mathbf{Q}_{\text{flex.}}^{(i)},$$

where $\mathbf{J}_c^{(i)}$ is the Jacobian matrix for the (i)th manipulator and $\widetilde{\mathbf{X}}$ can be considered as the generalized workspace variables of the assumed robot. We should note that $\widetilde{\mathbf{Q}}_{\text{flex.}}^{(i)}$ is the disturbance force resulting from the solar panel vibrations in the object manipulation maneuver. Also, $\widetilde{\mathbf{Q}}^{(i)}$ is the vector of generalized forces in the work space. Similarly, choosing the impedance law for each end-effector as:

$$\widetilde{\mathbf{M}}_{\text{des}}\ddot{\widetilde{\mathbf{e}}}^{(i)} + \widetilde{\mathbf{K}}_d\dot{\widetilde{\mathbf{e}}}^{(i)} + \widetilde{\mathbf{K}}_p\widetilde{\mathbf{e}}^{(i)} = -\mathbf{F}_c, \quad (24)$$

where $\widetilde{\mathbf{e}}^{(i)} = \widetilde{\mathbf{X}}_{\text{des}}^{(i)} - \widetilde{\mathbf{X}}^{(i)}$ is the system tracking error for each manipulator and $\widetilde{\mathbf{M}}_{\text{des}}$, $\widetilde{\mathbf{K}}_d$, $\widetilde{\mathbf{K}}_p$ are the gain matrices for the proposed controller of the robotic system. Thus, the required force for motion control of the end-effectors by using the MIC law is expressed as

$$\widetilde{\mathbf{Q}}_m^{(i)} = \widetilde{\mathbf{H}}^{(i)}\widetilde{\mathbf{M}}_{\text{des}}^{-1}\left[\widetilde{\mathbf{M}}_{\text{des}}\ddot{\widetilde{\mathbf{X}}}_{\text{des}}^{(i)} + \widetilde{\mathbf{K}}_d\dot{\widetilde{\mathbf{e}}}^{(i)} + \widetilde{\mathbf{K}}_p\widetilde{\mathbf{e}}^{(i)} + \mathbf{F}_c\right] + \widetilde{\mathbf{C}}^{(i)} - \widehat{\widetilde{\mathbf{Q}}}_{\text{flex.}}^{(i)}, \quad (25)$$

in which $\widehat{\widetilde{\mathbf{Q}}}_{\text{flex.}}^{(i)}$ is a model-based term to compensate the vibration of solar panels. Considering the stated approach in the dynamics modelling, this modification term seems more necessary. In fact, the MIC law is extended and it can be applied to the rigid-flexible multibody systems in addition to the rigid systems by this modification term. Also, it has been recommended that the same impedance characteristics for the manipulated object and end-effectors can be chosen [37]:

$$\mathbf{M}_{\text{des}} = \widetilde{\mathbf{M}}_{\text{des}}, \quad \mathbf{K}_d = \widetilde{\mathbf{K}}_d, \quad \mathbf{K}_p = \widetilde{\mathbf{K}}_p, \tag{26}$$

Finally, the required forces for object manipulation to be supplied by actuators are

$$\widetilde{\mathbf{Q}}^{(i)} = \widetilde{\mathbf{Q}}^{(i)}_{\text{app}} + \widetilde{\mathbf{Q}}^{(i)}_{\text{react}} = \widetilde{\mathbf{Q}}^{(i)}_{m} + \widetilde{\mathbf{Q}}^{(i)}_{f} + \widetilde{\mathbf{Q}}^{(i)}_{\text{react}}, \tag{27}$$

where $\widetilde{\mathbf{Q}}^{(i)}_{m}$ is the control forces for end-effector motion and $\widetilde{\mathbf{Q}}^{(i)}_{\text{react}}$ is the reaction load on the end-effectors and virtually cancelled $\widetilde{\mathbf{Q}}^{(i)}_{f}$ as

$$\widetilde{\mathbf{Q}}^{(i)}_{\text{react}} = -\mathbf{F}^{(i)}_{e}, \tag{28}$$

where $\mathbf{F}^{(i)}_{e}$ is the exerted forces from end-effectors. Next, substituting (28), (25), and (21) into (27) and then the results into (22) yields

$$\widetilde{\mathbf{H}}^{(i)}\left(\mathbf{q}^{(i)}\right)\left\{\ddot{\widetilde{\mathbf{X}}}^{(i)} - \mathbf{M}^{-1}_{\text{des}}\left[\mathbf{M}_{\text{des}}\ddot{\widetilde{\mathbf{X}}}^{(i)}_{d} + \mathbf{K}_d\dot{\widetilde{\mathbf{e}}}^{(i)} + \mathbf{K}_p\widetilde{\mathbf{e}}^{(i)}\right]\right\}$$
$$+ \mathbf{G}^{\#}\mathbf{M}\left\{\ddot{\mathbf{X}}_o - \mathbf{M}^{-1}_{\text{des}}\left[\mathbf{M}_{\text{des}}\ddot{\mathbf{X}}_{o_{\text{des}}} + \mathbf{K}_d\dot{\mathbf{e}}_o + \mathbf{K}_p\mathbf{e}_o\right]\right\} = \mathbf{0}. \tag{29}$$

Because (29) must hold for any $\widetilde{\mathbf{H}}^{(i)}$ and $\mathbf{M}$, it can be concluded that:

$$\widetilde{\mathbf{H}}^{(i)}\left(\mathbf{q}^{(i)}\right)\left\{\ddot{\widetilde{\mathbf{X}}}^{(i)} - \mathbf{M}^{-1}_{\text{des}}\left[\mathbf{M}_{\text{des}}\ddot{\widetilde{\mathbf{X}}}^{(i)}_{d} + \mathbf{K}_d\dot{\widetilde{\mathbf{e}}}^{(i)} + \mathbf{K}_p\widetilde{\mathbf{e}}^{(i)}\right]\right\} = \mathbf{0},$$
$$\mathbf{M}\left\{\ddot{\mathbf{X}}_o - \mathbf{M}^{-1}_{\text{des}}\left[\mathbf{M}_{\text{des}}\ddot{\mathbf{X}}_{o_{\text{des}}} + \mathbf{K}_d\dot{\mathbf{e}}_o + \mathbf{K}_p\mathbf{e}_o\right]\right\} = \mathbf{0}. \tag{30}$$

By notating the fact that $\widetilde{\mathbf{H}}^{(i)}$ and $\mathbf{M}$ are positive definite mass matrices, it can result in

$$\mathbf{M}_{\text{des}}\ddot{\widetilde{\mathbf{e}}}^{(i)} + \mathbf{K}_d\dot{\widetilde{\mathbf{e}}}^{(i)} + \mathbf{K}_p\widetilde{\mathbf{e}}^{(i)} = \mathbf{0},$$
$$\mathbf{M}_{\text{des}}\ddot{\mathbf{e}}_o + \mathbf{K}_d\dot{\mathbf{e}}_o + \mathbf{K}_p\mathbf{e}_o = \mathbf{0} \tag{31}$$

which means all participating manipulators and the manipulated object exhibit the same designated impedance behaviour. Next, we study the extended multiple impedance control for the assumed space robot to perform an object manipulation task on the designated trajectory. Also, the simulations are done for the original MIC and the extended one to show the advantage of the new control algorithm.

Table 1: The geometric and mass parameters of the assumed robotic system.

i, j ($i = 1$, $i = 2$) j	m_{ij} (kg)	l (m)	$I_{Z_{ij}}$ (kg·m^2)
0	50	1	10
1	4	1	0.5
2	3	1	0.5

Table 2: The specifications parameters of the flexible members [38].

$m = 9$ (Kg)	$\rho = 1.4$ (g/cm^3)
$L_b = 4$ (m)	$E = 60$ (GPa)
$a = 16$ (cm^2)	$EI = 20$ (N·m^2)

5. Simulation Results and Discussions

Considering the same geometric and mass parameters for the main rigid system as given in Table 1 as they are applied by [27], two solar panels each 4 m long whose specification parameters are included in Table 2, are appended to the base. The assumed space robot with its geometric parameters and the defined frames is represented in Figure 3. Also, MATLAB/SIMULINK program is used to simulate and implement the stated control algorithm on the assumed space robot. So, the initial condition of the simulation study is stated in Table 3 which is the same for the two cases. It should be noted that the extracted dynamics model of space robot was verified by [4, 5]. As mentioned before, two cases of implementation of the extended MIC law are studied in simulations based on the designed trajectories, that is, CASE-I for the designed MTT and CASE-II for the circular path which has the radius of 15 m. These maneuvers are important cases since flexible modes are stimulated due to on-off nature of actuating forces in the first one, and centripetal accelerations in the second one. In each case, the simulations are done for the space robot with flexible solar panels (FSP) and without those, while the controller gain matrices remain the same. It should be noted that if the robotic system does not have FSP, then the extended MIC law changes to the original form of the MIC law.

First, to compare the MIC algorithm with the extended MIC, the MIC algorithm is applied on the assumed systems without FSP, and the operation is studied. Thus, considering the rigid system assumption (without FSP) for CASE-I, the MIC law can perfectly control the object on the designed path (Figures 4 and 5), which is absolutely misleading. Therefore, in a realistic analysis for practical implementations, the flexible effects should be considered, while those severe vibrations may even lead to fracture of flexible solar panels. However, as shown in Figures 6 and 7 for CASE-I and using the same controller gains, the effects of the disturbance forces due to vibrations of the flexible solar panels could create some errors in the position and the velocity, whereas the disturbance forces that generate these errors are shown in Figure 8 and

TABLE 3: The initial condition of the simulation study.

$\mathbf{q}(0) = [0\ 0\ 0\ 0.6\ 4.2\ 5.6\ 2.1\ \mu/7\ \pi/36]^T$	$\dot{\mathbf{q}}(0) = [0.85\ 0\ 0\ 0\ 0\ 0\ 0\ 0\ 0]^T$
$\bar{\mathbf{q}}_f(0) = [0\ \ 0\ \ 0]^T$	$\dot{\bar{\mathbf{q}}}_f(0) = [0\ \ 0\ \ 0]^T$
$\mathbf{K}_p = \widetilde{\mathbf{K}}_p = 900$	$\mathbf{K}_d = \widetilde{\mathbf{K}}_d = 300$

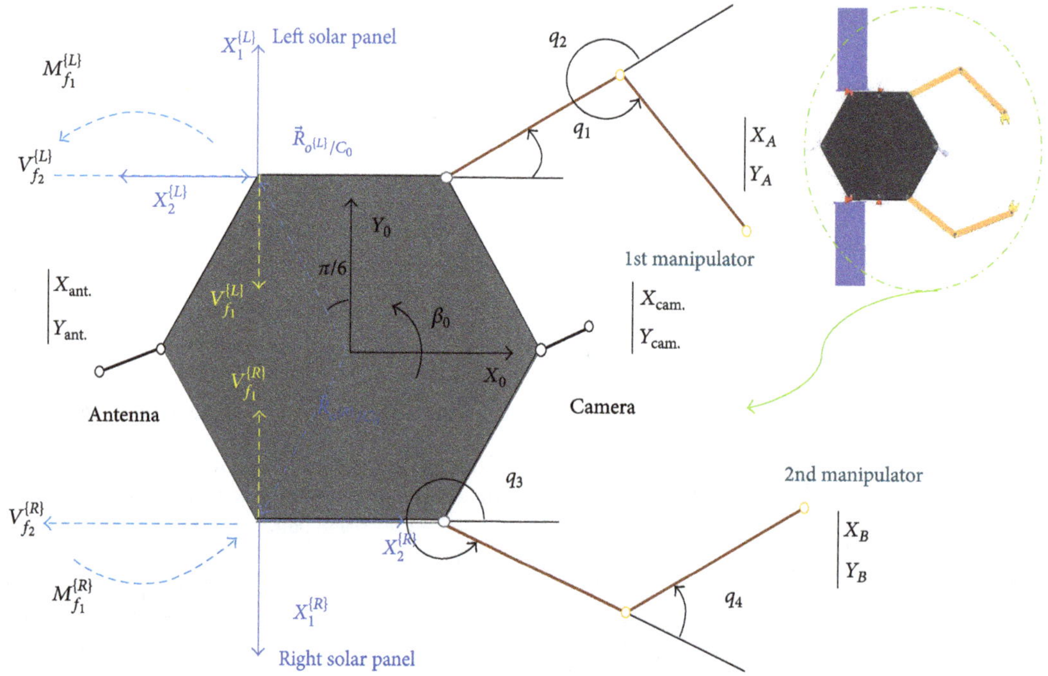

FIGURE 3: Schematic representation of assumed space robot system and its parameters and frames.

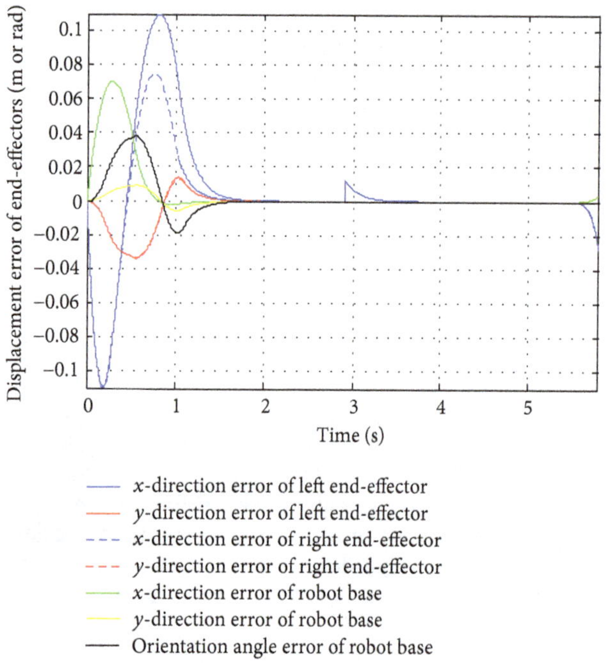

FIGURE 4: Error of the work space variables during object manipulation in CASE-I without FSP.

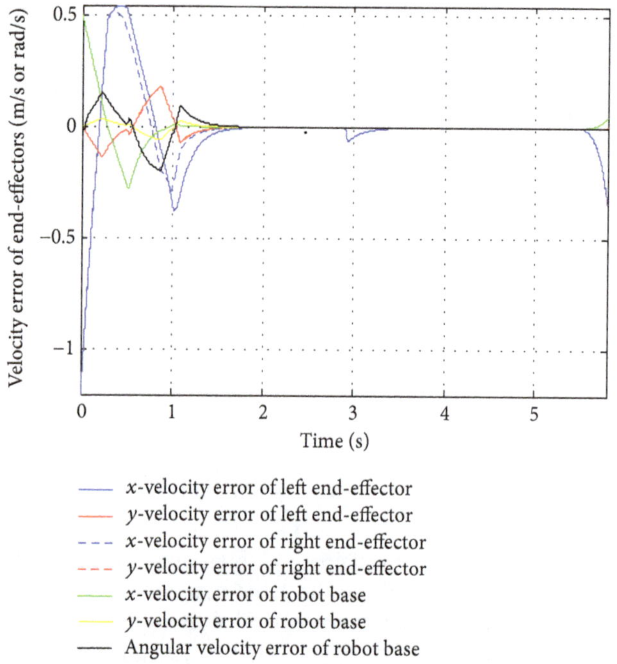

FIGURE 5: Error of the variable rates during object manipulation in CASE-I without FSP.

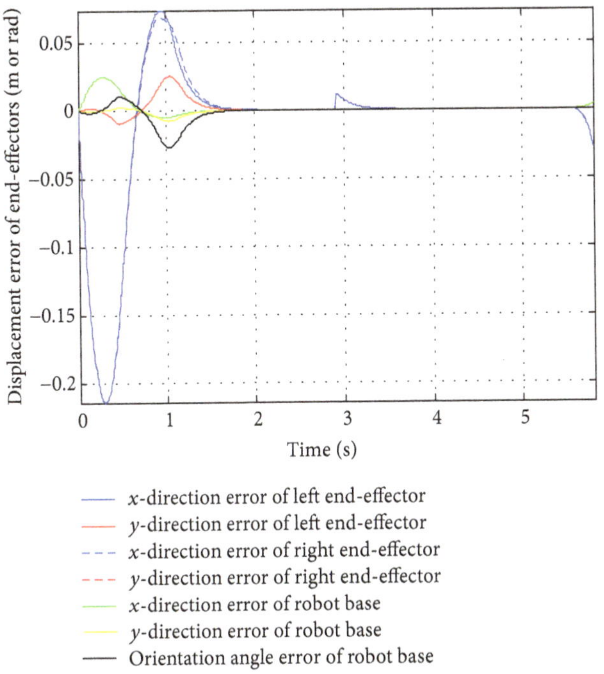

FIGURE 6: Error of the work space variables during object manipulation in CASE-I.

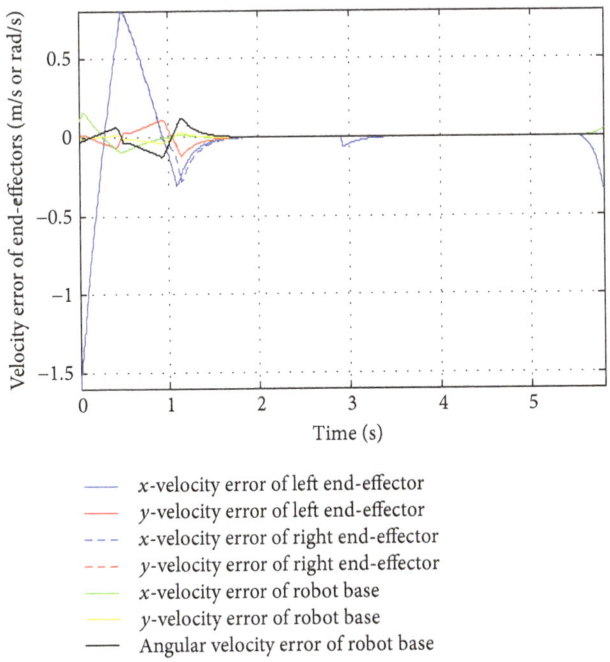

FIGURE 7: Error of the variable rates during object manipulation in CASE-I.

they could disturb the operation. These disturbing forces are caused by the deflection of the left and the right solar panels as shown in Figure 9. Although this is certainly undesirable, it should be controlled by this efficient strategy which is the property of the extended MIC controller than the original MIC algorithm. For instance, noting to some considerations

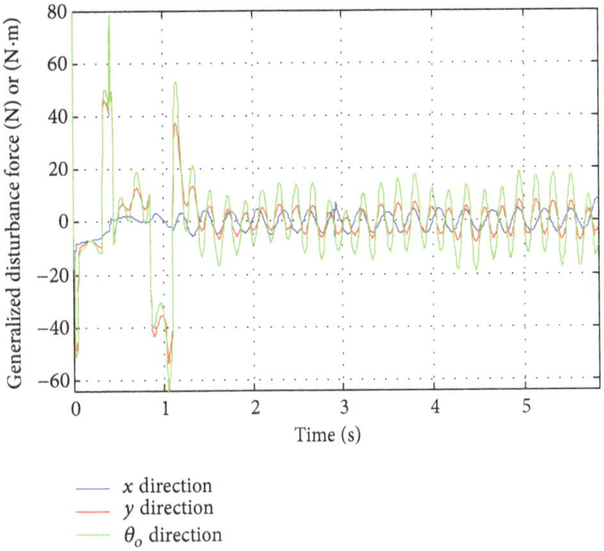

FIGURE 8: Generalized disturbance forces during object manipulation in CASE-I.

in path planning such as a constant speed for the robot base, or at least not inducing on-off impact disturbances, the extended MIC controller can successfully perform the object manipulation task. If the defined generalized variable of the robot base and the End-Effectors (the generalized workspace variables of the robot or $\widetilde{\mathbf{X}}$) are entirely controlled, the object would be on the desired path in which this is the basis of the MIC algorithm. It means that the object manipulation task is well done, while the flexible solar panels just stay true. This will be studied in the second case too. Also, an animated view of the system performing the object manipulation during the designed MTT of CASE-I is shown in Figure 10.

The implementation of the extended MIC law to move the object and accordingly the whole system on a circular path (CASE-II) is shown in Figures 11–15. This operation was done by the defined generalized variable of the robot base which is shown in Figure 3. Assuming a rigid system (without FSP), the original form of the MIC law can successfully complete the object manipulation task based on the planned path with negligible errors [4, 5]. Although the object manipulation task is successfully done in this case with FSP too, it would not be succeeded if the considerations in path designing such as a constant speed for the robot base are not taken into account, because the simulation was stopped after the flexible solar panels had been fractured. By using such considerations as discussed in CASE-I, the proposed extended MIC controller can effectively perform the task as shown in Figures 11 and 12. Deflection of solar panels is shown in Figure 13. Also, the disturbance forces of Figure 14 are produced due to these deflections along the designed circular path. An animated view of the system performing the task in CASE-II is shown in Figure 15. It should be noted that we have assumed that the space free-flying robot with the flexible solar panels possesses an initial velocity in all of these simulations. Also,

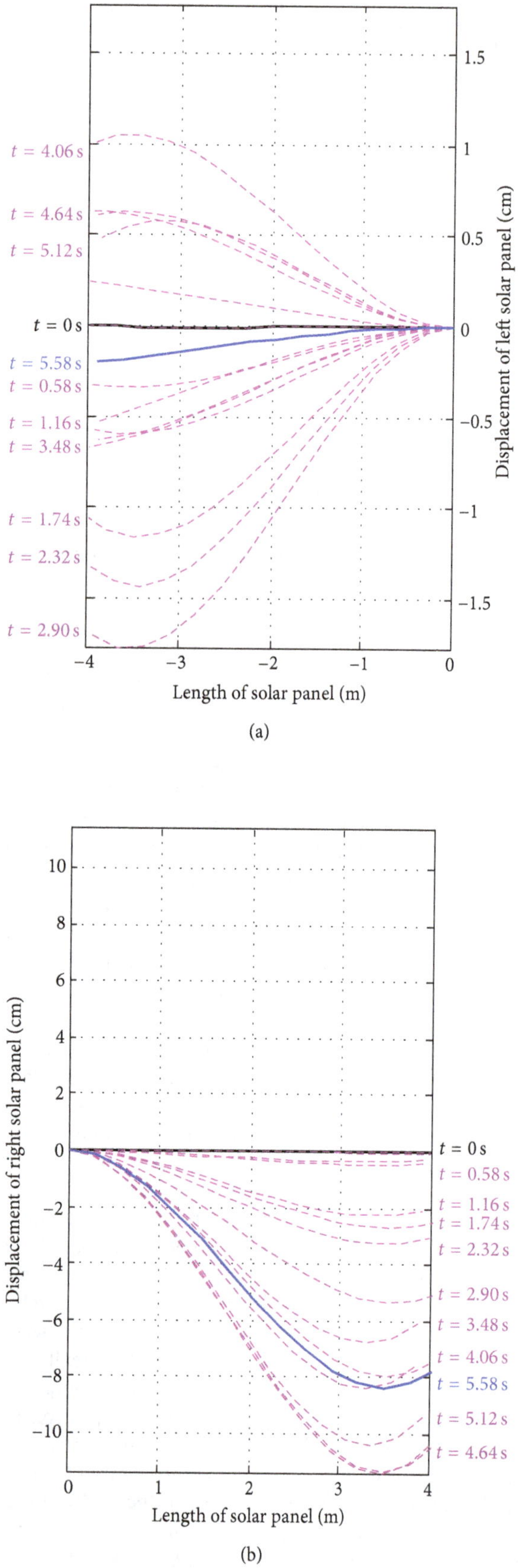

FIGURE 9: Deflection time history of the left and right flexible solar panels in CASE-I.

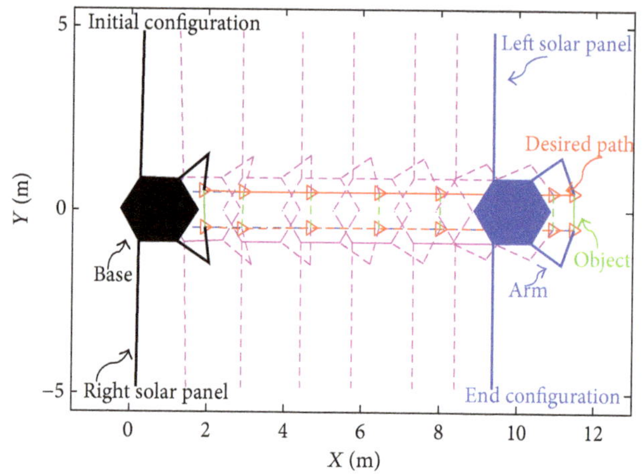

FIGURE 10: An animated view of the system during the object manipulation in CASE-I.

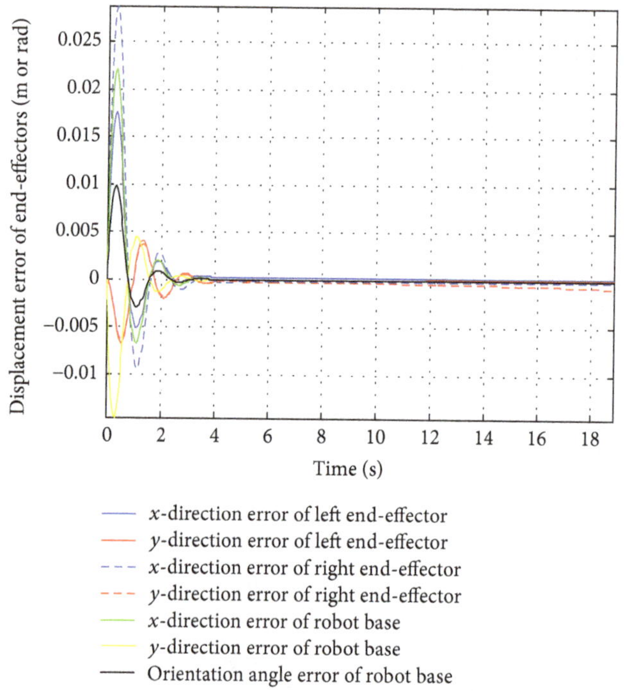

FIGURE 11: Error of the work space variables during object manipulation in CASE-II.

the flexible solar panels are at rest at the beginning of each operation as shown in Figures 9 and 14. It should be noted that by implementing the MIC algorithm on these two case studies, the simulation is broken since the flexible solar panels have more than the allowed displacement. This results in the fracture of these flexible appendages, and therefore, the object manipulation operation would not be perfectly done.

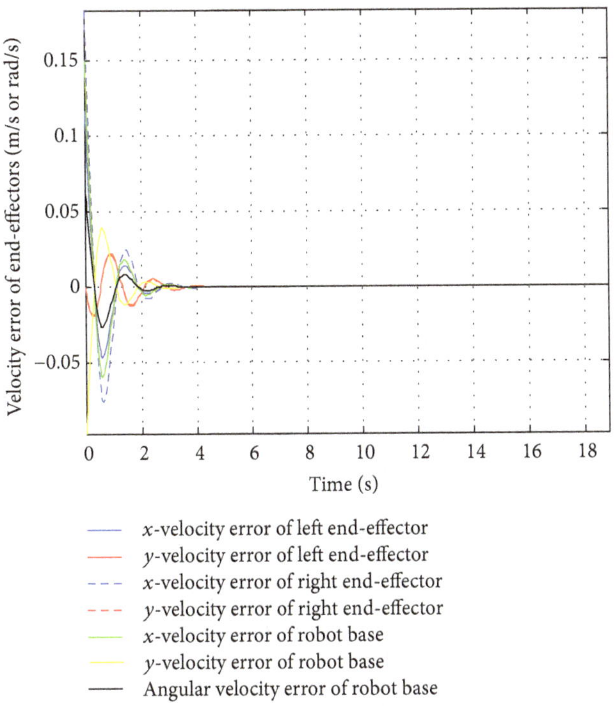

FIGURE 12: Error of the variable rates during object manipulation in CASE-II.

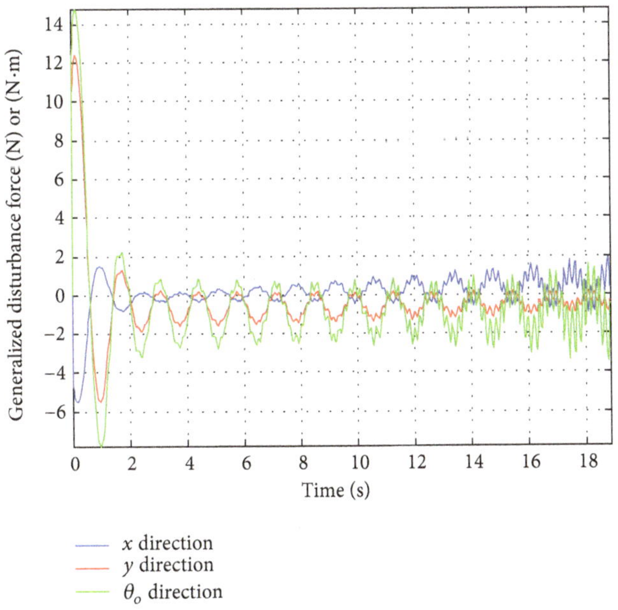

FIGURE 13: Generalized disturbance forces during object manipulation in CASE-II.

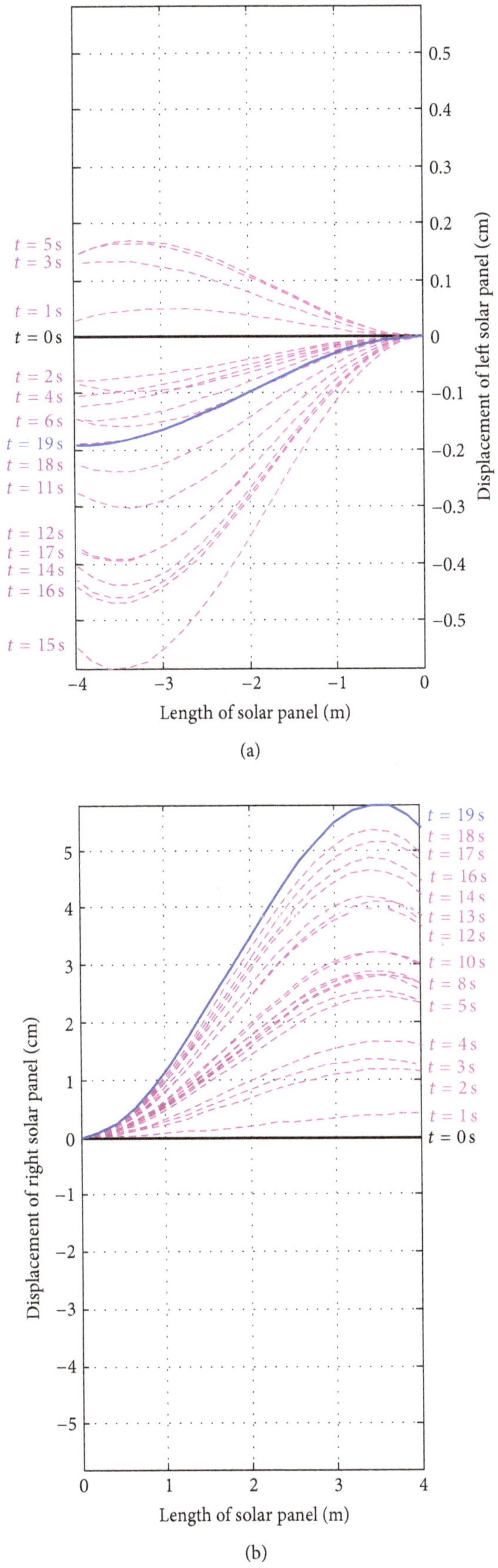

FIGURE 14: Deflection time history of the left and right flexible solar panels in CASE-II.

6. Conclusions

In this paper, dynamics modelling and control of a space robotic system with flexible appendages during a cooperative

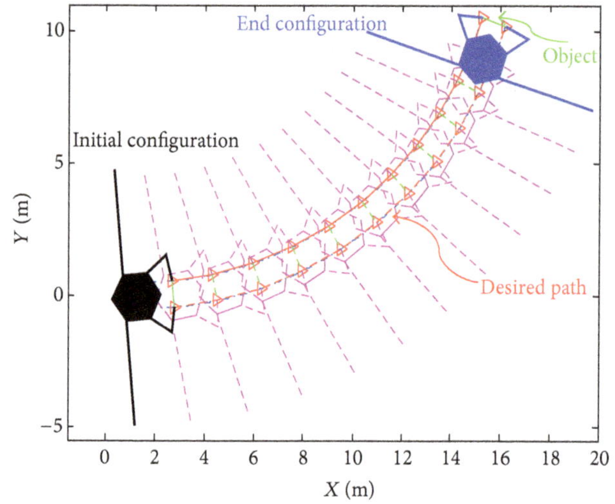

FIGURE 15: An animated view of the system during the object manipulation in CASE-II.

object manipulation task was discussed. This rigid-flexible multibody system necessitated delicate force exertion by numerous end-effectors to move the object along a specified path. After expressing the dynamic equations of a rigid space robot, the equations of the deformable bodies of this flexible multibody system were extracted. After that, the relationship between these two sets of equations was studied to obtain a practical precise dynamic model for designing the controller. So, the achieved dynamics model of a rigid-flexible multibody system decreased the computations of dynamics analysis, and it was useful in model-based control algorithms. Next, based on genetic algorithm approach using MATLAB/GATOOL, the trajectory was designed to move along a straight line based on a minimum time trajectory. This that the fast dynamics of the multibody system to be stimulated, and its effects could be studied. Then, an extended multiple impedance control was proposed and implemented on this system for object manipulation. As detailed, the multiple impedance control law was extended to compensate the disturbing forces due to vibrating motion of flexible appendages. Finally, by a comprehensive simulation routine, the obtained results were studied for the designed minimum time trajectory that complete the straight operation (CASE-I) and for the circular path (CASE-II). It was shown that vibration of the flexible solar panels results in generalized disturbance forces that are applied on the mobile base of the robot. These disturbance forces lead to undesirable errors of the end-effectors which were eliminated by the extended MIC controller. Moreover, these severe vibrations may even lead to fracture of flexible solar panels by applying the multiple impedance control. So, in this realistic analysis, the flexible effects could be considered and controlled by the stated efficient strategy. Furthermore, by noting some considerations in path planning to prevent on-off impact disturbances, the extended MIC controller could successfully perform the object manipulation task for these complicated rigid-flexible systems even if the straight path was designed based on a minimum time trajectory.

Nomenclature

$\mathbf{C}$:	Vector of quadratic nonlinear terms of velocity where $\widetilde{\mathbf{C}}$ defines in task space
$\mathbf{G}$:	Grasp matrix
$\mathbf{H}$:	Positive definition mass matrix of system where $\widetilde{\mathbf{H}}$ defines in task space
$\mathbf{I}$:	Second moment of area
$\mathbf{J}_c$:	The Jacobian matrix for the manipulators
$\mathbf{K}$:	Stiffness matrix of flexible member
$\mathbf{K}_p, \mathbf{K}_d, \mathbf{M}_{\text{des}}$:	Gain matrix of controller for object
$\widetilde{\mathbf{K}}_p, \widetilde{\mathbf{K}}_d, \widetilde{\mathbf{M}}_{\text{des}}$:	Gain matrix of controller for system in task space
$\mathbf{M}_f$:	Positive definition mass matrix of flexible member
M_{sys}:	Total mass of the rigid subsystem
$\mathbf{q}$:	Entity vector of generalized coordinate of rigid system
$\overline{\mathbf{q}}$:	Entity vector of generalized coordinate of flexible body
$\overline{\mathbf{q}}_f$:	Vector of elastic generalized coordinate of flexible body
$\overline{\mathbf{q}}_r$:	Vector of reference or rigid generalized coordinate of flexible body
$\mathbf{Q}$:	Vector of generalized forces
$\widetilde{\mathbf{Q}}_{\text{app}}$:	Vector of applied control forces
$\mathbf{Q}_e$:	Vector of generalized external forces of the flexible members
$\mathbf{Q}_{\text{flex.}}$:	Vector of generalized forces due to stimulation of the flexible members
$\widetilde{\mathbf{Q}}_m$:	Vector of control forces for end-effector motion
$\widetilde{\mathbf{Q}}_{\text{react}}$:	Vector of forces in task space that is exerted from object to end-effectors
$\mathbf{Q}_v$:	Quadratic velocity vector of flexible member
$\mathbf{R}_{C_0}, \dot{\mathbf{R}}_{C_0}, \ddot{\mathbf{R}}_{C_0}$:	Vector of position, velocity, and acceleration of robot bases in inertial frame
s:	Path parameter
$\mathbf{W}$:	Task weighting matrix
$\mathbf{X}_E^{(m)}, \dot{\mathbf{X}}_E^{(m)}$:	Vector of position and velocity of (m)th end-effectors
$\boldsymbol{\beta}_0$:	Generalized Euler angles variables of the robot base
$\boldsymbol{\theta}$:	Generalized variables of the robot joints
$\boldsymbol{\theta}_o$:	Object orientation angle
ω:	Angular velocity.

Superscript

$\{i\}$: Counter of flexible member
(i): Counter of rigid member of manipulators.

Subscript

f: Showing flexibility in a member
r: Showing rigidly in a member
0: Index of the base.

Conflict of Interests

The authors declare that they do not have any conflict of interests with others.

References

[1] S. A. A. Moosavian and A. Pourreza, "Heavy object manipulation by a hybrid serial-parallel mobile robot," *International Journal of Robotics and Automation*, vol. 25, no. 2, pp. 109–120, 2010.

[2] S. K. Dwivedy and P. Eberhard, "Dynamic analysis of flexible manipulators, a literature review," *Mechanism and Machine Theory*, vol. 41, no. 7, pp. 749–777, 2006.

[3] P. Zarafshan and S. A. A. Moosavian, "Manipulation control of a space robot with flexible solar panels," in *Proceedings of the IEEE/ASME International Conference on Advanced Intelligent Mechatronics (AIM '10)*, pp. 1099–1104, Montreal, Canada, July 2010.

[4] P. Zarafshan and S. A. A. Moosavian, "Control of a space robot with flexible members," in *Proceedings of the IEEE International Conference on Robotics and Automation (ICRA '11)*, Shanghai, China, 2011.

[5] P. Zarafshan and S. A. A. Moosavian, "Rigid-flexible interactive dynamics modelling approach," *Journal of Mathematical and Computer Modelling of Dynamics Systems*, vol. 18, no. 2, pp. 1–25, 2011.

[6] A. A. Ata and H. Johar, "Dynamic force/motion simulation of a rigid-flexible manipulator during task constrained," in *Proceedings of the IEEE International Conference on Mechatronics (ICM '04)*, pp. 268–273, June 2004.

[7] K. Yoshida, H. Nakanishi, H. Ueno, N. Inaba, T. Nishimaki, and M. Oda, "Dynamics, control and impedance matching for robotic capture of a non-cooperative satellite," *Advanced Robotics*, vol. 18, no. 2, pp. 175–198, 2004.

[8] B. Simeon, "On Lagrange multipliers in flexible multi-body dynamics," *Computer Methods Applied Mechanical Engineering*, vol. 19, no. 5, pp. 6993–7005, 2006.

[9] A. Fattah, J. Angeles, and A. K. Misra, "Dynamics of a 3-DOF spatial parallel manipulator with flexible links," in *Proceedings of the IEEE International Conference on Robotics and Automation*, pp. 627–632, May 1995.

[10] M. H. Korayem, H. Ghariblu, and A. Basu, "Dynamic load-carrying capacity of mobile-base flexible joint manipulators," *International Journal of Advanced Manufacturing Technology*, vol. 25, no. 1-2, pp. 62–70, 2005.

[11] T. M. Wasfy and A. K. Noor, "Computational strategies for flexible multibody systems," *Applied Mechanics Reviews*, vol. 56, no. 6, pp. 553–613, 2003.

[12] X. S. Ge and Y. Z. Liu, "The attitude stability of a spacecraft with two flexible solar arrays in the gravitational field," *Chaos, Solitons and Fractals*, vol. 37, no. 1, pp. 108–112, 2008.

[13] B. Pratiher and S. K. Dwivedy, "Non-linear dynamics of a flexible single link Cartesian manipulator," *International Journal of Non-Linear Mechanics*, vol. 42, no. 9, pp. 1062–1073, 2007.

[14] J. A. C. Ambrosio, "Dynamics of structures undergoing gross Motion and nonlinear deformations: a multi-body approach," *Computer Structure*, vol. 59, no. 6, pp. 1001–1012, 1996.

[15] C. Schmitke and J. McPhee, "Using linear graph theory and the principle of orthogonality to model multibody, multi-domain systems," *Advanced Engineering Informatics*, vol. 22, no. 2, pp. 147–160, 2008.

[16] M. J. Sadigh and A. K. Misra, "Stabilizing tethered satellite systems using space manipulators," in *Proceedings of the IEEE/RSJ/GI International Conference on Intelligent Robots and Systems*, pp. 1546–1553, September 1994.

[17] J. F. Deü, A. C. Galucio, and R. Ohayon, "Dynamic responses of flexible-link mechanisms with passive/active damping treatment," *Computers and Structures*, vol. 86, no. 3-5, pp. 258–265, 2008.

[18] P. Rocco and W. J. Book, "Modelling for two-time scale force/position control of flexible robots," in *Proceedings of the 13th IEEE International Conference on Robotics and Automation*, pp. 1941–1946, April 1996.

[19] A. Green and J. Z. Sasiadek, "Dynamics and trajectory tracking control of a two-link robot manipulator," *JVC/Journal of Vibration and Control*, vol. 10, no. 10, pp. 1415–1440, 2004.

[20] F. Landolsi, S. Choura, and A. H. Nayfeh, "Control of 2D flexible structures by confinement of vibrations and regulation of their energy flow," *Shock and Vibration*, vol. 16, no. 2, pp. 213–228, 2009.

[21] B. Subudhi and A. S. Morris, "Soft computing methods applied to the control of a flexible robot manipulator," *Applied Soft Computing Journal*, vol. 9, no. 1, pp. 149–158, 2009.

[22] M. Ouled Chtiba, S. Choura, S. El-Borgi, and A. H. Nayfeh, "Confinement of vibrations in flexible structures using supplementary absorbers: dynamic optimization," *JVC/Journal of Vibration and Control*, vol. 16, no. 3, pp. 357–376, 2010.

[23] K. Yoshida, D. N. Nenchev, P. Vichitkulsawat, H. Kobayashi, and M. Uchiyama, "Experiments on the point-to-point operations of a flexible structure mounted manipulator system," *Advanced Robotics*, vol. 11, no. 4, pp. 397–411, 1997.

[24] A. Ebrahimi, S. A. A. Moosavian, and M. Mirshams, "Comparison between minimum and near-minimum time optimal control of a flexible slewing spacecraft," *Journal of Aerospace Science and Technology*, vol. 3, no. 3, pp. 135–142, 2006.

[25] V. J. Modi, Y. Cao, C. W. De Silva, and A. K. Misra, "A class of novel space platform-based manipulators with slewing and deployable links: analyses and experiments," *JVC/Journal of Vibration and Control*, vol. 7, no. 8, pp. 1111–1161, 2001.

[26] A. Suleman, "Multibody dynamics and nonlinear control of flexible space structures," *JVC/Journal of Vibration and Control*, vol. 10, no. 11, pp. 1639–1661, 2004.

[27] S. A. A. Moosavian, R. Rastegari, and E. Papadopoulos, "Multiple impedance control for space free-flying robots," *Journal of Guidance, Control, and Dynamics*, vol. 28, no. 5, pp. 939–947, 2005.

[28] K. S. Chang, R. Holmberg, and O. Khatib, "Augmented object model: cooperative manipulation and parallel mechanism dynamics," in *Proceedings of the IEEE International Conference on Robotics and Automation (ICRA '00)*, pp. 470–475, April 2000.

[29] S. A. A. Moosavian and H. R. Ashtiani, "Cooperation of robotic manipulators using non-model-based multiple impedance control," *Journal of Industrial Robot*, vol. 35, no. 6, pp. 549–558, 2008.

[30] S. A. A. Moosavian and E. Papadopoulos, "Explicit dynamics of space free-flyers with multiple manipulators via SPACEMAPLE," *Advanced Robotics*, vol. 18, no. 2, pp. 223–244, 2004.

[31] A. A. Shabana, *Dynamics of Multi-Body Systems*, Cambridge University Press, 3rd edition, 2005.

[32] R. Jain and P. M. Pathak, "Trajectory planning of 2 DOF planar space robot without attitude controller," *World Journal of Modelling and Simulation*, vol. 4, no. 3, pp. 196–204, 2008.

[33] J. E. Bobrow, S. Dubowsky, and J. S. Gibson, "Time-optimal control of robotic manipulators," *International Journal of Robotics Research*, vol. 4, no. 3, pp. 244–258, 1985.

[34] Z. H. Jiang, "Impedance control of flexible robot arms with parametric uncertainties," *Journal of Intelligent and Robotic Systems*, vol. 42, no. 2, pp. 113–133, 2005.

[35] C. Ott, A. Albu-Schäffer, A. Kugi, and G. Hirzinger, "On the passivity-based impedance control of flexible joint robots," *IEEE Transactions on Robotics*, vol. 24, no. 2, pp. 416–429, 2008.

[36] T. Narikiyo and M. Ohmiya, "Control of a planar space robot: theory and experiments," *Control Engineering Practice*, vol. 14, no. 8, pp. 875–883, 2006.

[37] R. Rastegari and S. A. A. Moosavian, "Multiple impedance control of space free-flying robots via virtual linkages," *Acta Astronautica*, vol. 66, no. 5-6, pp. 748–759, 2010.

[38] V. R. Katti, K. Thyagarajan, K. N. Shankara, and A. S. K. Kumar, "Spacecraft technology," *Current Science*, vol. 93, no. 12, pp. 1715–1736, 2007.

Permissions

The contributors of this book come from diverse backgrounds, making this book a truly international effort. This book will bring forth new frontiers with its revolutionizing research information and detailed analysis of the nascent developments around the world.

We would like to thank all the contributing authors for lending their expertise to make the book truly unique. They have played a crucial role in the development of this book. Without their invaluable contributions this book wouldn't have been possible. They have made vital efforts to compile up to date information on the varied aspects of this subject to make this book a valuable addition to the collection of many professionals and students.

This book was conceptualized with the vision of imparting up-to-date information and advanced data in this field. To ensure the same, a matchless editorial board was set up. Every individual on the board went through rigorous rounds of assessment to prove their worth. After which they invested a large part of their time researching and compiling the most relevant data for our readers. Conferences and sessions were held from time to time between the editorial board and the contributing authors to present the data in the most comprehensible form. The editorial team has worked tirelessly to provide valuable and valid information to help people across the globe.

Every chapter published in this book has been scrutinized by our experts. Their significance has been extensively debated. The topics covered herein carry significant findings which will fuel the growth of the discipline. They may even be implemented as practical applications or may be referred to as a beginning point for another development. Chapters in this book were first published by Hindawi Publishing Corporation; hereby published with permission under the Creative Commons Attribution License or equivalent.

The editorial board has been involved in producing this book since its inception. They have spent rigorous hours researching and exploring the diverse topics which have resulted in the successful publishing of this book. They have passed on their knowledge of decades through this book. To expedite this challenging task, the publisher supported the team at every step. A small team of assistant editors was also appointed to further simplify the editing procedure and attain best results for the readers.

Our editorial team has been hand-picked from every corner of the world. Their multi-ethnicity adds dynamic inputs to the discussions which result in innovative outcomes. These outcomes are then further discussed with the researchers and contributors who give their valuable feedback and opinion regarding the same. The feedback is then collaborated with the researches and they are edited in a comprehensive manner to aid the understanding of the subject.

Apart from the editorial board, the designing team has also invested a significant amount of their time in understanding the subject and creating the most relevant covers. They scrutinized every image to scout for the most suitable representation of the subject and create an appropriate cover for the book.

The publishing team has been involved in this book since its early stages. They were actively engaged in every process, be it collecting the data, connecting with the contributors or procuring relevant information. The team has been an ardent support to the editorial, designing and production team. Their endless efforts to recruit the best for this project, has resulted in the accomplishment of this book. They are a veteran in the field of academics and their pool of knowledge is as vast as their experience in printing. Their expertise and guidance has proved useful at every step. Their uncompromising quality standards have made this book an exceptional effort. Their encouragement from time to time has been an inspiration for everyone.

The publisher and the editorial board hope that this book will prove to be a valuable piece of knowledge for researchers, students, practitioners and scholars across the globe.

List of Contributors

Stefan LeBel and Christopher J. Damaren
University of Toronto Institute for Aerospace Studies, 4925 Dufferin Street, Toronto, ON, Canada M3H 5T6

Morteza Shahravi and Milad Azimi
Space Research Institute, Tehran 15875-1774, Iran

J. Li
Shanghai Aircraft Manufacturing Co., Ltd., Shanghai 200436, China

T. Sreenuch
Integrated Vehicle Health Management Centre, Cranfield University, Bedford MK43 0AL, UK

A. Tsourdos
Division of Engineering Sciences, Cranfield University, Bedford MK43 0AL, UK

A. Nazarian Shahrbabaki, M. Bazazzadeh, A. Shahriari and M. Dehghan Manshadi
Department of Mechanical & Aerospace Engineering, Malek-Ashtar University of Technology, Shahin Shahr, Isfahan 83145/115, Iran

Mohamed Mostafa Y. B. Elshabasy
Department of Mechanical Engineering, Faculty of Engineering, Alexandria University, Alexandria 21544, Egypt

Yongki Yoon
Whirlpool Corporation, Saint Joseph Technology Center, 303 Upton Drive, Saint Joseph, MI 49085, USA

Ashraf Omran
CNH Industrial, 6900 Veterans Ave., Burr Ridge, IL 60527, USA

P. Ahmadi, M. Golestani and A. R. Vali
Department of Electrical and Electronic Engineering, Malek Ashtar University of Technology, Tehran, Iran

S. Nasrollahi
Department of Aerospace Engineering, Sharif University of Technology, Tehran, Iran

Mohammad Rasool Mojallizadeh
Department of Electrical Engineering, Najafabad Branch, Islamic Azad University, University Blvd., Najafabad, Isfahan, Iran

Bahram Karimi
Department of Electrical Engineering, Malek Ashtar University of Technology, Shahinshahr, Isfahan, Iran

A. Zanotti, G. Gibertini, D. Grassi and D. Spreafico
Politecnico di Milano, Dipartimento di Scienze e Tecnologie Aerospaziali, Campus Bovisa, Via La Masa 34, 20156 Milano, Italy

Mazyar Dawoodian
Department of Mechanical Engineering, I. A. University of Takestan, Takestan, Iran

Abdolrahman Dadvand
Department of Mechanical Engineering, Urmia University of Technology, Urmia, Iran

Amir Hassanzadeh
Department of Mechanical Engineering, Urmia University, Urmia, Iran

Antonio Andreini, Gianluca Caciolli, Bruno Facchini and Lorenzo Tarchi
DIEF, Department of Industrial Engineering Florence, University of Florence, Via S. Marta 3, 50139 Florence, Italy

Ali Ouacha, Ahmed Habbani and Jamal El Abbadi
Laboratoire d'Electronique et de Communications (LEC), Ecole Mohammadia d'Ing´enieurs (EMI), Universit´e Mohammed V-Agdal (UM5A), BP 765, Avenue Ibn Sina, Agdal, 10000 Rabat, Morocco

Bachir Bouamoud and Ahmed Habbani
Laboratoire SIME, Ecole Nationale Sup´erieure d'Informatique et d'Analyse des Syst`emes (ENSIAS), Universit´e Mohammed V-SOUISSI (UM5S), BP 713, Madinat Al Irfane, Avenue Mohammed ben Abdallah Regragui, Agdal, 10000 Rabat, Morocco

Adel A. Ghandakly
Department of ECE, California State University, Chico, Chico, CA 95929-0888, USA

Jason A. Reed
Principal Engineer AREVA NP Inc., 1345 Ridgeland Pkwy, Alpharetta, GA 30004, USA

Mohamed Abdullah, F. Gholamian and A. R. Zarei
Faculty of Chemistry, Malek Ashtar University, P.O. BOX 16705-3454, Tehran, Iran

R. Samar and M. Zamurad Shah
Mohammad Ali Jinnah University, Islamabad 44000, Pakistan

M. Nzar
Centres of Excellence in Science & Applied Technologies (CESAT), Islamabad, Pakistan

Francesco Topputo and Franco Bernelli-Zazzera
Department of Aerospace Science and Technology, Politecnico di Milano, Via La Masa 34, 20156 Milano, Italy

P. Zarafshan and S. Ali A. Moosavian
Center of Excellence in Robotics and Control, Advanced Robotics and Automated Systems Laboratory, Department of Mechanical Engineering, K.N. Toosi University of Technology, P.O. Box 19395-1999, Tehran, Iran

www.ingramcontent.com/pod-product-compliance
Lightning Source LLC
Chambersburg PA
CBHW050500200326
41458CB00014B/5254
9781632400574